CRC Handbook of EPR Spectra from Quinones and Quinols

Editor

Jens A. Pedersen, Ph.D.
Department of Chemistry
Physical Chemistry Division
Aarhus University
Aarhus, Denmark

CRC Press
Taylor & Francis Group
Boca Raton London New York

CRC Press is an imprint of the
Taylor & Francis Group, an **informa** business

First published 1985 by CRC Press
Taylor & Francis Group
6000 Broken Sound Parkway NW, Suite 300
Boca Raton, FL 33487-2742

Reissued 2018 by CRC Press

Library of Congress in Publication Data (Revised)
Main entry under title:

Handbook of EPR spectra from quinones and quinols.

Bibliography: p.
Includes index.
1. Quinone—Spectra—Handbooks, manuals. etc.
2. Hydroquinone—Spectra—Handbooks, manuals, etc.
3. Electron paramagnetic resonance spectroscopy—
Handbooks, manuals, etc. I. Pedersen, Jens A.
II. Title: Handbook of E.P.R. spectra from natural
and synthetic quinones and quinols.
QC463.Q54H36 1985 547'.596 84-12102
ISBN 0-8493-2955-8

A Library of Congress record exists under LC control number: 84012102

ISBN 13: 978-1-315-89393-8 (hbk)
ISBN 13: 978-1-351-07303-5 (ebk)

Visit the Taylor & Francis Web site at http://www.taylorandfrancis.com and the
CRC Press Web site at http://www.crcpress.com

PREFACE

This Handbook contains a compilation of Electron Paramagnetic Resonance (EPR) data of numerous semiquinones and related radicals. The Handbook is intended as a general reference guide to the structural studies of the radicals obtained by redox processes from natural and synthetic quinones and quinols. A chief difficulty in the general utility of EPR spectroscopy as an analytical tool is that even closely related radicals may furnish widely different spectra. When the spectra are analyzed in terms of splitting constants, however, the similarities between related radicals become apparent immediately. In organizing the Tables, therefore, effort has been made to attach importance to these similarities. Furthermore, variations in the splitting constants due to solvent, pH, or temperature effects have been located with the particular radical in order to make the data readily accessible; hopefully the clarity of presentation is not, therefore, compromised.

For the benefit of the newcomers to the field a short introduction to the principles of EPR is given, together with instructive examples which show how to correlate a radical structure with its EPR spectrum.

The Handbook should serve workers in a number of fields, especially since there is a growing interest in a variety of aspects of semiquinone chemistry. Besides the use in the structural characterization of numerous quinones/quinols the semiquinones are also useful for the following: as effective probes for studying intermolecular effects originating from counter ion and solvent association; for examining electronic theories of free radicals; and also for the insight one might get with systems where quinones/semiquinones participate, including blood clotting (menaquinones), aging (tocoquinones), energy conserving systems (plasto- and ubiquinones), and systems where the compounds act as insect and microbial controlling agents.

The editor is most grateful to the members of the Advisory Board for their useful criticisms and expert help. Special thanks are due to the many colleagues who generously provided compounds for analysis; these EPR analyses have been incorporated whenever possible.

The editor is pleased to acknowledge the expeditious and skillful help of several members of the technical staff at the Chemistry Department, Aarhus University, and owes a particular debt to the members of the Physical Chemistry Division for their valuable suggestions during the preparation of the manuscript.

Jens A. Pedersen
Aarhus
July 1983

THE EDITOR

Jens A. Pedersen, Ph.D., is Associate Professor at the Chemistry Department, Aarhus University.

Dr. Pedersen graduated from the University of Aarhus in 1968 with the M.Sc. degree in chemistry and in 1973 received the Ph.D. degree in physical chemistry. He served as a Research Associate at the University of Western Ontario (London, Ontario) from 1970 to 1971. He assumed his present position in 1974.

Dr. Pedersen is a member of the Phytochemical Society of Europe and the Danish Botanical Society. He has served as Vice Chairman of the Chemistry Department, the University of Aarhus, from 1976 to 1977.

He has published some 30 research papers and several national and international presentations. His major current research interests relate to the chemistry of free radicals as studied by magnetic resonance spectroscopy and advanced computer technique.

ADVISORY BOARD

TABLE OF CONTENTS

INTRODUCTION

Semiquinones are relatively stable free radicals formed by one-electron reduction of quinones or by one-electron oxidation of quinols. Thus, in alkaline media, the 1,4-benzosemiquinone radical anion ($\dot{Q}^-$) appears as an intermediate in the process

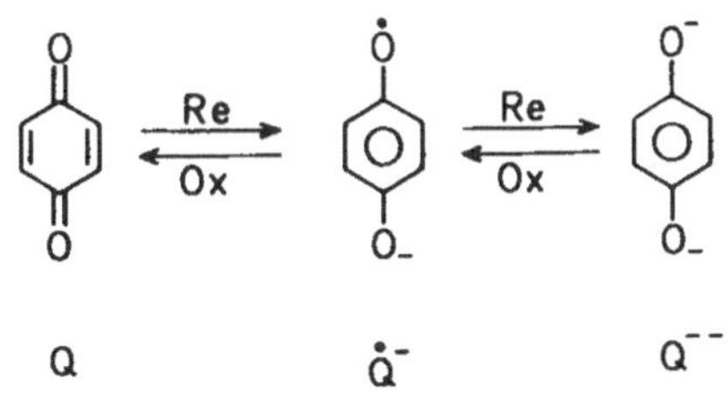

(1)

A semiquinone is by virtue of the odd number of electrons amenable to electron paramagnetic resonance (EPR) spectroscopy. In general, the protons in the molecule split the EPR signal from the unpaired electron into a unique splitting pattern, called the hyperfine structure. Although the pattern may consist of hundreds of individual lines, it can be completely described in terms of a small number of splitting constants, in fact constituting a "structural fingerprint" of the semiquinone.

Moreover, since redox processes of the type (1) occur without significant structural changes, the EPR spectrum of a semiquinone identifies the parent quinone or quinol as well. This, in connection with the inherent high sensitivity of EPR spectroscopy, has made EPR spectra of semiquinones an important analytical tool in the study of naturally occurring quinones and quinols. The analytical application rests on the substantial and rapidly growing body of published semiquinone spectra, obtained from chemically well-defined compounds.

The present compilation of splitting constants was undertaken primarily to give ready access to the available reference data, thus enhancing the analytical potential of the EPR method in studies involving quinones and related compounds.

The tables comprise representative results for *all* reported semiquinone-radicals derived from quinones or quinols having six-membered quinone rings without heteroatoms, supplemented by data for other radicals derived from mono- and polyhydric phenols and naphthols. Five- and seven-membered quinone rings are represented by acenaphthenequinones and tropolones, respectively.

The compilation is based on the literature as covered by *Chemical Abstracts* from 1958 to the beginning of 1983. As a rule, all EPR data concerning the classes of radicals mentioned are included. Omissions are made, however, where the description or interpretation of the results was found inadequate and where new, accurate data supersede older ones. Data from some 200 unpublished spectra run in the editors laboratory are included to fill "gaps" in some of the tables and to augment the very scanty information on certain groups of radicals, e.g., the anthrasemiquinones.

ASSIGNMENTS

The assignment of observed hyperfine splitting constants to nuclei at specific sites is crucial for any interpretation in terms of molecular properties. Most splitting constants included in the tables were originally assigned by the authors. However, conflicting or incomplete assignments in the literature are not uncommon. All assignments entered into the tables have been critically revised in the light of the newest evidence, empirical rules for substituent effects, and, finally, trends within a family of related radicals, e.g., an anion radical and its complexes with various cations. The assignments thus obtained are internally consistent and agree with the available evidence.

Where an assignment deviates from that given by the authors, the reference is marked with a superscript plus sign (e.g., 27^+).

The data, comprising hyperfine splitting constants, g-factor, solvent system, and method of radical generation, are organized in five sections under the headings **Anions, Neutral Species, Cations, Semiquinone-Cation Complexes, and Isotopes (^{13}C and ^{17}O).**

To facilitate the use of the tabulated data, a brief introduction to the analysis of EPR spectra is given. Moreover, a collection of semiquinone spectra is shown in order to illustrate characteristic features. The analytical application of EPR spectroscopy is shown with examples of determinations of quinones/quinols in crude plant extracts.

PRINCIPLES OF EPR*

A free electron placed in a magnetic field, H, will be found in either of two Zeeman states having an energy separation proportional to the strength of the field. This is accounted for by assigning to the electron a spin angular momentum with quantum number $S = {}^1/_2$, and an associated magnetic moment. It follows from the quantum mechanics of angular momentum that the electron can be in either of $2S + 1 = 2$ states, associated with the values ($\pm{}^1/_2$) of the "magnetic" quantum number m_S, which specifies the orientation of the magnetic moment in the external field.

The electronic energy is given by

$$E(m_S) = E_o + \beta g_e H m_S \qquad (2)$$

where E_o represents the energy in the absence of the magnetic field, β is the Bohr magneton (9.274×10^{-21} erg G^{-1}), and g_e a dimensionless constant, the electronic g-factor with the value 2.002319. In an EPR experiment transitions are induced between Zeeman states in a strong magnetic field, by means of microwave radiation of proper frequency ν_o, determined by the *resonance condition*

$$h\nu_o = \Delta E = E(m_S = {}^1/_2) - E(m_S = -{}^1/_2)$$

or with Equation 2

$$h\nu_o = \beta g_e H \qquad (3)$$

The level diagram and the EPR spectrum of a (hypothetical) sample of free electrons are depicted in Figure 1.

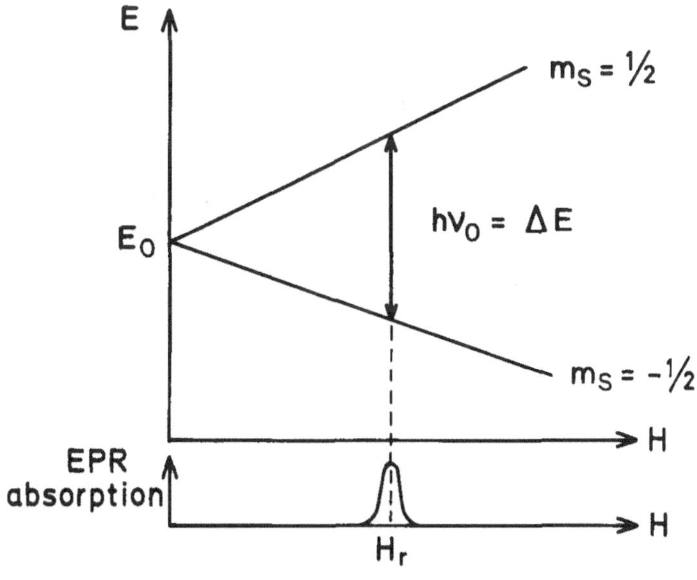

FIGURE 1

* The following outline is intended to serve as an introduction to the EPR studies in solution. For more detailed information, one of the authoritative reference texts listed under EPR monographs should be consulted.

In Figure 1 the EPR spectrum is represented as absorption vs. field strength, since in the normal experimental procedure the microwave frequency, ν_0, is held constant whereas H is varied until the resonance condition (Equation 3) is met.The field at which the absorption line occurs, is called the *resonance field*, H_r. Note the slope of the energy levels is determined by g_e.

RADICALS

The spins of electrons bound in molecules are usually paired, so that most molecules with an even number of electrons have no net magnetic moments due to the electrons. Such molecules do not respond to magnetic fields in the way required to yield an EPR spectrum. On the other hand, radicals having an odd number of electrons behave in the EPR experiment much like free electrons. In molecular quantum mechanics the electronic ground state of radicals is assigned an electron spin $S = {}^1/_2$. Strictly speaking, this quantum number characterizes the full electronic state of the radical rather than a single electron. However, in accordance with molecular orbital theory and the Aufbau principle, it has proved convenient to attribute the magnetic properties of the radical to the spin of a single electron occupying a specific molecular orbital. This so-called ***unpaired electron*** represents the radical as far as the EPR spectrum is concerned. Hence the description given above for a sample of free electrons applies to a sample of radicals, as well, if g_e in Equation 3 is substituted by the g-factor of the unpaired electron, i.e., the value required to represent the magnetic moment of the radical.

HYPERFINE SPLITTINGS

Semiquinone radicals give rise to spectra exhibiting from one to hundreds of lines disposed in a centrosymmetric pattern of total spread up to 30 G. The line pattern or hyperfine structure arises from interaction between the magnetic moment of the unpaired electron and one or more atomic nuclei. Many commonly occurring nuclei possess magnetic moments associated with spins.* The spin quantum number assigned to nuclei is denoted I and has one of the values 0, $^1/_2$, 1, 3/2 . . . depending on the nucleus in question. A nucleus with spin I has 2I + 1 substates labeled by the magnetic quantum number m_I, m_I = I, I − 1, I − 2 . . . −I. The states correspond to different orientations of the nuclear moment in an external magnetic field.

In a radical containing a nucleus with spin I ≠ 0, the magnetic moment of the unpaired electron may interact with the local magnetic field set up by the nuclear moment. Since the local field experienced by the unpaired electron depends on the quantum number m_I, so do the energy levels of the electron. In fact, for each value of m_S there are 2I + 1 closely spaced levels (Figure 2). If the external field is much stronger than the local fields and the radicals are dissolved in a low-viscosity solvent so that they tumble rapidly, the possible EPR transitions are simply obtained from the resonance condition

$$h\nu = g\beta H + a'm_I \tag{4}$$

where the term $a'm_I$ appears as a result of the interaction with the local field. For fixed frequency, ν_o, 2I + 1 equidistant EPR lines of equal intensity, each associated with a particular value of m_I, are observed at the resonance fields

$$H(m_I) = H_o - am_I \tag{5}$$

where $H_o = h\nu_o/g\beta$ is the resonance field in the absence of magnetic nuclei, and $a = a'/g\beta$ is the hyperfine splitting constant measured in Gauss.

Identical nuclei in identical chemical environments within a radical will give rise to the same hyperfine splitting. Such nuclei are called *equivalent.* A radical containing n equivalent nuclei of spin I yields equidistant EPR lines at the resonance fields

$$H(m_I^1, m_I^2, \ldots, m_I^n) = H_o - a\sum_{j=1}^{n} m_I^j \tag{6}$$

There are $(2I + 1)^n$ different sets $(m_I^1, m_I^2, \ldots, m_I^n)$, each specifying a transition in the EPR spectrum. However, the right-hand side of Equation 6 yields only 2nI + 1 different resonance fields, or distinct EPR lines. Hence all but the two outermost EPR lines are degenerate, so that the observed relative intensities increase towards the center of the spectrum. For nuclei with spin I = $^1/_2$ the relative intensities are given by the binomial coefficients of order n (Pascal's triangle, see Scheme 1) and for spin I = 1 by the trinomial coefficients and so on.

These results may be generalized to radicals containing r groups of equivalent nuclei. If the i'th group contains n^i equivalent nuclei with spin I^i and splitting constant a^i, then the number of lines in the spectrum is given by the product

$$N = \prod_{i=1}^{r} (2n^iI^i + 1) \tag{7}$$

These lines are positioned at the resonance fields given as

$$H(m_I^{i,j}) = H_o - \sum_{i=1}^{r} a^i \sum_{j=1}^{n^i} m_I^{i,j} \tag{8}$$

* A table of nuclear moments may be found in the *CRC Handbook of Chemistry and Physics*, 64th ed., Weast, R. C., Ed., CRC Press, Boca Raton, Fla., 1983-1984, E-65.

where i labels the groups of equivalent nuclei and j labels the nuclei within the groups. The number of individual transitions is given by the product

$$D = \prod_{i=1}^{r} (2I^i + 1)^{n^i} \tag{9}$$

Since N < D if there are equivalent nuclei, a number of transitions will coincide to produce degenerate lines in the spectrum. The outermost lines are always nondegenerate, however. This implies that D may be expressed in terms of the height h(k) of the k'th line as

$$D = \sum_{k=1}^{N} h(k)/h(1) \tag{10}$$

The linear field separation between the two outermost lines in the spectrum called the total width (TW), may be expressed by means of Equation 8 as

$$TW = 2 \sum_{i=1}^{r} n^i I^i a^i \tag{11}$$

Examples include the following:

1. *Radical with a single proton with splitting constant a*: $r = n = 1$, $I = {}^1/_2$, and $m_I = \pm {}^1/_2$, and therefore two resonance fields are obtained from Equation 6

$$H_1 = H_0 - {}^1/_2 a$$

$$H_2 = H_0 + {}^1/_2 a$$

The spectrum consists of two lines of equal intensity, separated by a as shown in Figure 2. As usual in EPR spectroscopy the spectrum is represented as the first derivative of the absorption curve.

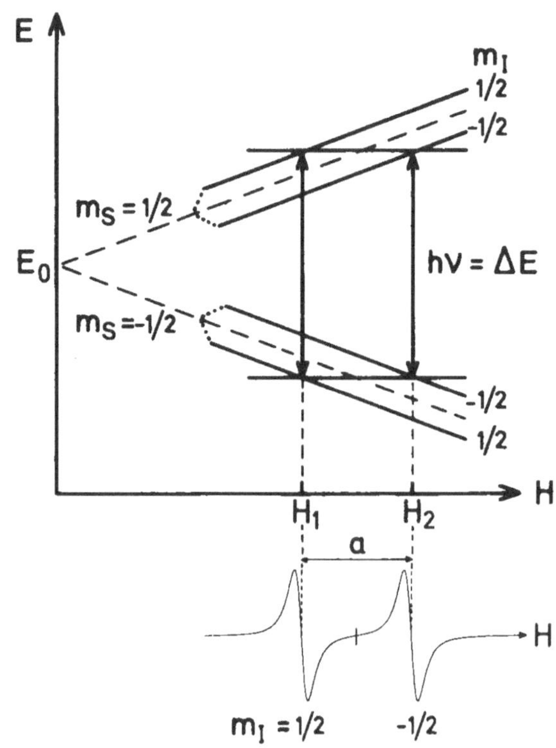

FIGURE 2.

Scheme 1
BINOMIAL, TRINOMIAL, AND TETRANOMIAL COEFFICIENTS GIVING THE RELATIVE LINE INTENSITIES OBSERVED FROM n NUCLEI OF SPIN I = $^1/_2$, 1, AND $^3/_2$, RESPECTIVELY

n	Relative intensities	N	D
$I = {}^1/_2 \mid N = n + 1 \mid D = 2^n$:			
0	1	1	1
1	1 1	2	2
2	1 2 1	3	4
3	1 3 3 1	4	8
4	1 4 6 4 1	5	16
5	1 5 10 10 5 1	6	32
6	1 6 15 20 15 6 1	7	64
7	1 7 21 35 35 21 7 1	8	128
8	1 8 28 56 70 56 28 8 1	9	256
9	1 9 36 84 126 126 84 36 9 1	10	512
$I = 1 \mid N = 2n + 1 \mid D = 3^n$:			
1	1 1 1	3	3
2	1 2 3 2 1	5	9
3	1 3 6 7 6 3 1	7	27
4	1 4 10 16 19 16 10 4 1	9	81
5	1 5 15 30 45 51 45 30 15 5 1	11	243
6	1 6 21 50 90 126 141 125 90 50 21 6 3	13	729
$I = {}^3/_2 \mid N = 3n + 1 \mid D = 4^n$:			
1	1 1 1 1	4	4
2	1 2 3 4 3 2 1	7	16
3	1 3 6 10 12 12 10 6 3 1	10	64
4	1 4 10 20 31 40 44 40 31 20 10 4 1	13	256

2. *Radical with two equivalent protons*: $r = 1$, $n = 2$, $I = {}^1/_2$, $m_I = \pm {}^1/_2$. The spectrum contains three lines (Equation 7). The resonance fields obtained from Equation 6 are

$$H(^1/_2, ^1/_2) = H_o - a$$

$$H(^1/_2, -^1/_2) = H_o$$

$$H(-^1/_2, ^1/_2) = H_o$$

$$H(-^1/_2, -^1/_2) = H_o + a$$

The spectrum is a triplet with the relative intensities 1:2:1 and line separation equal to a as shown in Figure 3.

1. The smallest splitting constant is equal to the separation H(2) − H(1) (or H(N) − H(N − 1)).
2. If h(1) ≈ h(2), the smallest splitting stems from a single proton (cf. Figure 2), and the second-smallest splitting constant then equals H(3) − H(1). If on the other hand, h(2) ≈ 2h(1) or h(2) ≈ 3h(1), then the smallest splitting stems from two or three equivalent nuclei, respectively. This rule applies to $I = {}^1/_2$ nuclei only.
3. When all but one splitting constant, a^r, have been determined, a^r may be obtained as

$$a^r = \frac{1}{2n^rI^r}\left(TW - 2\sum_{i=1}^{r-1} n^iI^ia^i\right) \tag{12}$$

Example 1 — 2-Methoxy-1,4-benzosemiquinone (No. 446).* As shown in Figure 5, 28 lines are observed. First the total number P of protons associated with splittings is determined *via* Equation 9, which for $I = {}^1/_2$ becomes

$$D = 2^P$$

or, by means of Equation 10

$$2\sum_{k=1}^{28} h(k)/(h(1) + h(28)) = 2^P$$

yielding

$$2^P \sim 62 \text{ or } P = 6$$

By Rule 1, $a^1 = H(2) - H(1)$. Furthermore, since h(1) ≈ h(2), Rule 2 yields $a^2 = H(3) - H(1)$. Since h(3) is ≈ 3h(1), a 1:3:3:1 intensity pattern is searched for. This pattern is exhibited by the lines 1, 3, 5, and 8, indicating the presence of three equivalent protons with splitting constant a^2. The 8th-line spectrum (Figure 5b) constructed from a^1(1H) and a^2(3H) can be superimposed on the lines 1 through 5, 7, and 8, and 11, so that the first line not accounted for is line number 6. Hence H(6) − H(1) is equal to a^3 (Figure 5c) and, since h(6) ≈ h(1), only one proton is associated with a^3. Hence the splitting constant of the 6th proton remains to be determined by Rule 3, $a^4 = TW - (a^1 + 3a^2 + a^3)$, with TW = H(28) − H(1).

According to Equation 7 the spectrum should contain 32 lines rather than 28. However, a reconstruction of the spectrum by means of Equation 8 shows an accidental overlap of lines in the center, so that the lines 13 to 16 each contain two components of intensity 1 and 3, respectively.

Example 2 — 5-(2-Carboxylatoethyl)-4-oxido-1,2-benzosemiquinone (No. 402). This experimental spectrum (Figure 6) contains some lines (marked by arrows) from an unknown radical. Such spurious EPR lines are frequently encountered. They should be clearly marked before an analysis of the main spectrum is attempted. Usually spurious lines may be recognized through a different rate of decay and a different spectral centerpoint (g-factor). The number of protons involved in the spectrum is determined, as in the first example, from

$$D = 2^P = 2\sum_{k=1}^{33} h(k)/(h(1) + h(33)) = 76$$

* Number from the chapter on Explanation of Tables.

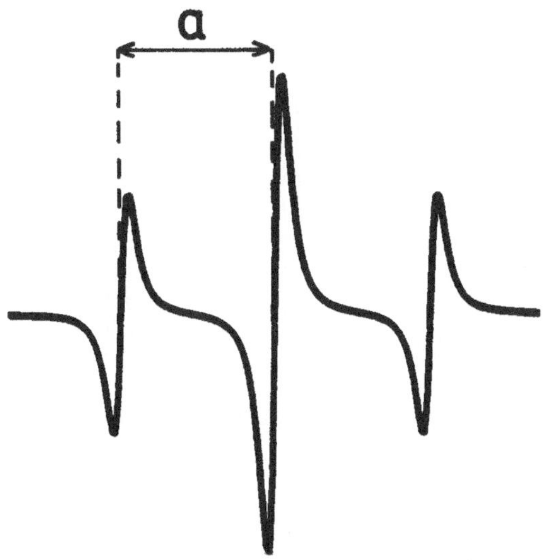

FIGURE 3.

3. *Radical with three protons, two being equivalent*: $r = 2$, $n^1 = 1$, $n^2 = 2$, with splitting constants a^1 and a^2. The resonance fields are obtained from Equation 8. The appearance of the spectrum depends on the relative magnitude of a^1 and a^2 as illustrated in Figure 4.

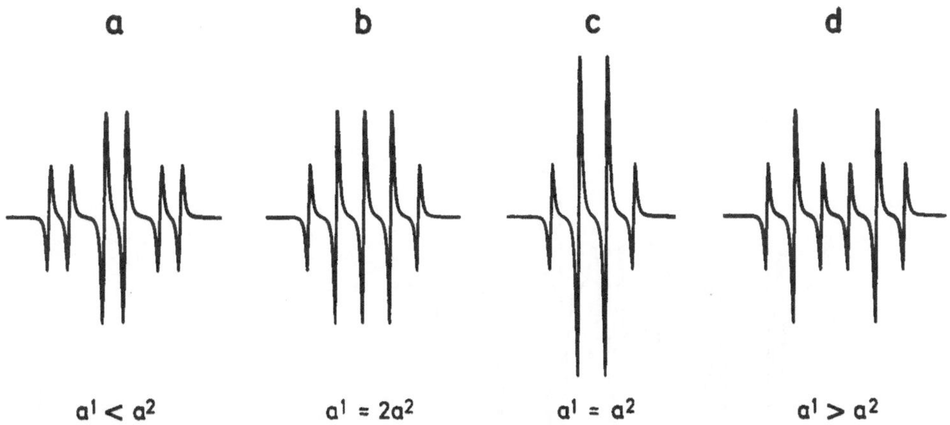

FIGURE 4.

The number of lines derived from Equation 7 is six (four if $a^1 = a^2$). For $a^1 = 2a^2$ only five lines appear due to *accidental* degeneracy. The spectrum (Figure 4c) is characteristic of three equivalent protons, e.g., from a methyl group.

Determination of Splitting Constants from Experimental Spectra

The analysis of reasonably well-resolved EPR spectra from the type of radicals considered in this handbook is illustrated through two examples, for simplicity chosen so that all hyperfine splittings arise from protons ($I = {}^1/_2$). A few simple rules are often helpful in the analysis. Assume that the spectrum to be analyzed consists of N lines which are numbered consecutively from left to right, and that the resonance field and height of line k are H(k) and h(k), respectively. Then the rules may be expressed:

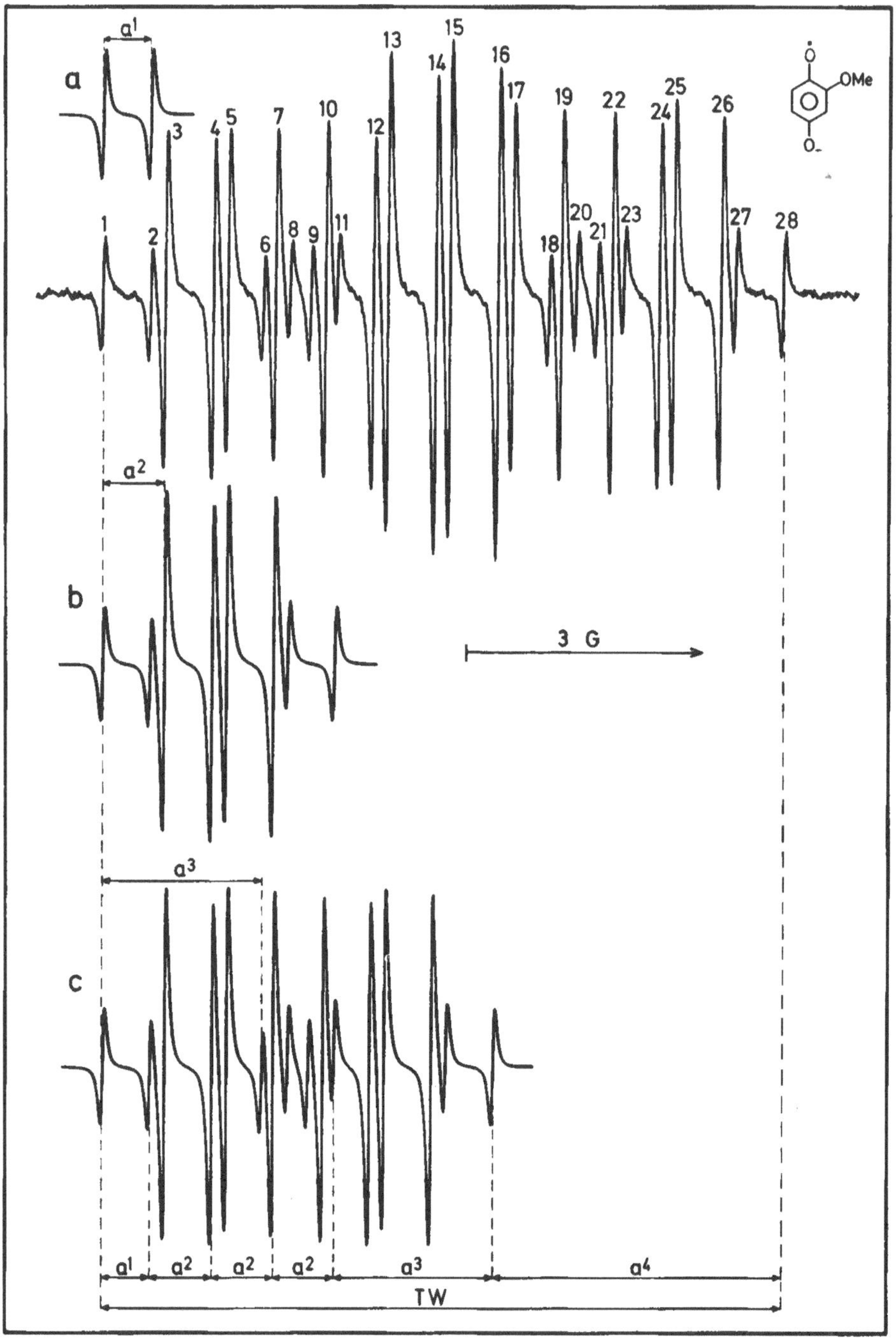

FIGURE 5.

Radical:	2-Methoxy-1,4-benzosemiquinone (no. 447)
Solvent:	w/EtOH
Method:	Oxidation in an alkaline medium
a_H/Gauss:	0.602(a^1), 0.788(a^2), 2.029(a^3), 3.601(a^4)
Lw/Gauss:	0.054

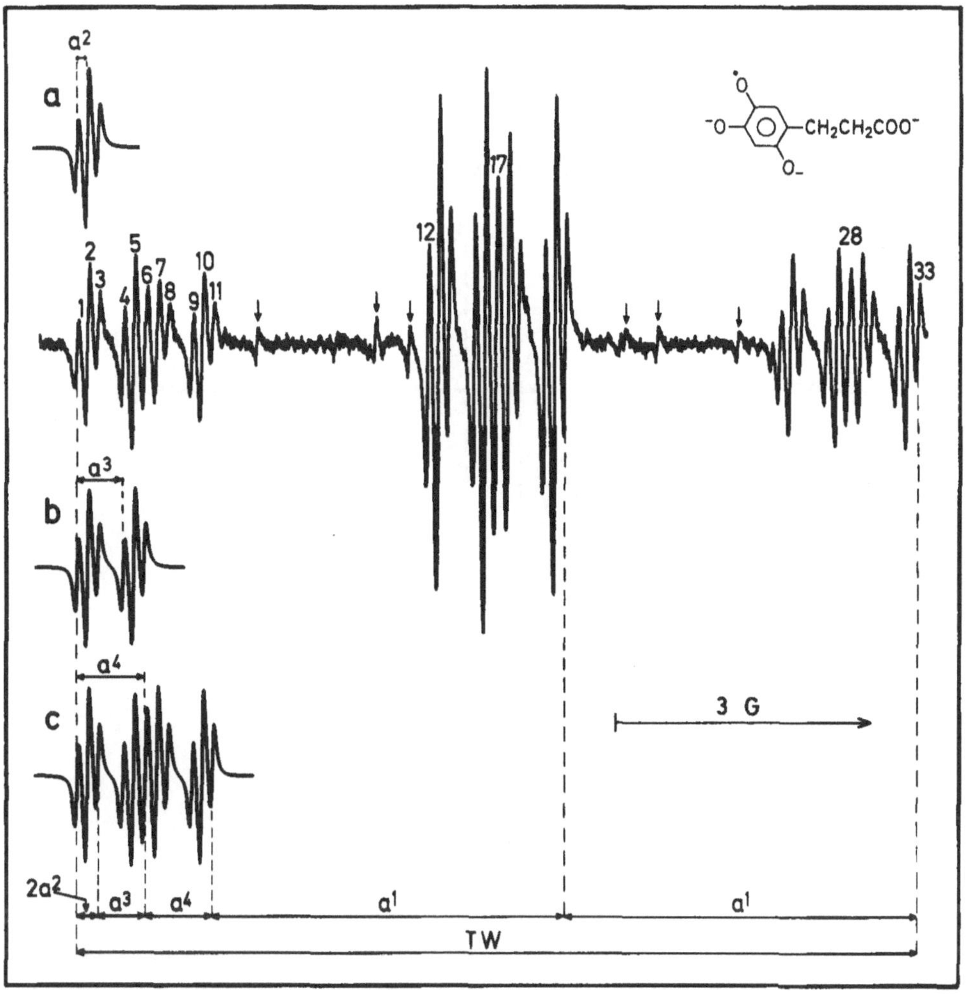

FIGURE 6.

Radical:	5-(2-Carboxylatoethyl)-4-oxido-1,2-benzosemiquinone (no. 402)
Solvent:	w/EtOH
Method:	Oxidation in alkaline medium
a_H/Gauss:	0.123(a^2), 0.538(a^3), 0.810(a^4), 4.116(a^1)
Lw/Gauss:	0.045

Again six protons are involved since the value of D is closer to 2^6 than to either 2^5 or 2^7. The spectrum consists of three 11-line subspectra with the same hyperfine structure and with relative intensities close to the ratios 1:2:1. This indicates a large splitting from two equivalent protons, the splitting constant a^1 being immediately obtainable as H(12) − H(1). By Rule 1 a^2 is found as H(2) − H(1) and, since h(2) ≈ 2h(1), two equivalent protons are associated with a^2 (Figure 6a). This accounts for the lines 1, 2, and 3 so that a^3 is equal to H(4) − H(1). As shown in Figure 6b, a^3 stems from a single proton. By Rule 3, a^4 is finally determined as $a^4 = TW - (2a^1 + 2a^2 + a^3)$. A reconstruction of the spectrum from Equation 8 shows that the lines 6, 17, and 28 are composite lines formed by accidental overlap of two components of equal intensity.

In order to check the analysis, a reconstruction of the spectrum is usually made. For simple, well-resolved spectra like the foregoing examples, a stick-plot diagram calculated from Equation 8 suffices. However, when the number of lines increases and overlap between lines becomes frequent, a computer simulation taking the lineshape into account is required.

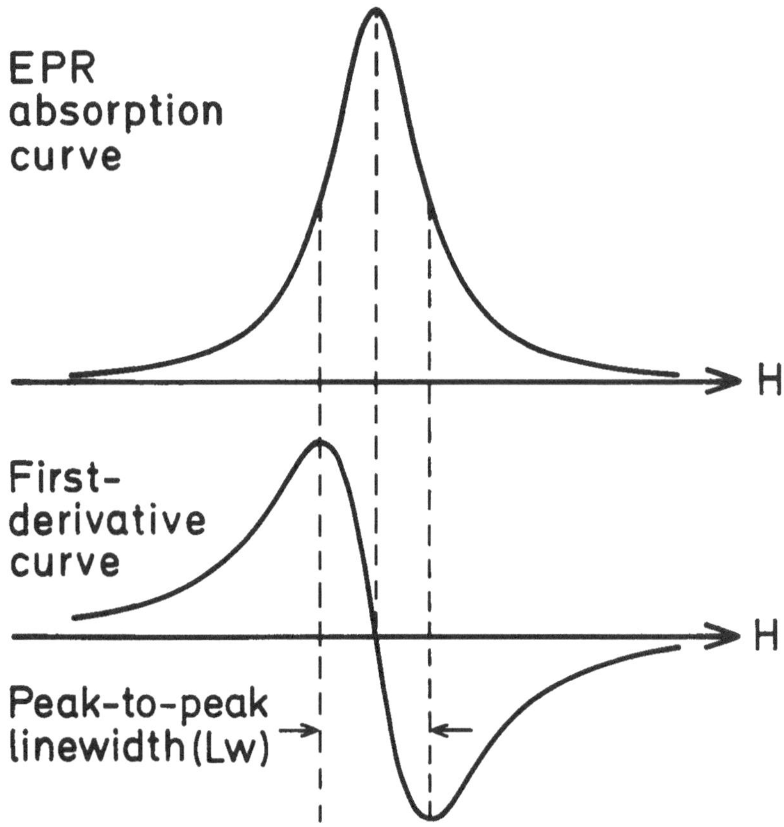

FIGURE 7.

The first derivative of a Lorentzian curve (Figure 7) closely matches the shape of EPR lines from radicals in diluted liquid solutions ($<10^{-4}M$), as illustrated in Figures 17 and 18b. As shown in Figure 18a a stick-plot diagram may be insufficient to confirm the analysis of the spectrum, since a detailed comparison with the experimental curve is not feasible. The derivative signal in Figure 7 is increasing and the one in Figure 17 decreasing on the left-hand side. This phase difference is totally irrelevant for the spectral analysis, and recorded spectra of either phase occur throughout in the literature and in this handbook.

If the EPR spectrum is recorded on paper, the set of splitting constants and the linewidth are adjusted by trial and error until the calculated spectrum closely matches the experimental spectrum (cf. Figure 61). If, on the other hand, the EPR spectrometer is interfaced to a computer, the analysis of the digitally recorded spectrum may be refined by an iterative least-squares parameter adjustment. The experimental spectra shown in this handbook have been analyzed digitally.

Assignment of Splitting Constants

The assignment of splitting constants to nuclei at specific sites in the radical requires some care. Splittings due to nuclei in substituents may often be recognized immediately, as illustrated by Examples 1 and 2 above. Thus in Example 1, the splitting constant a^2 associated with three equivalent protons evidently belongs to the methoxyl group. Similarly, in Example 2 the constants a^1 and a^2 belong to two pairs of equivalent methylene protons. From the general experience that the unpaired electron density decreases with distance from the radical nucleus, the large constant a^1 is assigned to the α-methylene protons and a^2 to the β-methylene protons.

Scheme 2
SUBSTITUTED 4-OXIDO-1,2-NAPHTHOSEMIQUINONES (NOS. 590 TO 593 AND 602 TO 604) WITH SPLITTING CONSTANTS STATED/GAUSS

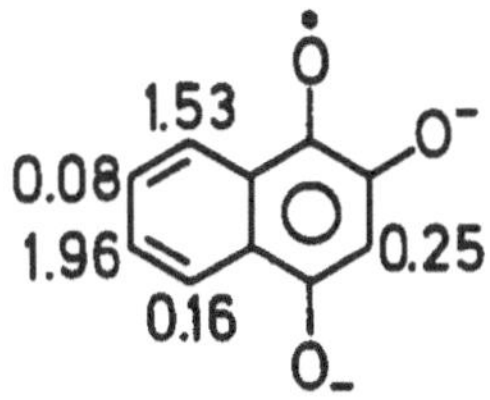

No. 590

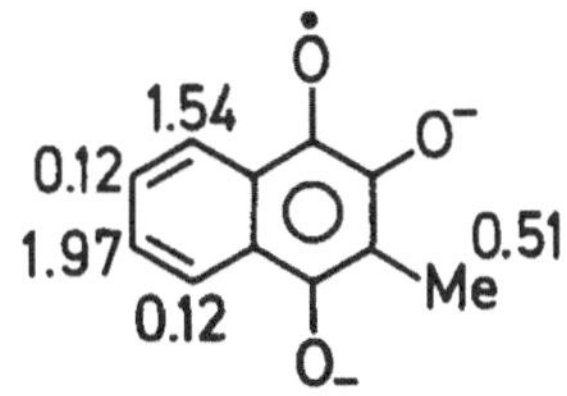

No. 591

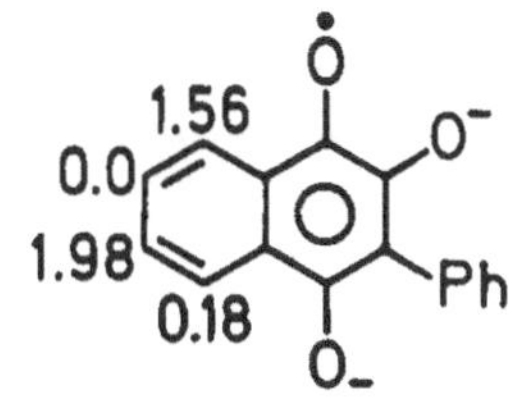

No. 593

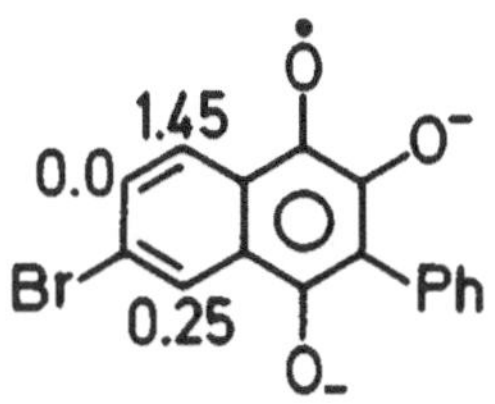

No. 602

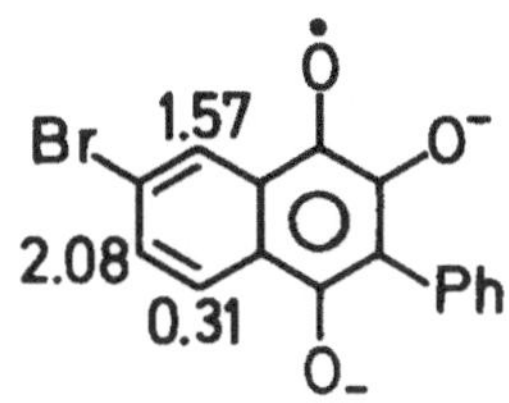

No. 603

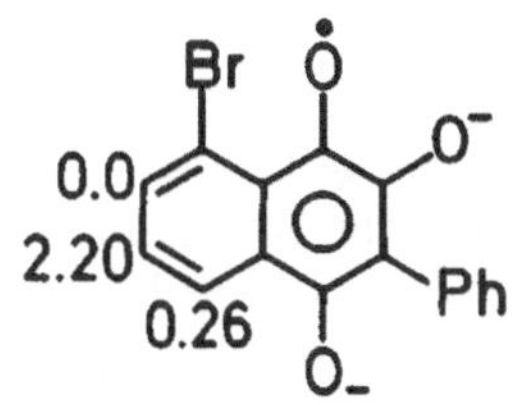

No. 604

Data from Hewgill and Mullings, Reference 105.

Assignment of splitting constants to the aromatic protons of the radical nucleus frequently requires comparison of data for a family of radicals obtained from a basic structure through substitution. The following discussion applies to the aromatic protons only. The most direct assignment results from site-specific substitution with deuterium, which affects only the splitting constant a_H belonging to the substituted atom. Taking the spin and magnetic moment of the deuteron into account, a_H may be expressed

$$a_H = 1.44\,(TW_H - TW_D) \qquad (13)$$

where TW_H and TW_D are the total widths of the fully protonated and monodeuterated radicals, respectively. This procedure has been widely used to identify splitting constants associated with labile protons (such as hydroxyl protons) where TW_D may be obtained simply by preparing the radical sample in a deuterated solvent.

All substituents other than deuterium change the distribution of the unpaired electron in the radical, and the splitting constants of the aromatic protons change accordingly. For certain substituents, such as alkyl groups or halogen atoms, the changes are often quite small, so that the splitting constants of the aromatic protons may still be recognized. Thus the assignments for a family of substituted radicals may sometimes be made without taking recourse to deuterium substitution.

The assignment of the five splitting constants of 4-oxido-1,2-naphthosemiquinone (No. 590, Scheme 2) illustrates the procedure. Consider first the derived radicals Nos. 593, 602, 603, and 604, where the 3-position has been blocked by a phenyl group from which no hyperfine splitting is observed. In each of the bromosubstituted species, the three remaining proton splittings are practically unaffected, allowing an unambiguous assignment at the positions 5 to 8. Comparison of the splittings of the radicals Nos. 591 and 593 indicates that the phenyl group has no specific influence on the splitting constants. Hence the assignment made for radical No. 593 applies to the unsubstituted radical No. 590 as well, suggesting in turn that the splitting from the proton at position 3 has the value 0.25 G.

The changes of the splitting constants of the aromatic protons induced by neighboring substituents may be used to assist the assignments, especially in monosubstituted radicals. The basic assumption is that the effects of substituents on the splitting constants of the protons are *additive*.

If a_i^o is the splitting constant of the proton at position i in the unsubstituted radical and a_i^j is the value of that constant observed when a given substituent, R, is introduced at position j, the *additivity principle* may be expressed

$$a_i^{j,k} = a_i^j + a_i^k - a_i^o \qquad (14)$$

where $a_i^{j,k}$, is the value of the constant belonging to position i, measured when the substituents R_1 and R_2 are introduced at position j and k, respectively. The width of the spectrum of the unsubstituted radical is

$$T(o) = \sum_{\ell=1}^{n} a_\ell^o$$

Similarly, for the radicals substituted at position j, and at the positions j and k, the widths may be expressed

$$T(j) = \sum_{\ell \neq j} a_\ell^j \text{ and } T(j,k) = \sum_{\ell \neq j,k} a_\ell^{j,k}$$

where any contribution from the substituents has been subtracted. With these expressions, Equation 14 leads to

$$\sum_{i \neq j,k} a_i^{j,k} = T(j,k) = (T(j) - a_k^j) + (T(k) - a_j^k) - (T(o) - a_j^o - a_k^o) \qquad (15)$$

If only a single set of equivalent positions in the radical is considered, and R_1 is taken equal to R_2, it follows that

$$T(j) = T(k),\ a_k^j = a_j^k,\text{ and } a_j^o = a_k^o$$

Hence Equation 15 may be written

$$a_k^j = a^o + T(j) - {}^1/_2(T(o) + T(j,k)) \qquad (16)$$

Thus the splitting constant at position k may be predicted from three appropriate spectral widths, T, and the splitting constant of the equivalent protons in the unsubstituted radical. The following chart illustrates the assignment of the three splittings from the aromatic protons in monosubstituted 1,4-benzosemiquinones by means of Equation 16.

It should be emphasized that the additivity principle only applies to set of spectra obtained under strictly similar conditions with respect to solvent system, temperature, and method of radical generation. Moreover, any instrument drift during the experiment may lead to incorrect and highly misleading T-parameters. Violation of the additivity principle may often be attributed to conformational differences between the radicals in the set.

The g-Factors

Determination of g-factors (Equations 4 and 5) involves measurement of H_o and the microwave frequency ν_o. For a group of related radicals such as the semiquinones the variation

Chart 1
ASSIGNMENT OF OBSERVED SPLITTING CONSTANTS (GAUSS) OF SUBSTITUTED 1,4-BENZOSEMIQUINONE ANIONS BY ADDITIVITY RULES

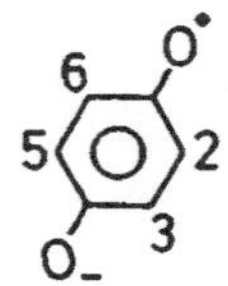

Substituents	a^o	a^2_3	a^2_5	a^2_6	T(2)	T(2,3)	T(2,5)	T(2,6)	Ref.
Me	2.33[a]	1.70 (1.75)[b]	2.55 (2.59)	2.37 (2.42)	6.62	5.08	3.40	3.74	1
t–Bu	2.35	1.55 —	2.07 (2.02)	2.82 (2.92)	6.44	[c] (5.08)	4.14	2.34	11
CH_2COO^-	2.39	1.87	2.40 (2.43)	2.56	6.83	[c]	4.02	[c]	8, 78
COO^-	2.36	2.20	2.01	2.57 (2.56)	6.78	[c]	[c]	3.72	18
COOEt	2.39	3.57	1.40 (1.53)	2.45	7.42	[c]	7.00	[c]	8, 78
OMe	2.41	0.93 (1.52)	3.58 (3.63)	2.28 (2.55)	6.79	5.72	1.50	3.66	78, 95
S–CH_2–Ph	2.36	1.28 —	2.77 (2.88)	2.10 (2.02)	6.15	[c] (2.51)	1.82	3.54	74
Ph	2.37	1.99	2.12 (2.10)	2.56	6.67	[c]	4.40	[c]	48
Ph–*p*–Cl	2.37	2.10	2.54 (2.59)	2.10	6.74	[c]	3.56	[c]	48
F	2.40	1.85	3.15 (3.14)	1.95	6.95	[c]	2.42	[c]	65
$SiMe_3$	2.40	2.68	2.10 (2.11)	2.52	7.30	[c]	5.58	[c]	73
Me (cations)	2.36	1.36 (1.77)	3.05 (3.20)	2.00 (2.00)	6.41	5.16	1.92	3.62	241

[a] Variations of a^o are due to different compositions of solvent.
[b] Values in parenthesis are calculated by means of Equation 16, simplified to $a^j_k = T(j) - a^o - {}^1/_2T(j,k)$.
[c] Experimental value not available.

of the values of H_0 at constant ν_0 is of the order 10^{-4}. Hence the g-factors are most conveniently determined by comparing the positions of the spectra on the magnetic field scale with that of a standard radical with known g-factor, thus bypassing the absolute determination of H_0 and ν_0. The spectrum of the radical U, for which the g-factor is to be determined, is run simultaneously with that of the standard radical S. Denoting the resonant fields H_U and H_S, the g-factor of U is related to the one of the standard g_S by the equation

$$g_U = g_S \left(1 - \frac{\Delta H}{H_S + \Delta H}\right) \tag{17}$$

where $\Delta H = H_U - H_S$.

A convenient standard for the determination of g-factors of semiquinones is tetrahydroxy-1,4-benzoquinone, which easily converts to a radical (No. 520) with a single narrow line. The recommended experimental conditions and the radical characteristics for the g-factor standard (No. 520)* are

1. Solvent: Alkaline water, 10^{-1} *M* NaOH
2. Concentration: 10^{-3} *M* in above solvent
3. Radical formation: Flushing the solution with atmospheric oxygen
4. Spectrum: One narrow line; peak-to-peak linewidth 30 mG (minimum observed) Line surrounded by two ^{13}C satellites $a(^{13}C) = 0.66$ G (see Figure 21)
5. g-Factor: 2.00487 at 25°C
6. Stability: Several hours at best; radical decays continuously

* For a list of other standards, see Wertz and Bolton, loc. cit. EPR Monographs.

AN ANALYTICAL APPLICATION

Many quinones and quinols occurring in plants may be detected and identified by EPR spectroscopy in crude alcoholic extracts of the plant material, since these compounds can be simply converted to the relatively stable semiquinone radicals in alkaline media by an appropriate redox process. Most other radicals that might arise are too short lived to be detected, so that no chemical purification of the extracts is required. As an example, the results of a semiquantitative determination of juglone, 2-methyljuglone, and hydroquinone in crude plant extracts are shown in Scheme 3. For details of this and other applications, see References 15, 16, and 456.

Scheme 3
NATURALLY OCCURRING QUINONES/QUINOLS DETECTED AND IDENTIFIED BY EPR

a) Juglone (observed as 5-hydroxy-1,4-naphthosemiquinone, no. 613)[a]

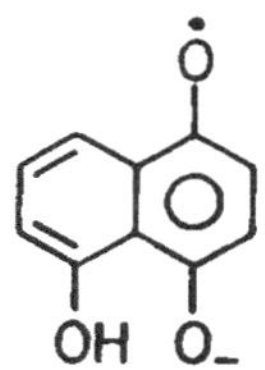

Observed in:
- *Juglandaceae* (leaves)
  - ***Carya alba*** **(0.14)**[b]
  - *C. cordiformis* **(blank)**
  - *C. glabra* (blank)
  - *C. laciniosa* (0.10)
  - *C. ovalis* **(blank)**
  - *C. ovata* (blank)
  - ***Juglans cathayensis*** (0.31)
  - *J. cinerea* (0.31)
  - *J. draconis* (0.44)
  - ***J. hindsii*** **(0.09)**
  - *J. mandshurica* (0.28)
  - *J. nigra* (0.42)
  - ***J. regia*** **(0.44)**
  - ***J. rupestris*** (0.29)
  - ***J. sieboldiana*** (0.57)
  - ***J. stenocarpa*** **(0.28)**
  - *Pterocarya fraxinifolia* (0.59)
  - ***P. hubehensis*** **(0.38)**
  - ***P. rhoifolia*** **(0.33)**
  - *P. stenoptera* (0.37)

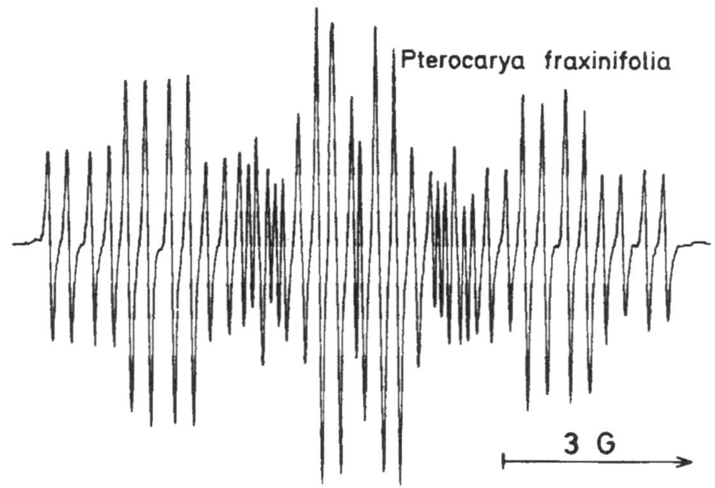

FIGURE 8. Spectrum observed from leaf extract of *Pterocarya fraxinifolia*. For experimental details, see no. 613, Table I.15.2 (Reference 16).

b) 2-Methyljuglone (observed as 5-hydroxy-2-methyl-1,4-naphthosemiquinone, no. 614)[a]

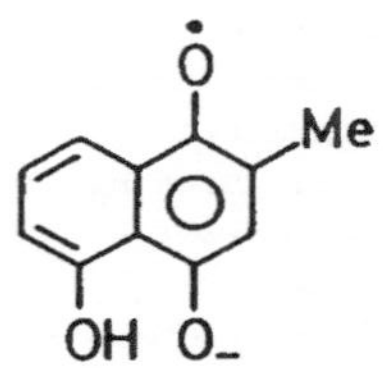

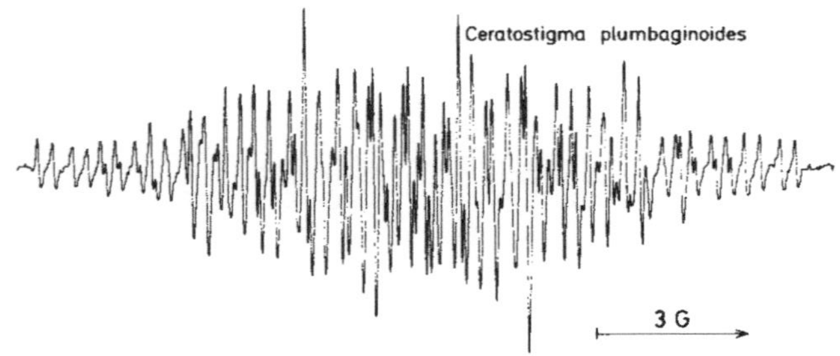

FIGURE 9. Spectrum observed from root extract of *Cerotostigma plumbaginoides*. For experimental details, see no. 614, Table I.15.2 (Reference 16).

Observed in:
- *Droceraceae* (leaves)
  - *Drocera binata*
- *Plumbaginaceae* (leaves, roots)
  - *Ceratostigma plumbaginoides*

c) Hydroquinone (observed as 1,4-benzosemiquinone, no. 107)[a]

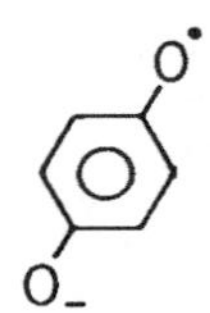

Observed in:
- *Asteraceae* (leaves)
  - *Senecio elegans*
- *Ericaceae*
  - *Arctostaphylos alpina*
  - *Vaccinum vitis-idaea*
- *Lamiaceae* (leaves)
  - *Dracocephalum ruyschiana*
  - *Origanum vulgare*

Scheme 3 (continued)
NATURALLY OCCURRING QUINONES/QUINOLS DETECTED AND IDENTIFIED BY EPR

Lycopodiaceae (leaves, dried)
- *Lycopodium cumingii*
- *L. hastatum*
- ***L. hippurideum***
- *L. sp. aff. crassum*
- *L. sp. aff. rufescens*
- ***L. tetragonum***

Rosaceae (leaves)
- *Pyrus amygdaliformis* (0.31)[b]
- *P. amygdaliformis* var. *persica* (0.17)
- *P. betulaefolia* (0.18)
- *P. communis* (0.35)
- ***P. elaeagrifolia* (0.18)**
- *P. elaeagrifolia* var. *kotschyana* (0.22)
- *P. nivalis* (0.17)
- ***P. pashia* (<0.02)**
- ***P. salicifolia* (0.06)**
- *P. salicifolia* var. *pendula* (<0.02)
- *P. serrulata* (0.11)
- ***P. ussuriensis* (0.20)**
- ***P. ussuriensis* var. *hondoensis* (0.19)**
- ***P. communis* × *nivalis* (0.17)**
- *Sorbusaria* × *P. communis* (<0.02) (*sorbopyrus auricularis*)
- ***P. calleryana* var. *faurei* (blank)**
- ***Cyndonia oblonga* (blank)**

Rubiaceae (flowers)
- *Galium molluga*

***Rutaceae* (flowers)**
- ***Phellodendron amurense***
- *P. japonicum*
- *P. sachaliense*

Saxifragaceae (leaves)
Bergenia bifolia (0.10)
B. busiana (0.09)
B. ciliata (blank)
B. cordifolia (0.22, red leaves)
B. cordifolia (0.06, green leaves)
B. crassifolia (<0.02)
B. ligulata (blank)
B. subsiliata (0.18)

[a] All semiquinones observed in w/EtOH (alkaline).
[b] Figure shows milligrams of compound per grams of dry weight.

SEMIQUINONE SPECTRA

The following collection contains 50 high-resolution EPR spectra analyzed digitally with the simplifying assumption of constant linewidth. The experimental conditions used in generating radicals are given in abbreviated form (see **Explanation to Tables**). The instrumental settings (Instr. sett.) applied are stated in the following order:

1. Scan width (4 G, 10 G, etc.)/scan time (sec, min)
2. Filter time constant (sec)
3. Microwave power (mW)
4. Number of accumulated scans (acc)

The spectra were run on a Bruker® ER 200 EPR spectrometer or a Varian® E-15 EPR spectrometer with a modulation frequency of 12.5 and 10 kHz, respectively. The spectra were stored digitally as 4000 discrete signals (resolution, 2.5 mG per interval for a 10-G spectrum).

The spectra illustrate characteristic features for each example mentioned below (numbers in brackets are figure numbers):

1. Spectral changes from linewidth (Lw) variations (10)
2. Smallest a_H < Lw (40, 44, 47, 49, 50)
3. Smallest $a_H \simeq$ Lw (20, 25, 29, 34)
4. Line-broadening towards higher field (48, 56, 58, 59)
5. Line-broadening towards lower field (31)
6. Solvent dependence of splitting constants (60)
7. Simulations by computer (18, 38, 47—49)
8. Data subtractions by computer (12, 23)
9. Stick-plot diagram (18)
10. ^{13}C splitting analysis (11)
11. Deuterium substitution (46)
12. Dissociation of hydroxyl proton (27, 52—55)

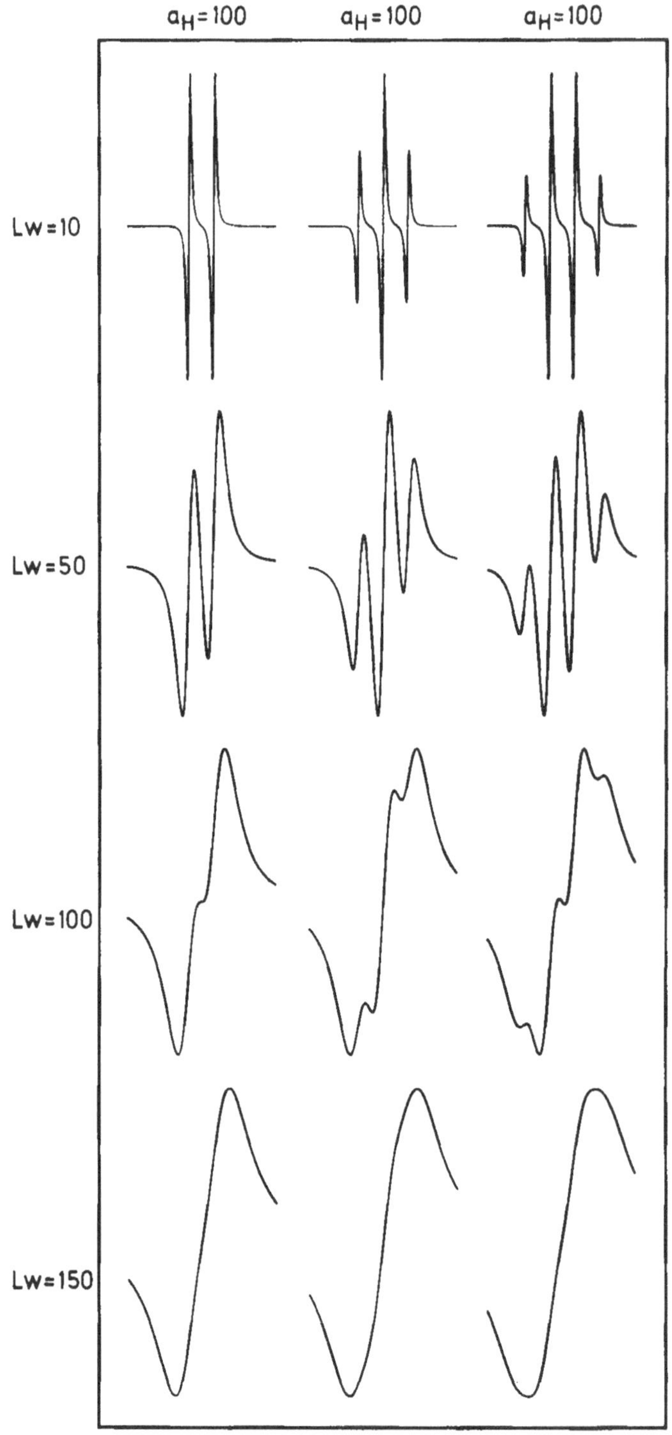

FIGURE 10. Simulations of doublet (1:1), triplet (1:2:1), and quartet (1:3:3:1) spectra with constant splitting (a_H) and linewidth (Lw) varied, or vice versa: a_H and Lw in arbitrary unit.

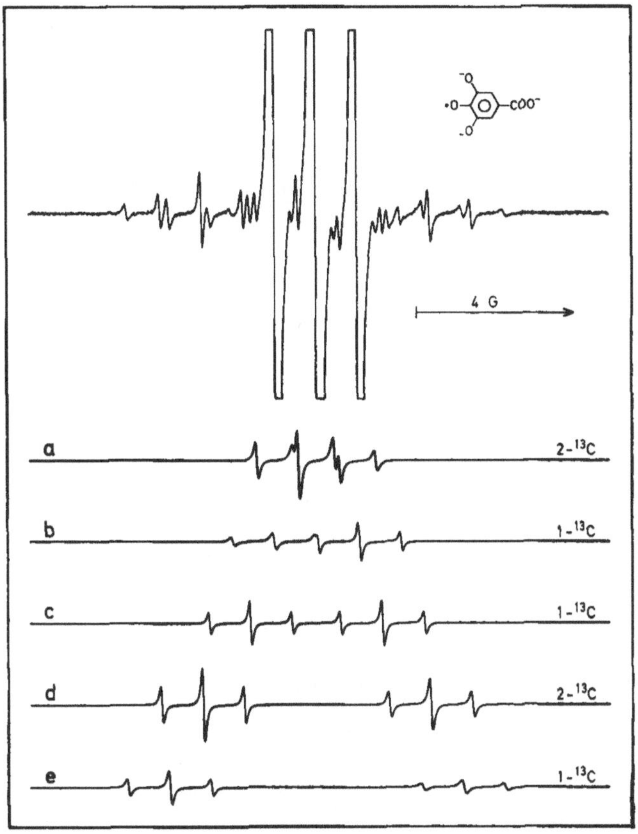

FIGURE 11.

Radical:	5-Carboxylatopyrogallolsemiquinone (no. 288)
Solvent:	w/EtOH
Method:	Ox alk
a_H/G:	1.088(2H), main triplet, lines out of scale
Lw/G:	0.190
Instr. sett.:	20 G/5 min, 0.03 sec, 0.1 mW

PARAMETERS USED IN SIMULATIONS

Curve	$a(^{13}C)$/G	Linewidth/G Left (1:2:1)	Linewidth/G Right (1:2:1)	Intensity (relative)
a	1.172	0.084	0.097	2.37
b	1.735	0.072	0.080	1.07
c	2.259	0.108	0.076	0.98
d	5.999	0.075	0.089	2.16
e	7.733	0.087	0.123	1.10

Note: Top curve — Experimental spectrum consisting of signals from the radical containing no ^{13}C nuclei (main triplet [2H, 1:2:1] at center) and from seven radicals each containing one ^{13}C nucleus at a different position (satellite lines in pairs of 1:2:1 triplets). Since four of the seven carbon atoms are pairwise equivalent, only five distinct ^{13}C spectra are observed with relative intensities 1:2:1:1:2:1 (cf. a to e in scheme). Curves a to e are simulated spectra of the radicals containing one ^{13}C nucleus. Note that the linewidths of the left- and right-hand triplets are different.

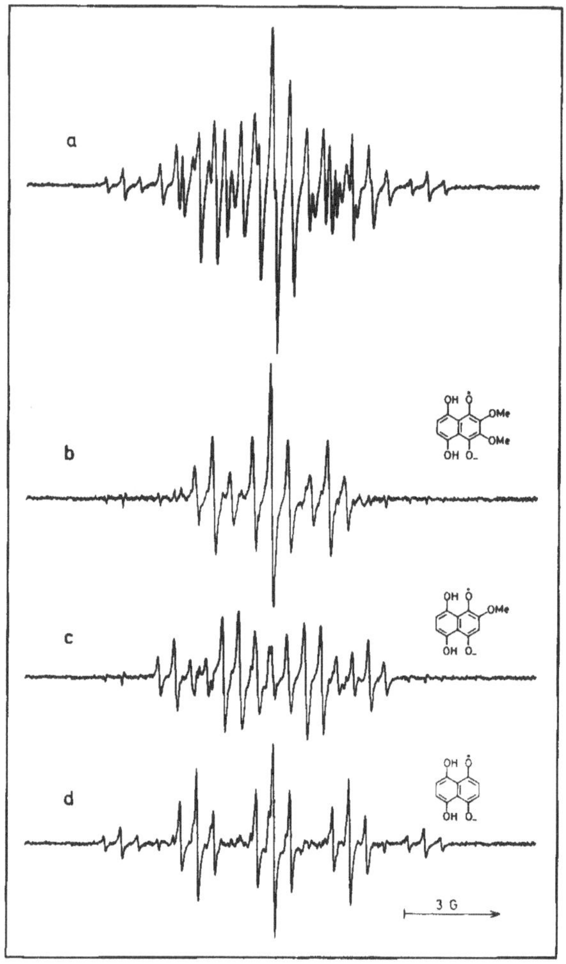

FIGURE 12.

Radicals a: Mixture of three naphthazarin radicals (nos. 637, 649, and 655)
Solvent: DMF
Method: Re el
Instr. sett.: 20 G/2 min, 0.03 sec, 0.1 mW

Radical b: 2,3-Dimethoxynaphthazarin semiquinone (No. 655)
a_H/G: 0.548(2H), 1.806(2H)
Offset/G: −0.104 (relative to no. 637)
Lw/G: 0.104
Spectrum obtained by data subtraction: simulations of nos. 637 and 649 subtracted from spectra of Figure 12a

Radical c: 2-Methoxynaphthazarin semiquinone (no. 649)
a_H/G: 0.489(1H), 0.522(1H), 1.513(1H), 2.017(1H), 2.585(1H)
Offset/G: −0.037 (relative to no. 637)
Lw/G: 0.089
Spectrum obtained by data subtraction: simulations of nos. 637 and 655 subtracted from spectra of Figure 12a

Radical d: Naphthazarine semiquinone (no. 637)
a_H/G: 0.527(2H), 2.397(4H)
Lw/G: 0.075
Spectrum obtained by data subtraction: simulations of nos. 649 and 655 subtracted from spectra of Figure 12a

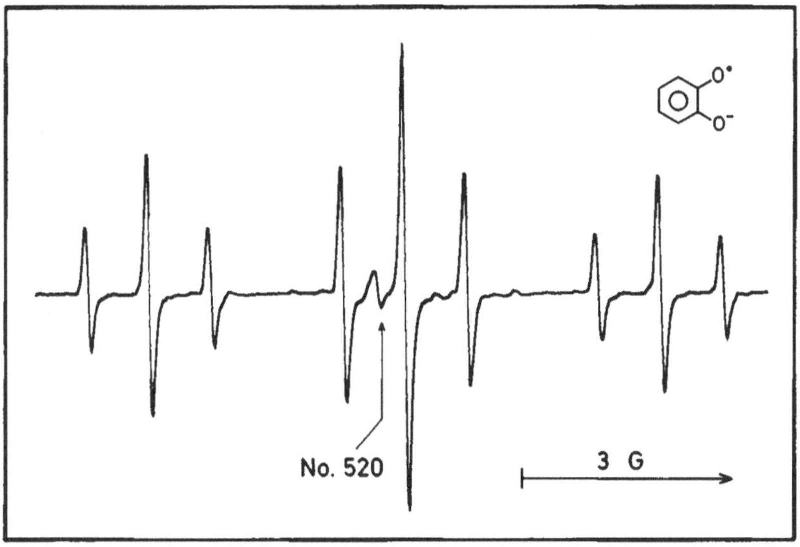

FIGURE 13.

Radical:	1,2-Benzosemiquinone (no. 1)
Solvent:	w/EtOH
Method:	Ox alk
a_H/G:	0.88(2H), 3.65(2H)
Lw/G:	~0.05
Instr. sett.:	20 G/4 min, 0.03 sec, 0.5 mW

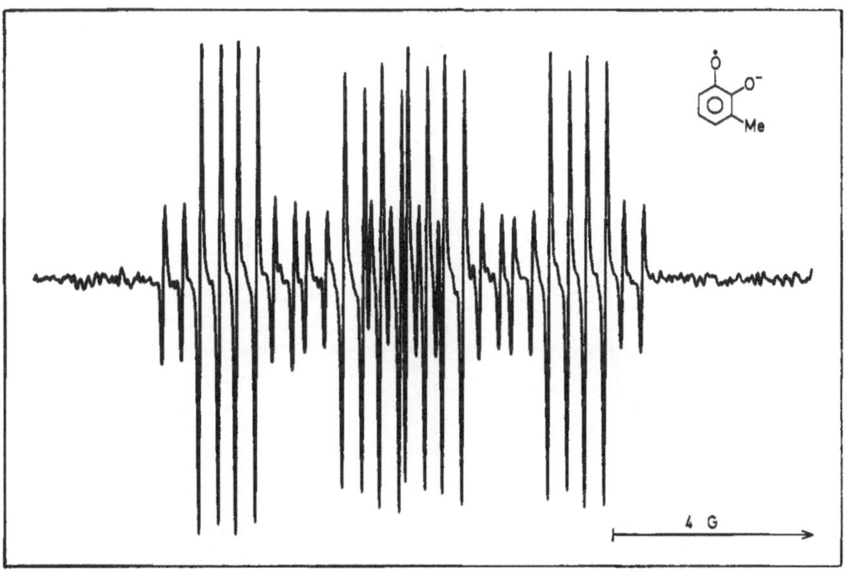

FIGURE 14.

Radical:	3-Methyl-1,2-benzosemiquinone (no. 2)
Solvent:	w/EtOH
Method:	Ox alk
a_H/G:	0.396(1H), 0.743(3H), 2.880(1H), 4.146(1H)
Lw/G:	0.045
Instr. sett.:	20 G/5 min, 0.1 sec, 0.1 mW

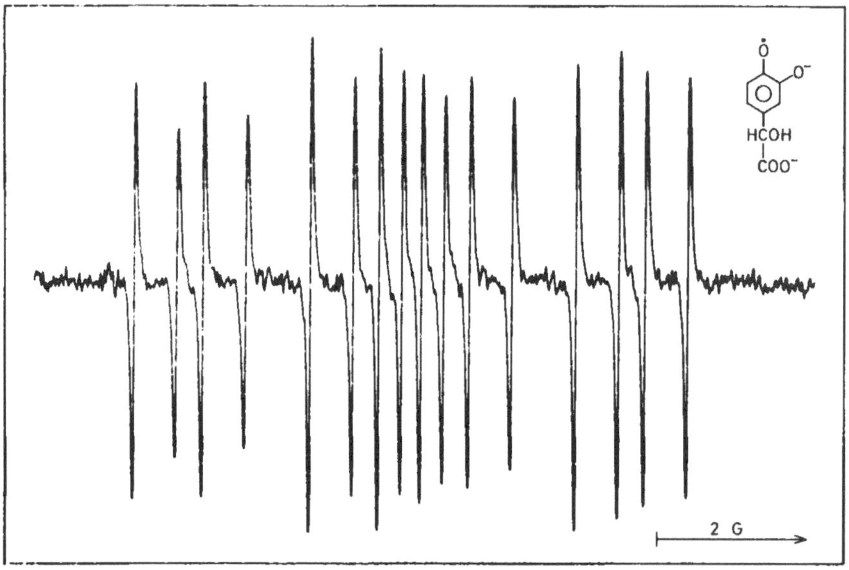

FIGURE 15.

Radical:	4-Hydroxycarboxylatomethyl-1,2-benzosemiquinone (no. 56)
Solvent:	w/EtOH
Method:	Ox alk (pH ~13.2)
a_H/G:	0.568(1H), 0.911(1H), 2.329(1H), 3.545(1H)
Lw/G:	0.043
Instr. sett.:	10 G/2 min, 0.03 sec, 0.05 mW, 8 acc

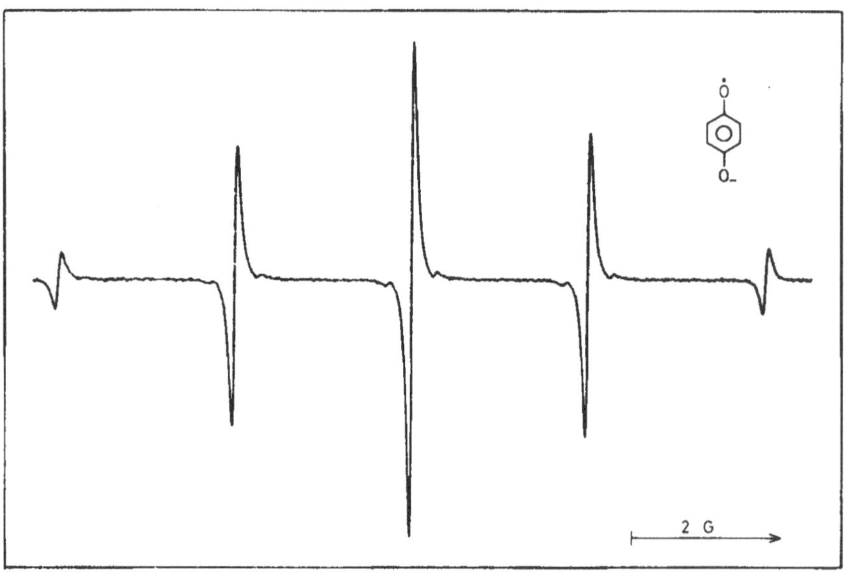

FIGURE 16.

Radical:	1,4-Benzosemiquinone (no. 107)
Solvent:	w/EtOH
Method:	Ox alk
a_H/G:	2.357(4H)
$a(^{13}C)$/G:	0.630 (positions 2, 3, 5, 6)
Lw/G:	0.074
Instr. sett.:	10 G/1 min, 0.03 sec

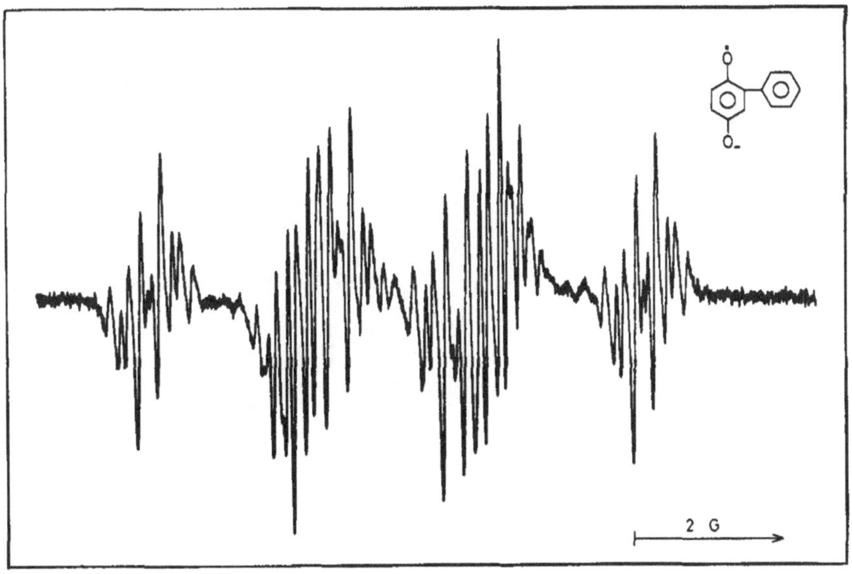

FIGURE 17.

Radical:	2-Phenyl-1,4-benzosemiquinone (no. 116)
Solvent:	w/EtOH
Method:	Ox alk
a_H/G:	0.164(2H), 0.263(3H), 1.971(1H), 2.077(1H), 2.539(1H)
Lw/G:	0.064
Instr. sett.:	10 G/2 min, 0.03 sec, 0.05 mW, 4 acc

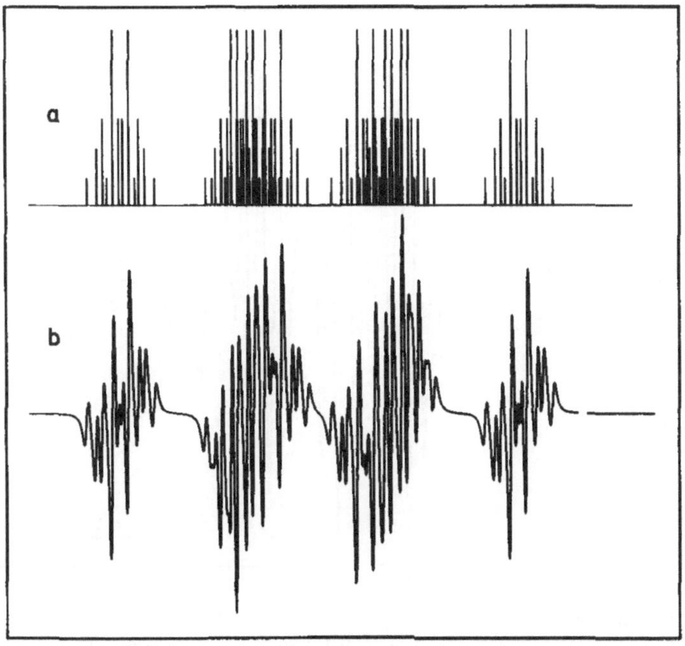

FIGURE 18. (a) Stick plot of No. 116; (b) computer simulation of no. 116.

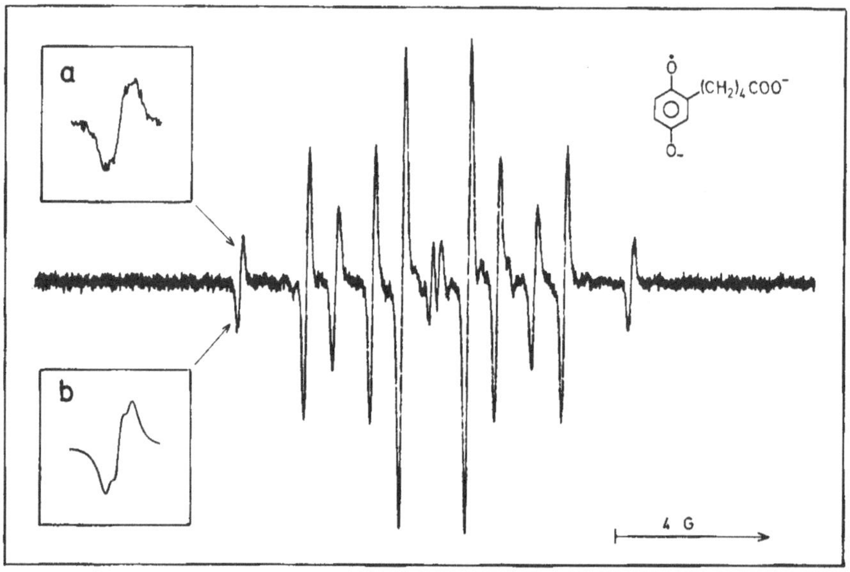

FIGURE 19.

Radical:	6-Isopropyl-2-methyl-1,4-benzosemiquinone (no. 168)
Solvent:	w/EtOH
Method:	Ox alk
a_H/G:	1.356(1H), 1.760(1H), 1.833(1H), 2.237(3H)
Lw/G:	0.146
Instr. sett.:	20 G/2 min, 0.01 sec, 0.05 mW, 8 acc

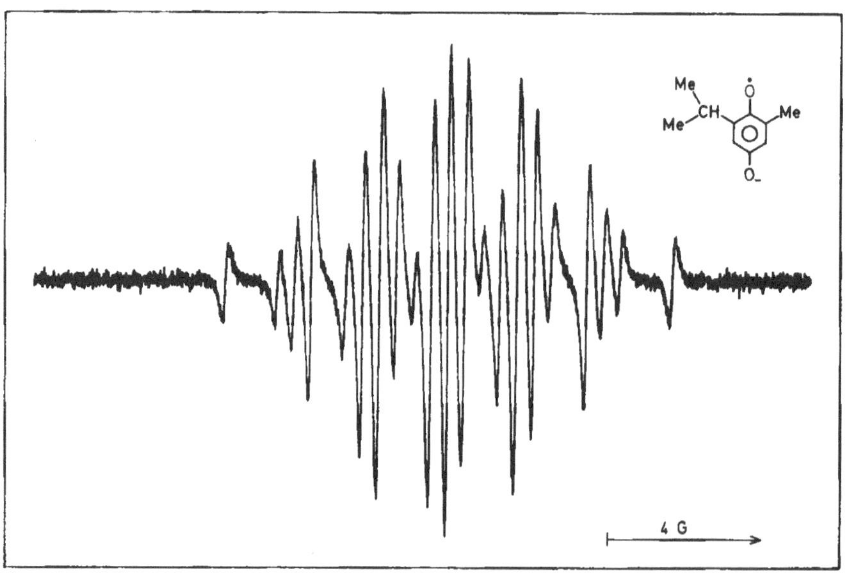

FIGURE 20.

Radical:	2-(4-Carboxylatobutyl)-1,4-benzosemiquinone (no. 178)
Solvent:	w/EtOH
Method:	Ox alk
a_H/G:	0.055(2H), 1.708(1H), 1.733(2H), 2.462(1H), 2.515(1H)
Lw/G:	0.059
Instr. sett.:	20 G/2 min, 0.01 sec, 0.04 mW, 4 acc
Inset:	(a) Line 1 enlarged; (b) line 1 simulated

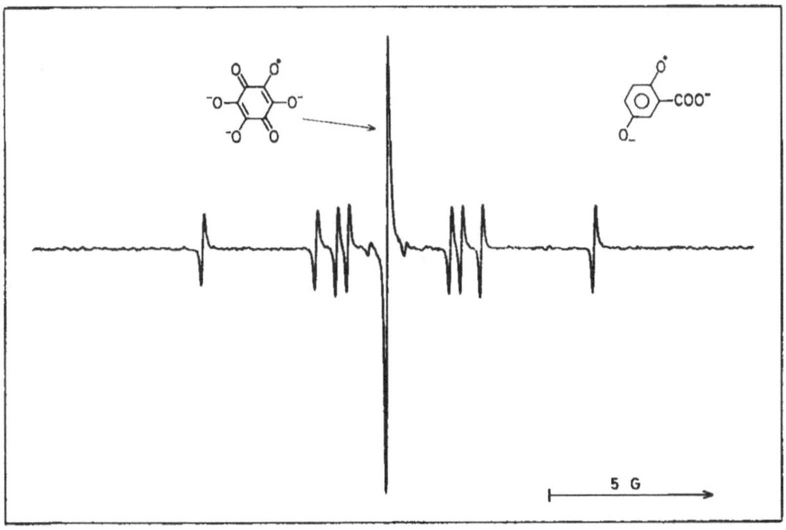

FIGURE 21.

Radical 1: 2-Carboxylato-1,4-benzosemiquinone (no. 192)
Solvent: w/EtOH
Method: Ox alk (pH ~13.2)
a_H/G: 2.061(1H), 2.430(1H), 2.630(1H)
Lw/G: 0.047

Radical 2: Tetraoxido-1,4-benzosemiquinone (no. 520; used as a g-factor standard)
Solvent: w
Method: Ox alk (pH ~13)
$a(^{13}C)$/G: 0.663
Lw/G: 0.051
Instr. sett.: 20 G/2 min, 0.1 sec, 0.1 mW

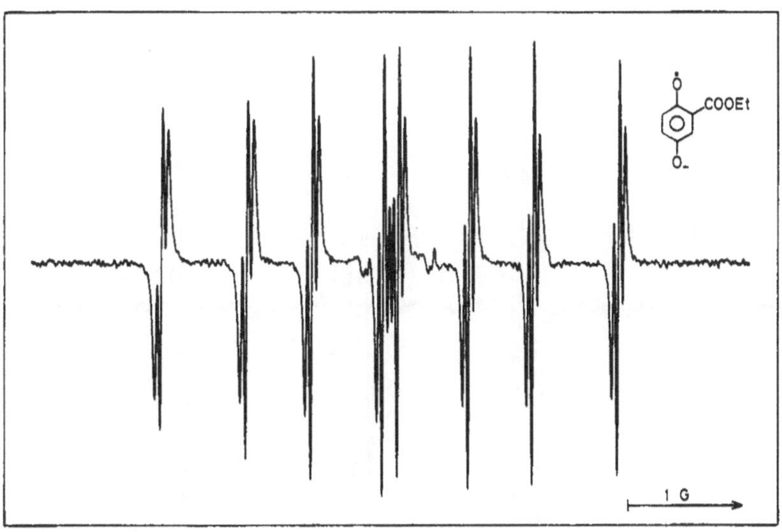

FIGURE 22.

Radical: 2-Ethoxycarbonyl-1,4-benzosemiquinone (no. 196)
Solvent: w/EtOH
Method: Ox alk
a_H/G: 0.092(2H), 1.445(1H), 2.537(1H), 3.727(1H)
Lw/G: 0.049
Instr. sett.: 10 G/2 min, 0.03 sec, 0.25 mW

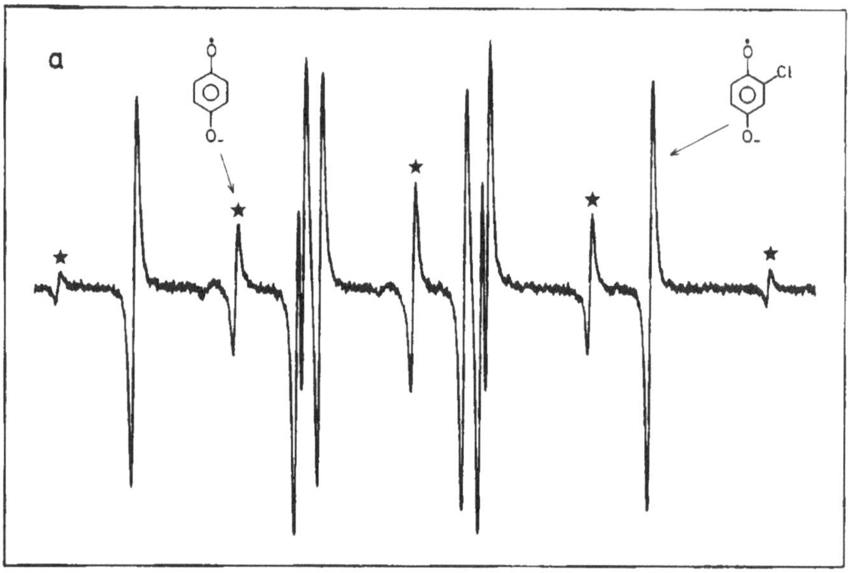

FIGURE 23a.

Radical 1:	2-Choloro-1,4-benzosemiquinone (no. 260)
a_H/G:	2.147(1H), 2.241(1H), 2.459(1H)
Lw/G:	0.059
Radical 2:	1,4-Benzosemiquinone (no. 107)
a_H/G:	2.359(4H)
Lw/G:	0.058
Solvent:	w/EtOH
Method:	Ox alk
Instr. sett.:	10 G/2 min, 0.1 sec, 0.06 mW, 4 acc

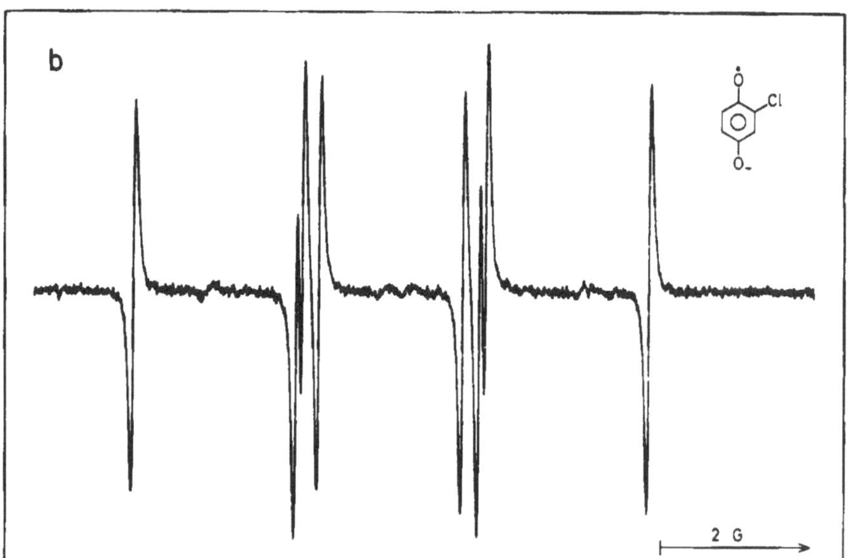

FIGURE 23b.
Data subtraction: simulation of no. 107 subtracted from spectra of Figure 23a

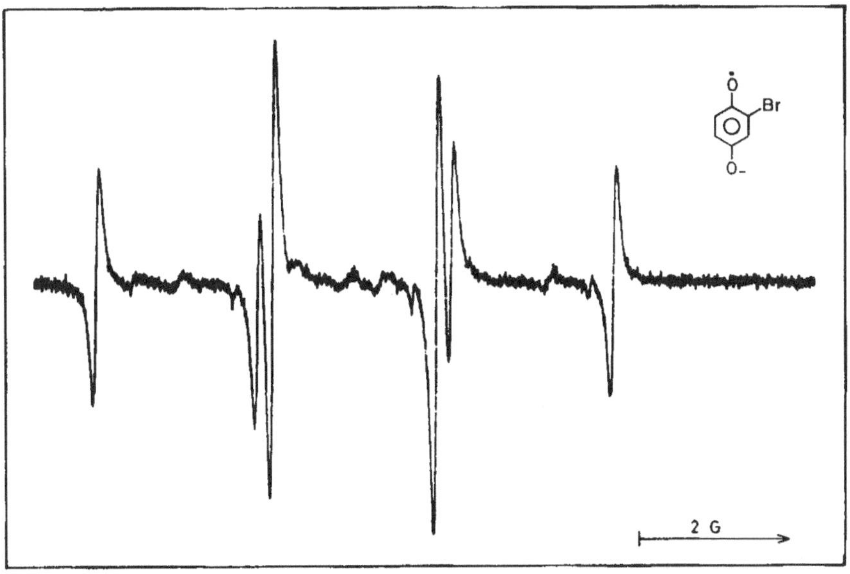

FIGURE 24.

Radical:	2-Bromo-1,4-benzosemiquinone (no. 261)
Solvent:	w/EtOH
Method:	Ox alk
a_H/G:	2.159(1H), 2.355(1H), 2.367(1H)
Lw/G:	0.086
Instr. sett.:	10 G/2 min, 0.1 sec, 0.06 mW, 4 acc

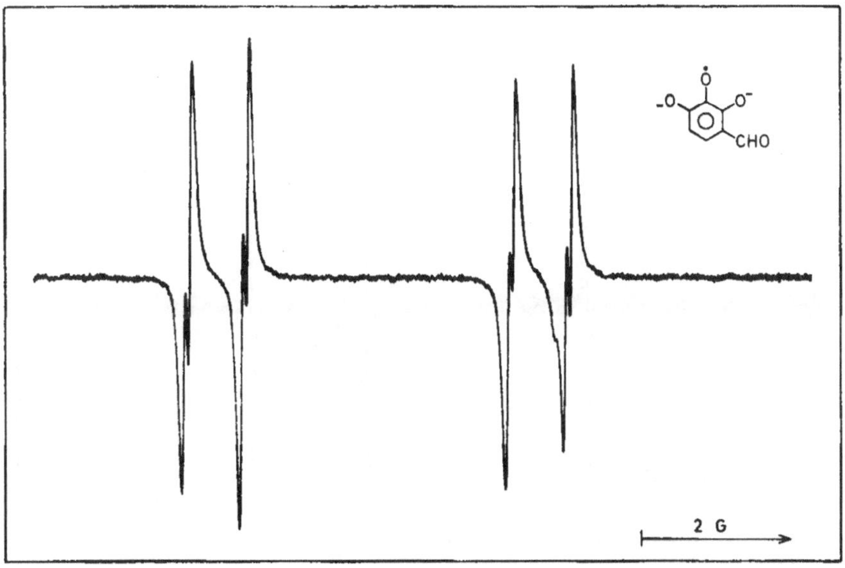

FIGURE 25.

Radical:	4-Formylpyrogallol semiquinone anion (no. 278)
Solvent:	w/EtOH
Method:	Ox alk (pH ~13)
a_H/G:	0.088(1H), 0.772(1H), 4.308(1H)
Lw/G:	0.071
Instr. sett.:	10 G/2 min, 0.03 sec, 0.06 mW, 2 acc

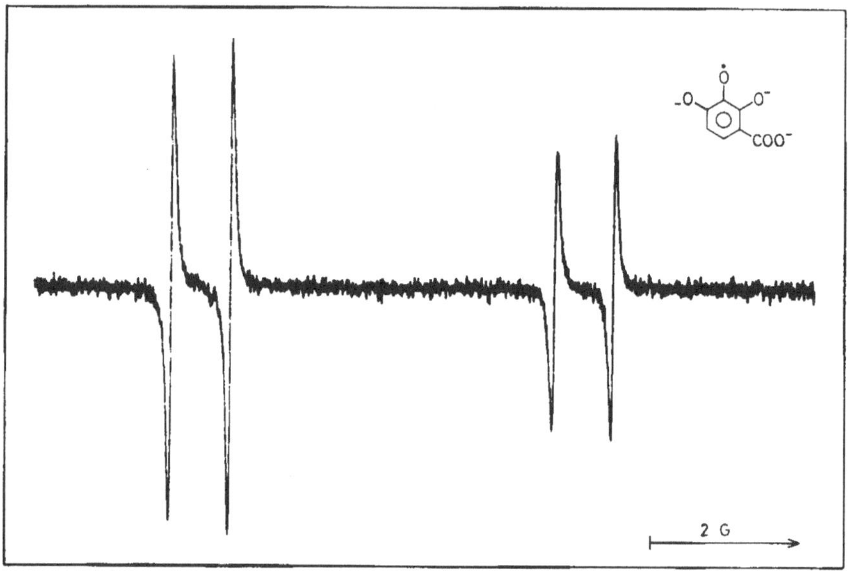

FIGURE 26.

Radical:	4-Carboxylatopyrogallol semiquinone anion (no. 283)
Solvent:	w/EtOH
Method:	Ox alk (pH ~13.7)
a_H/G:	0.781(1H), 5.095(1H)
Lw/G:	0.063
Instr. sett.:	10 G/2 min, 0.03 sec, 0.1 mW, 2 acc

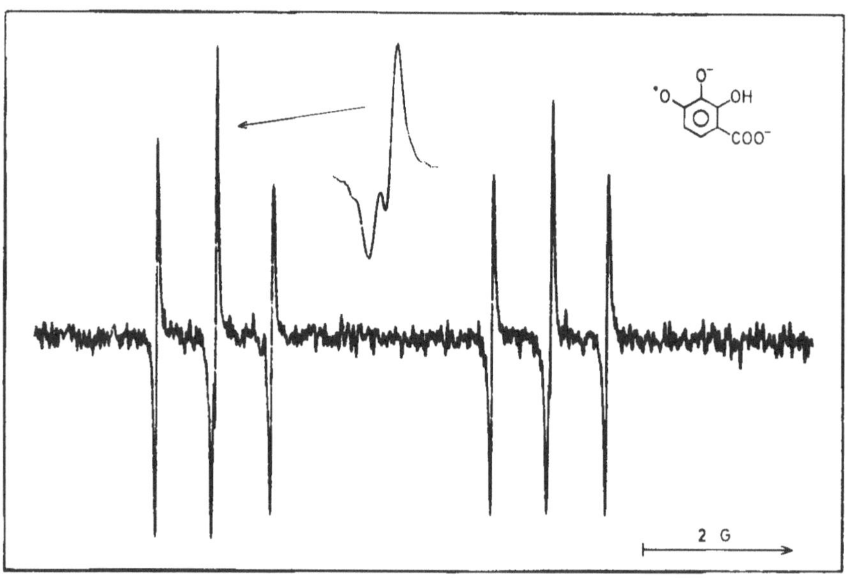

FIGURE 27.

Radical:	4-Carboxylatopyrogallol semiquinone (no. 302)
Solvent:	w/EtOH
Method:	Ox alk (pH ~13)
a_H/G:	0.740(1H), 0.782(1H), 4.472(1H)
Lw/G:	0.036
Instr. sett.:	10 G/2 min, 0.03 sec, 0.1 mW

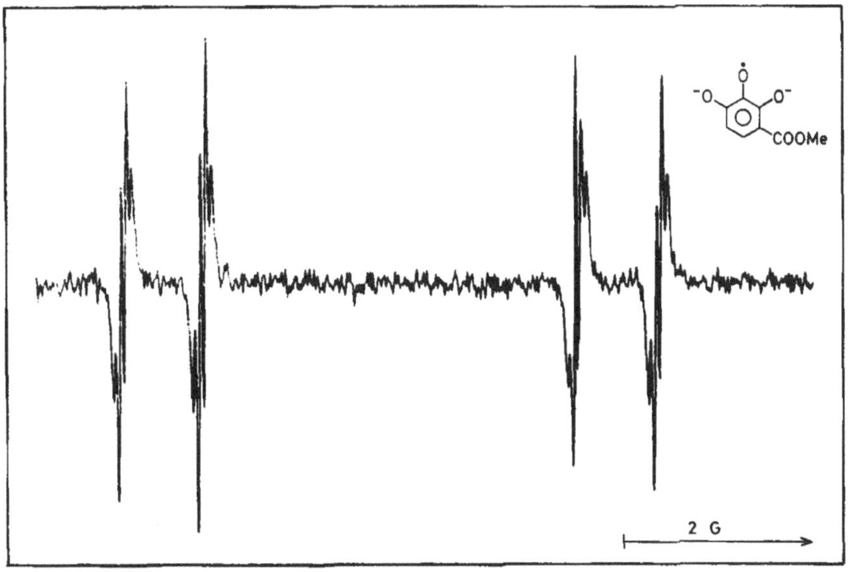

FIGURE 28.

Radical:	4-Methoxycarbonylpyrogallol semiquinone anion (no. 284)
Solvent:	w/EtOH
Method:	Ox alk (pH ~13.2)
a_H/G:	0.058(3H), 0.859(1H), 4.860(1H)
Lw/G:	0.036
Instr. sett.:	10 G/2 min, 0.03 sec, 0.04 mW, 4 acc

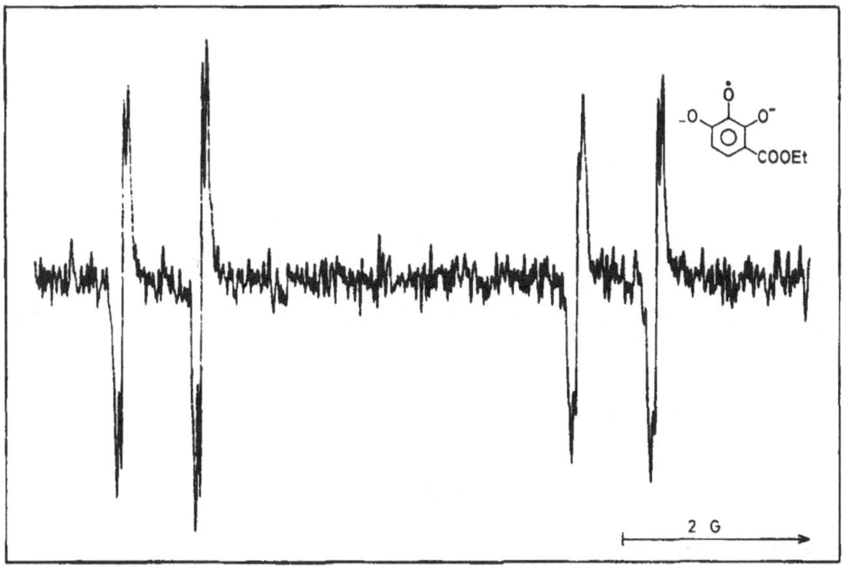

FIGURE 29.

Radical:	4-Ethoxycarbonylpyrogallol semiquinone anion (no. 285)
Solvent:	w/EtOH
Method:	Ox alk
a_H/G:	0.048(2H), 0.847(1H), 4.865(1H)
Lw/G:	0.040
Instr. sett.:	10 G/2 min, 0.03 sec, 0.04 mW, 7 acc

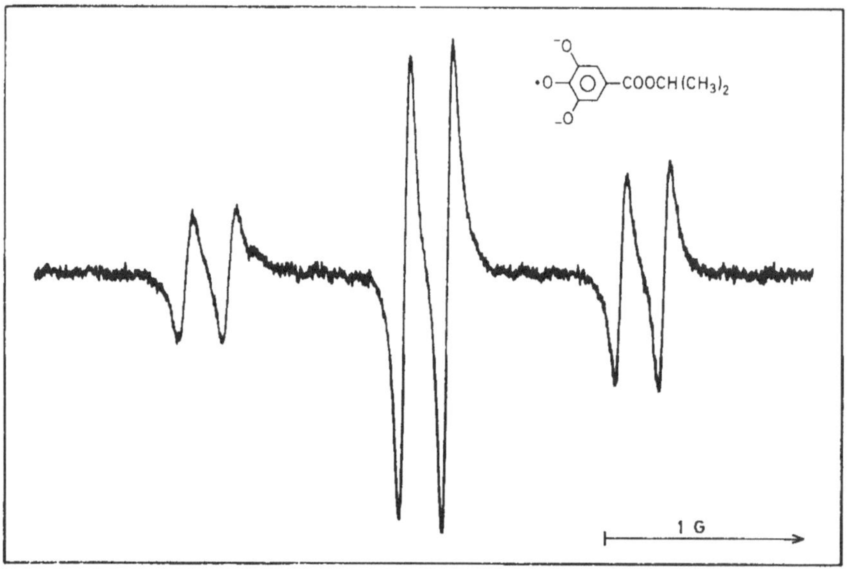

FIGURE 30.

Radical:	5-Propoxycarbonylpyrogallol semiquinone anion (no. 291)
Solvent:	w/EtOH
Method:	Ox alk
a_H/G:	0.379(2H), 1.105(2H)
Lw/G:	0.089
Instr. sett.:	4 G/20 sec, 0.01 sec, 0.1 mW, 4 acc

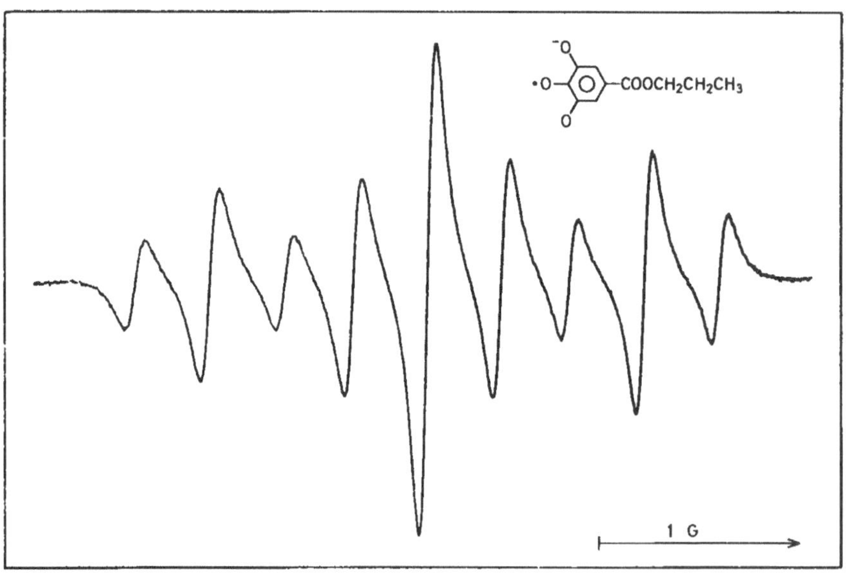

FIGURE 31.

Radical:	5-Isopropoxycarbonylpyrogallol semiquinone anion (no. 298)
Solvent:	w/EtOH
Method:	Ox alk
a_H/G:	0.220(1H), 1.104(2H)
Lw/G:	0.054
Instr. sett.:	4 G/20 sec, 0.01 sec, 0.1 mW, 8 acc

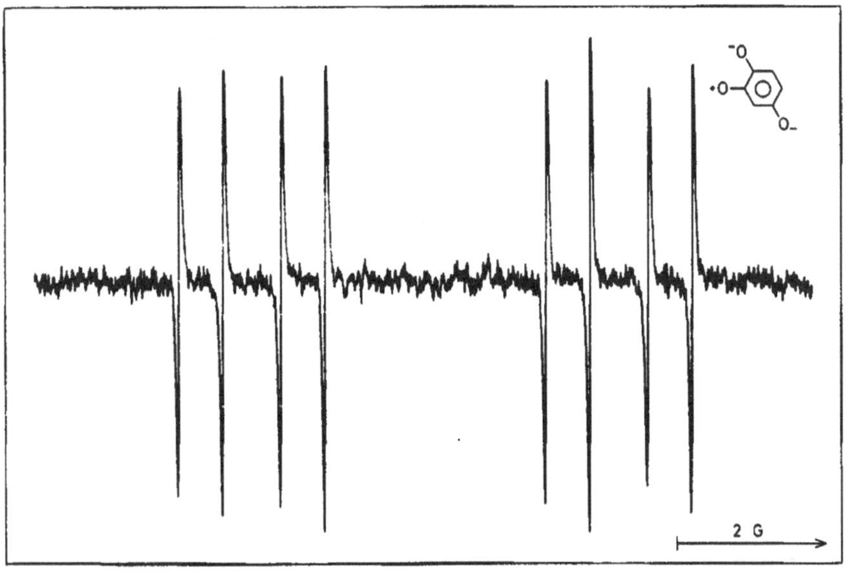

FIGURE 32.

Radical:	4-Oxido-1,2-benzosemiquinone (no. 308)
Solvent:	w/EtOH
Method:	Ox alk
a_H/G:	0.585(1H), 1.350(1H), 4.895(1H)
Lw/G:	0.035
Instr. sett.:	10 G/2 min, 0.03 sec, 0.036 mW, 5 acc

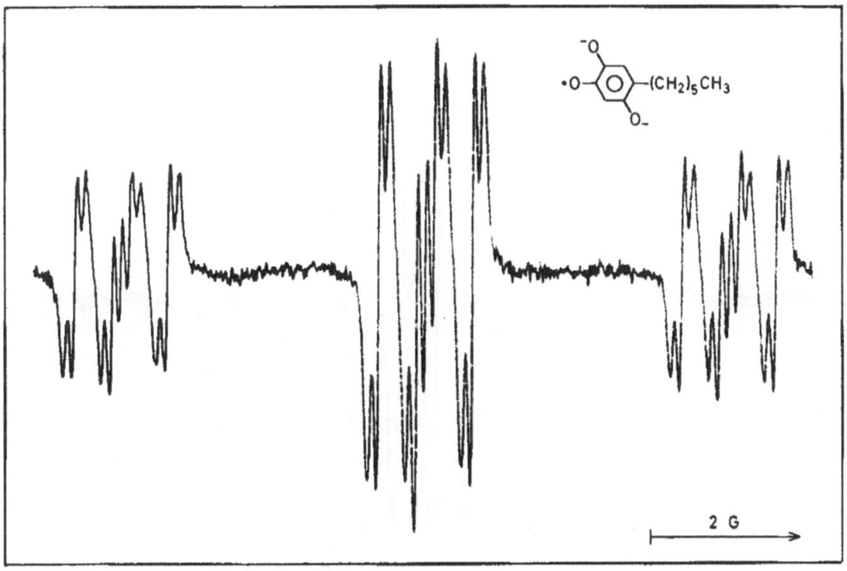

FIGURE 33.

Radical:	5-Hexyl-4-oxido-1,2-benzosemiquinone (no. 312)
Solvent:	w/EtOH
Method:	Ox alk
a_H/G:	0.117(2H), 0.503(1H), 0.747(1H), 4.044(2H)
Lw/G:	0.090
Instr. sett.:	10 G/2 min, 0.03 sec, 0.04 mW, 4 acc

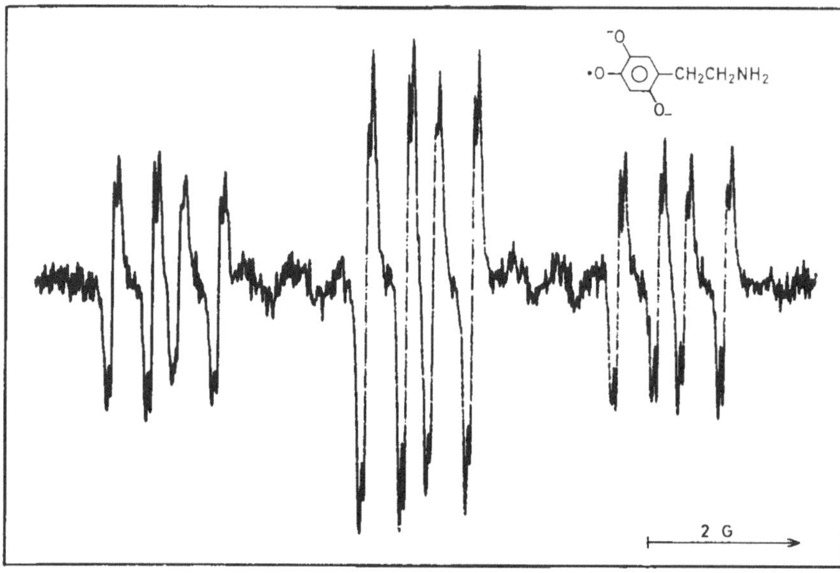

FIGURE 34.

Radical:	5-(2-Aminoethyl)-4-oxido-1,2-benzosemiquinone (no. 415)
Solvent:	w/EtOH
Method:	Ox alk
a_H/G:	0.054(2H), 0.542(1H), 0.896(1H), 3.332(2H)
Lw/G:	0.060
Instr. sett.:	10 G/2 min, 0.03 sec, 0.04 mW, 4 acc

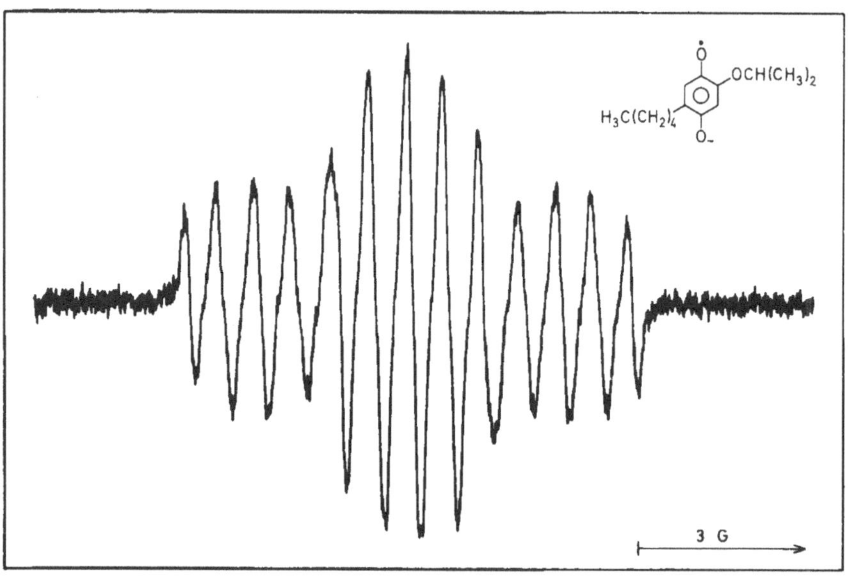

FIGURE 35.

Radical:	2-Isopropoxy-5-pentyl-1,4-benzosemiquinone (no. 465)
Solvent:	w/EtOH
Method:	Re DTN/alk
a_H/G:	0.540(1H), 0.687(1H), 1.344(1H), 2.739(2H)
Lw/G:	0.180
Instr. sett.:	20 G/2 min, 0.03 sec, 0.1 mW, 18 acc

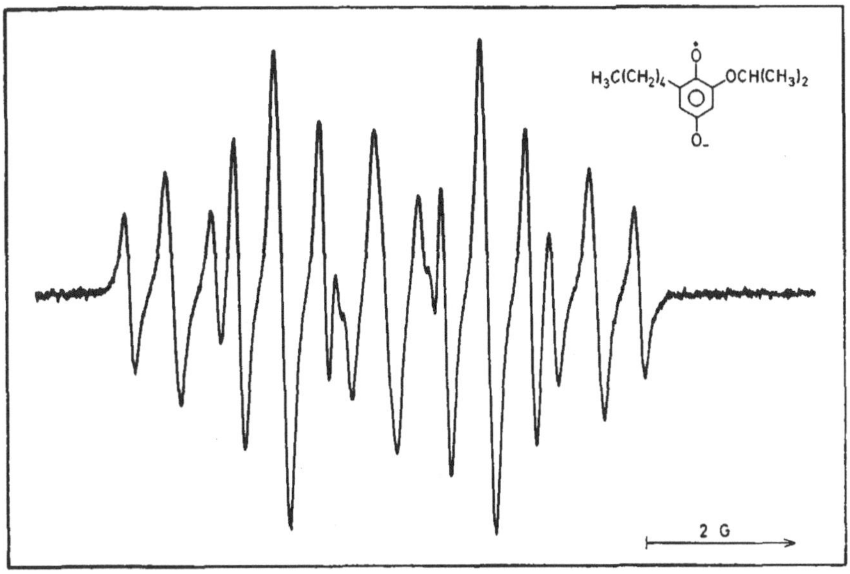

FIGURE 36.

Radical:	2-Isopropoxy-6-pentyl-1,4-benzosemiquinone (no. 468)
Solvent:	w/EtOH
Method:	Re DTN/alk
a_H/G:	0.528(1H), 0.608(1H), 1.440(2H), 2.753(1H)
Lw/G:	0.121
Instr. sett.:	10 G/2 min, 0.1 sec, 0.1 mW, 19 acc

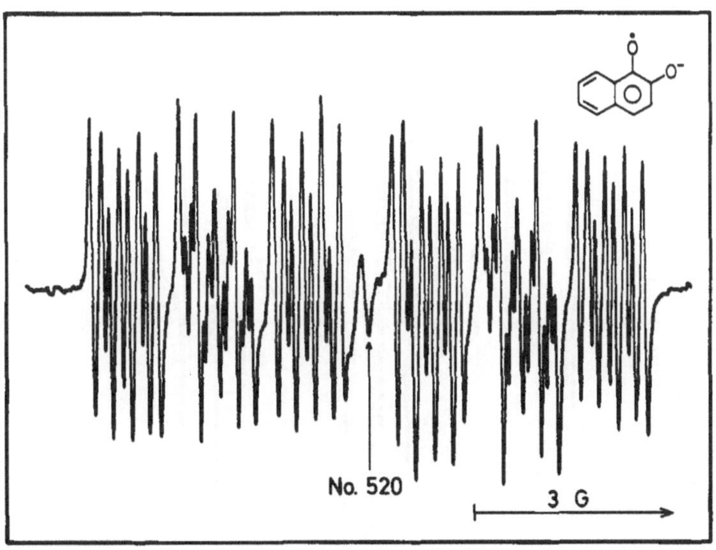

FIGURE 37.

Radical:	1,2-Naphthosemiquinone (no. 557)
Solvent:	w/EtOH
Method:	Ox alk/H_2O_2
a_H/G:	0.17(1H), 0.27(1H), 0.56(1H), 1.36(1H), 1.45(1H), 4.61(1H)
Lw/G:	~0.045
Instr. sett.:	20 G/8 min, 0.3 sec, 0.5 mW

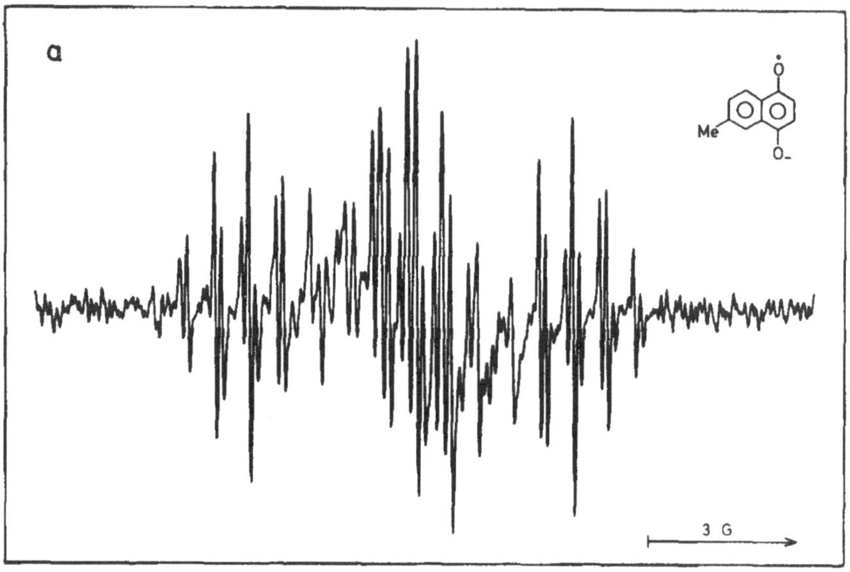

FIGURE 38a.

Radical:	6-Methyl-1,4-naphthosemiquinone (no. 564)
Solvent:	w/EtOH
Method:	Ox alk
a_H/G:	0.546(1H), 0.553(1H), 0.686(1H), 0.691(3H), 3.165(1H), 3.347(1H)
Lw/G:	0.051
Instr. sett.:	20 G/2 min, 0.03 sec, 0.04 mW, 16 acc.

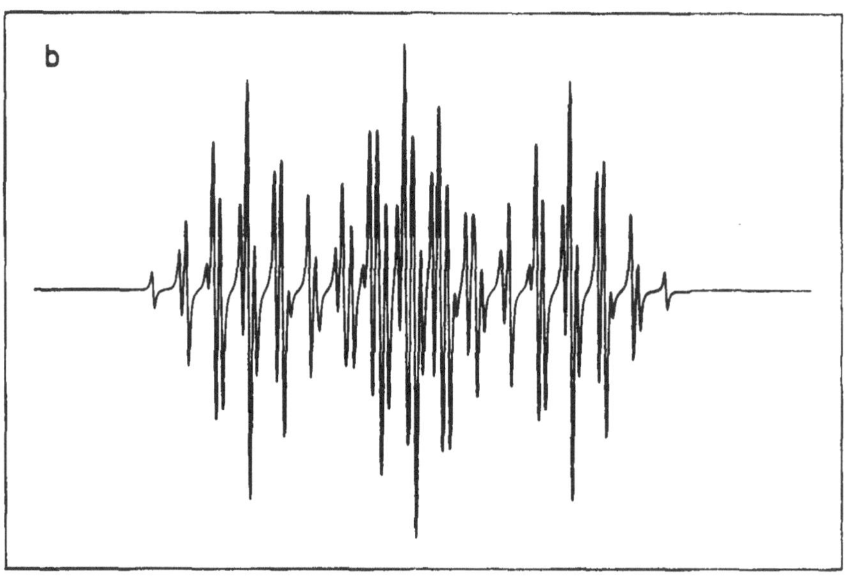

FIGURE 38b. Computer simulation of No. 564.

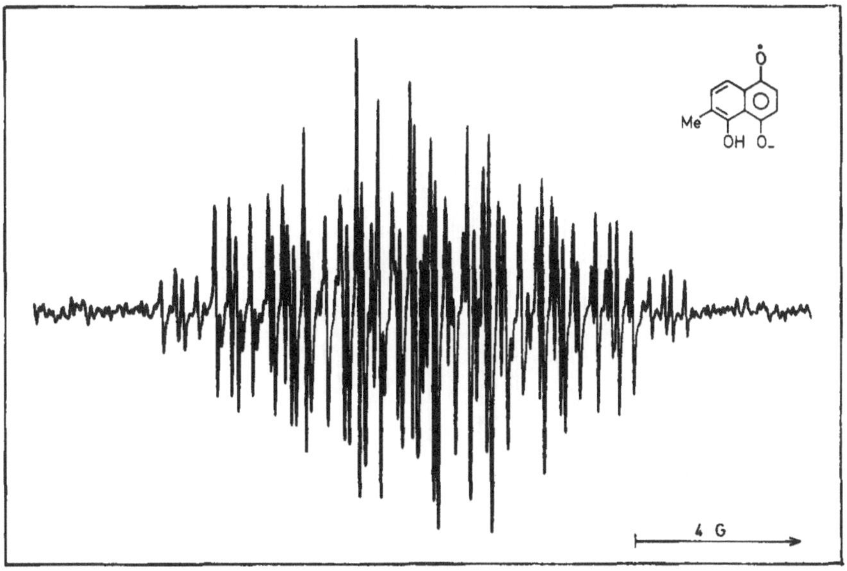

FIGURE 39.

Radical:	6-Methyl-5-hydroxy-1,4-naphthosemiquinone (no. 617)
Solvent:	w/EtOH
Method:	Re DTN/alk
a_H/G:	0.349(1H), 0.509(1H), 1.288(3H), 1.387(1H), 3.066(1H), 3.422(1H)
Lw/G:	0.044
Instr. sett.:	20 G/2 min, 0.03 sec, 0.04 mW, 15 acc

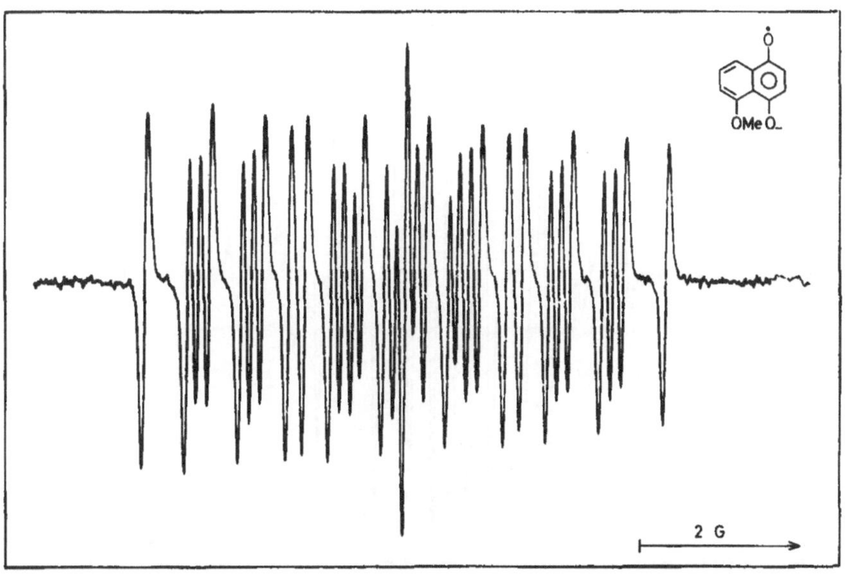

FIGURE 40.

Radical:	5-Methoxy-1,4-naphthosemiquinone (no. 619)
Solvent:	w/EtOH
Method:	Re DTN/alk
a_H/G:	0.037(3H), 0.727(1H), 0.913(1H), 1.102(1H), 2.464(1H), 3.727(1H)
Lw/G:	0.058
Instr. sett.:	10 G/1 min, 0.03 sec, 0.05 mW

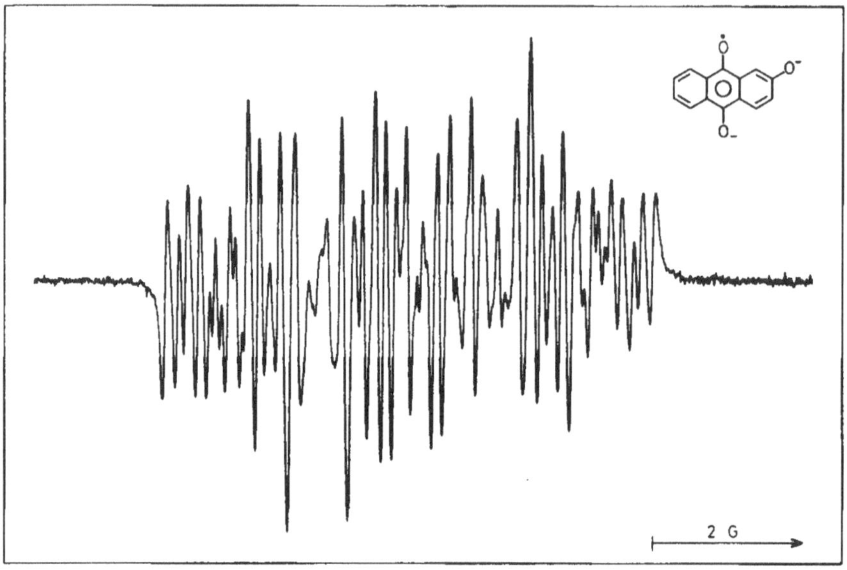

FIGURE 41.

Radical:	2-Oxidoanthrasemiquinone (no. 710)
Solvent:	w/EtOH
Method:	Re DTN/alk (pH ~13.2)
a_H/G:	0.154(1H), 0.261(1H), 0.548(1H), 0.621(1H), 1.021(1H), 1.608(1H), 1.96(1H)
Lw/G:	0.071
Instr. sett.:	10 G/2 min, 0.03 sec, 0.3 mW

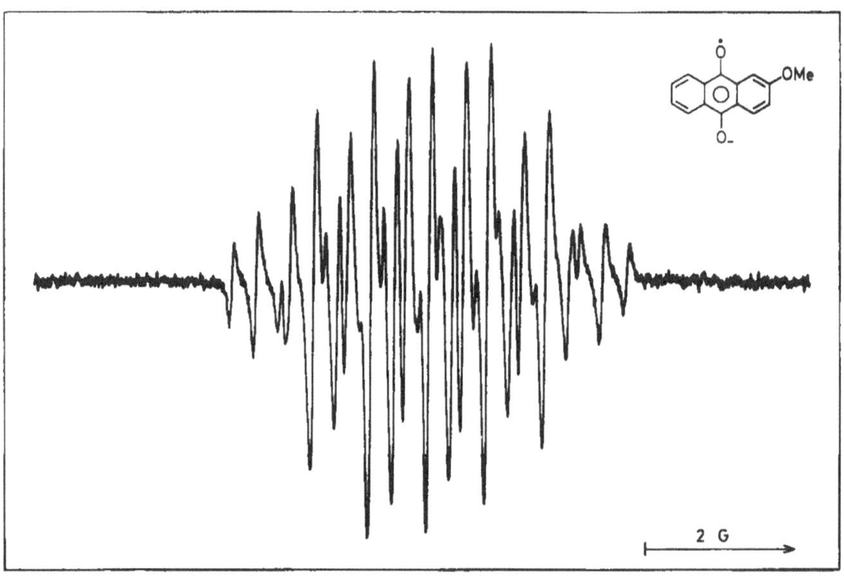

FIGURE 42.

Radical:	2-Methoxyanthrasemiquinone (no. 711)
Solvent:	w/EtOH
Method:	Re DTN/alk
a_H/G:	<0.015(3H), 0.310(1H), 0.338(1H), 0.737(1H), 0.785(1H), 0.788(1H), 1.081(1H), 1.237(1H)
Lw/G:	0.054
Instr. sett.:	10 G/2 min, 0.03 sec, 0.05 mW

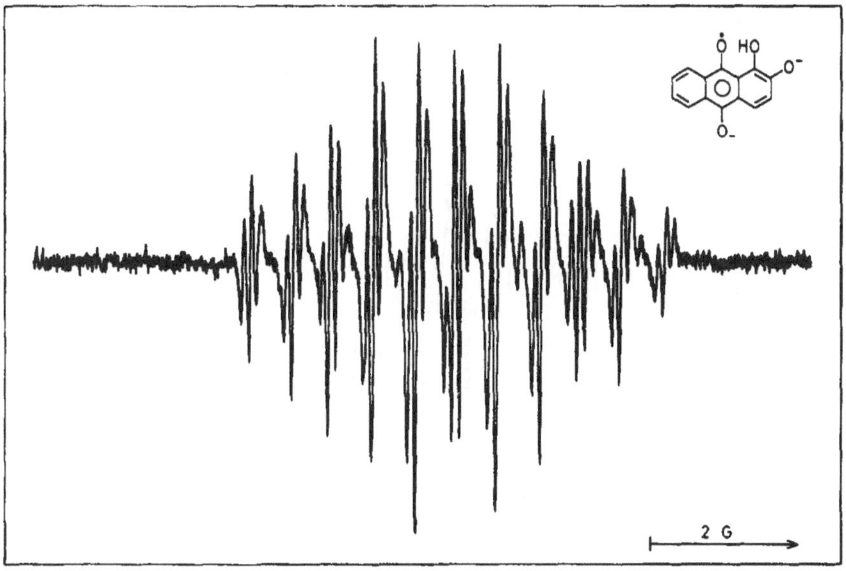

FIGURE 43.

Radical:	1-Hydroxy-2-oxidoanthrasemiquinone (no. 713)
Solvent:	w/EtOH
Method:	Re DTN/alk (pH ~13.6)
a_H/G:	0.098(1H), 0.108(1H), 0.580(1H), 0.593(1H), 1.054(1H), 1.075(1H), 2.234(1H)
Lw/G:	0.043
Instr. sett.:	10 G/5 min, 0.1 sec, 4 acc

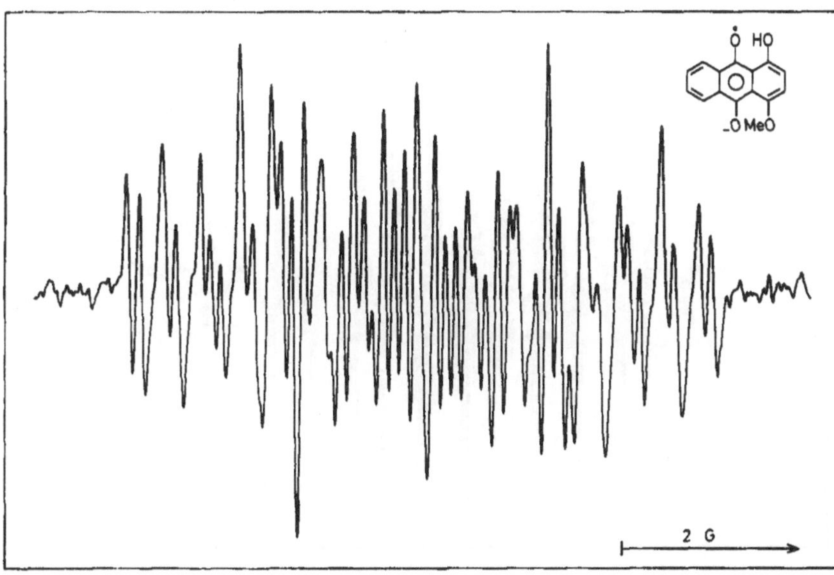

FIGURE 44.

Radical:	1-Hydroxy-4-methoxyanthrasemiquinone (no. 718)
Solvent:	w/EtOH
Method:	Re DTN/alk (pH ~13.2)
a_H/G:	0.026(3H), 0.142(1H), 0.396(1H), 0.454(1H), 0.891(1H), 0.958(1H), 1.954(1H), 2.177(1H)
Lw/G:	0.036
Instr. sett.:	10 G/2 min, 0.1 sec, 0.05 mW, 16 acc

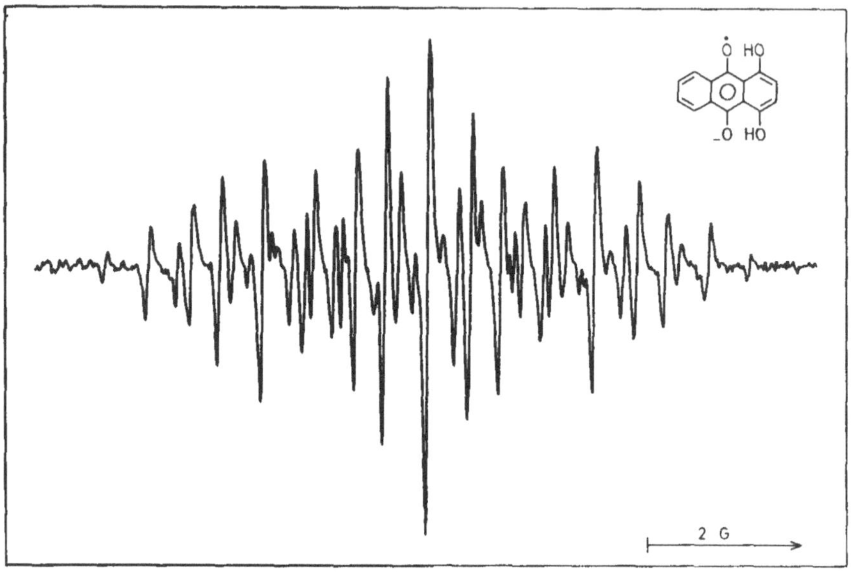

FIGURE 45.

Radical:	1,4-Dihydroxyanthrasemiquinone (no. 715)
Solvent:	w/EtOH
Method:	Re DTN/alk
a_H/G:	0.520(2H), 0.554(2H), 0.905(2H), 2.084(2H)
Lw/G:	0.045
Instr. sett.:	10 G/2 min, 0.03 sec, 0.3 mW

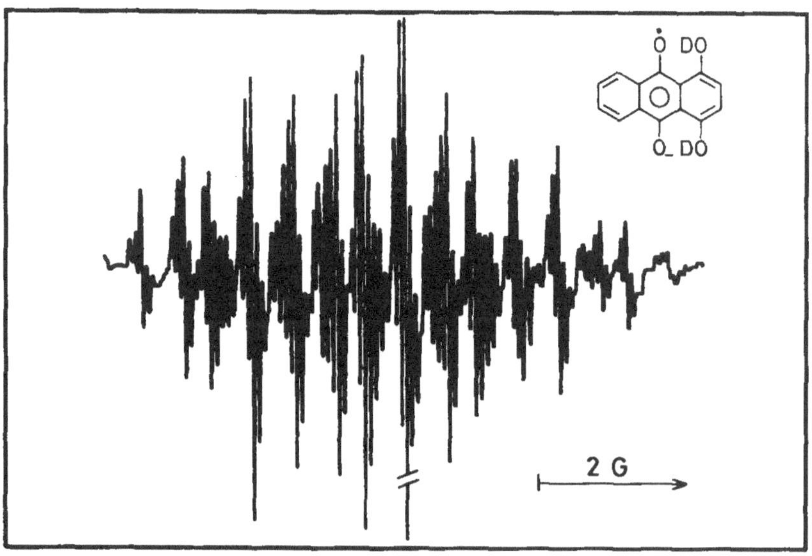

FIGURE 46.

Radical:	1,4-Dihydroxyanthrasemiquinone (no. 715D)
Solvent:	D_2O/EtOD
Method:	Re DTN/NaOD
a_H/G:	0.07(2D), 0.55(2H), 0.90(2H), 2.08(2H)
Lw/G:	~0.05
Instr. sett.:	20 G/8 min, 0.3 sec, 0.2 mW

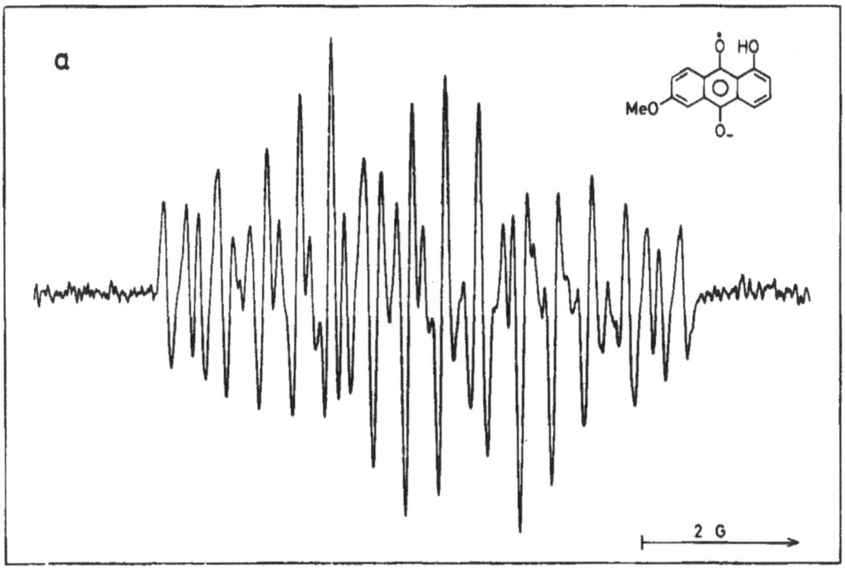

FIGURE 47a.

Radical:	1-Hydroxy-6-methoxyanthrasemiquinone (no. 721)
Solvent:	w/EtOH
Method:	Re DTN/alk (pH ~13.2)
a_H/G:	0.028(3H), 0.282(1H), 0.433(1H), 0.666(1H), 0.862(1H), 1.039(1H), 1.440(1H), 1.829(1H)
Lw/G:	0.043
Instr. sett.:	10 G/2 min, 0.03 sec, 0.04 mW, 23 acc

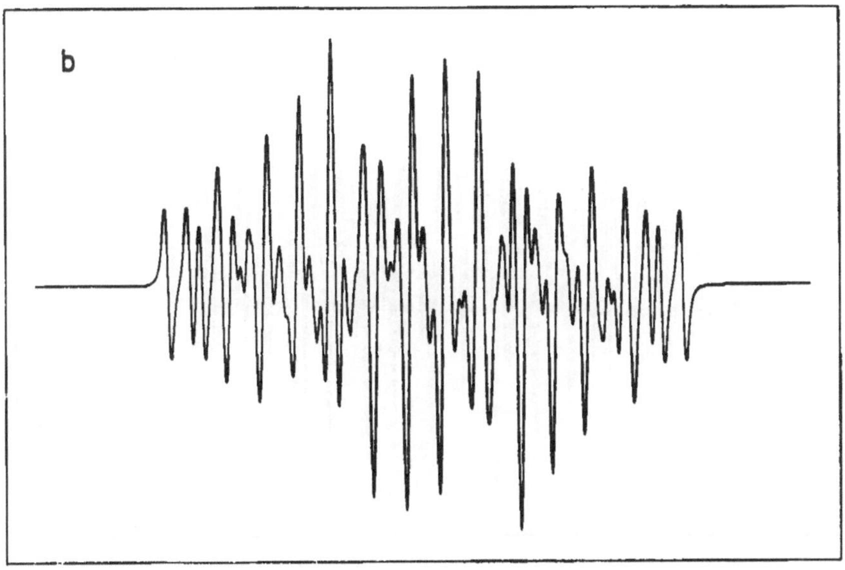

FIGURE 47b. Computer simulation of no. 721.

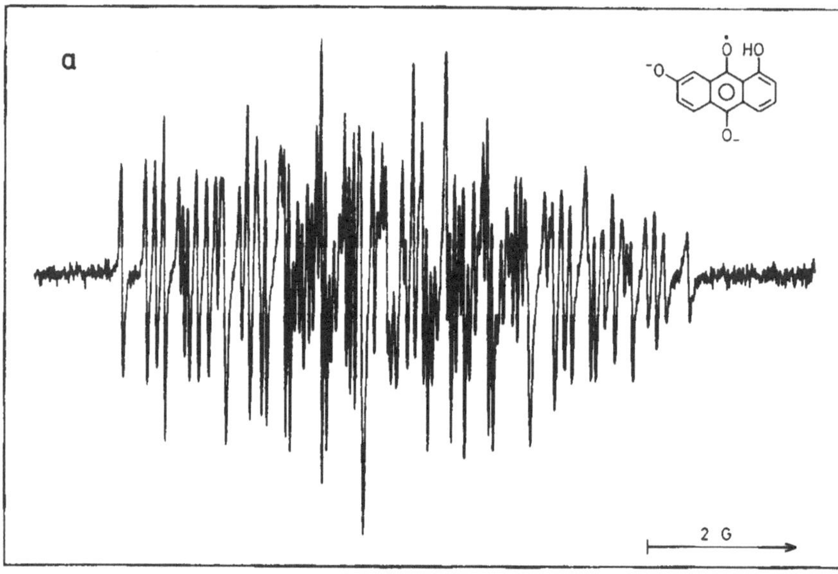

FIGURE 48a.

Radical:	1-Hydroxy-7-oxidoanthrasemiquinone (no. 722)
Solvent:	w/EtOH
Method:	Re DTN/alk (pH ~12.7)
a_H/G:	0.301(1H), 0.416(1H), 0.526(1H), 0.766(1H), 1.272(1H), 1.740(1H), 2.092(1H)
Lw/G:	0.033
Instr. sett.:	10 G/2 min, 0.03 sec, 0.035 mW, 36 acc

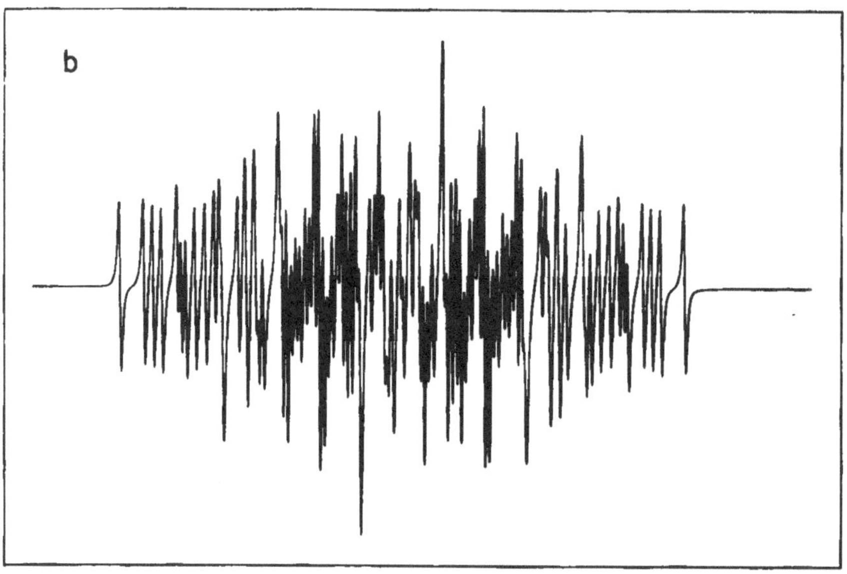

FIGURE 48b. Computer simulation of no. 722.

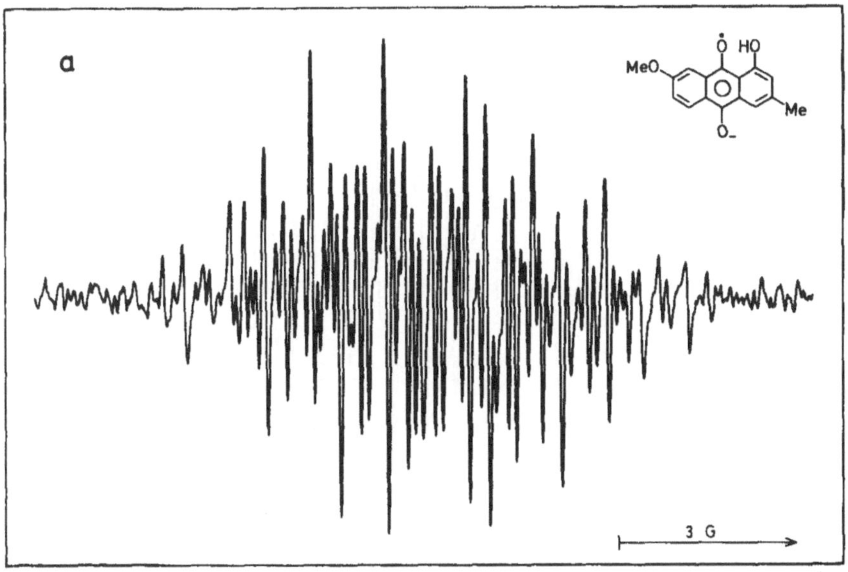

FIGURE 49a.

Radical:	1-Hydroxy-3-methyl-7-methoxyanthrasemiquinone (no. 724)
Solvent:	w/EtOH
Method:	Re DTN/alk
a_H/G:	0.026(3H), 0.333(1H), 0.344(1H), 0.800(1H), 1.124(1H), 1.208(1H), 1.247(1H), 1.370(3H)
Lw/G:	0.039
Instr. sett.:	20 G/5 min, 0.1 sec, 0.04 mW, 16 acc

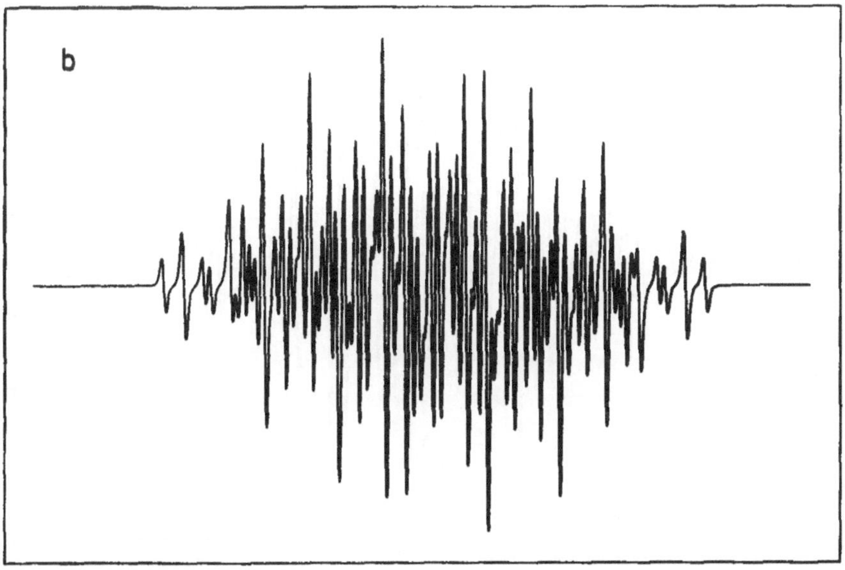

FIGURE 49b. Computer simulation of no. 724.

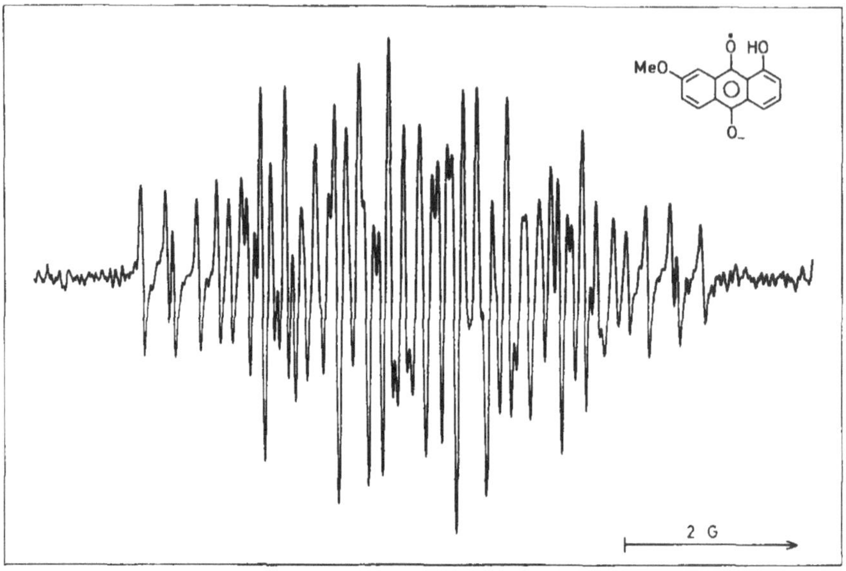

FIGURE 50.

Radical:	1-Hydroxy-7-methoxyanthrasemiquinone (no. 723)
Solvent:	w/EtOH
Method:	Re DTN/alk
a_H/G:	0.015(3H), 0.280(1H), 0.352(1H), 0.859(1H), 0.999(1H), 1.188(1H), 1.352(1H), 1.374(1H)
Lw/G:	0.031
Instr. sett.:	10 G/2 min, 0.03 sec, 0.04 mW, 8 acc

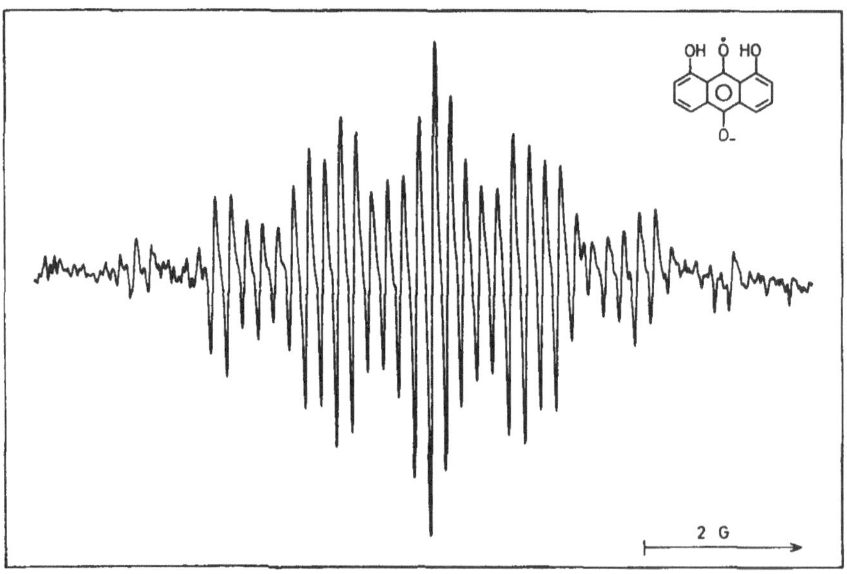

FIGURE 51.

Radical:	1,8-Dihydroxyanthrasemiquinone (no. 725)
Solvent:	w/EtOH
Method:	Re DTN/alk
a_H/G:	0.196(2H), 0.989(2H), 1.202(2H), 1.591(2H)
Lw/G:	0.049
Instr. sett.:	10 G/2 min, 0.01 sec, 25 acc

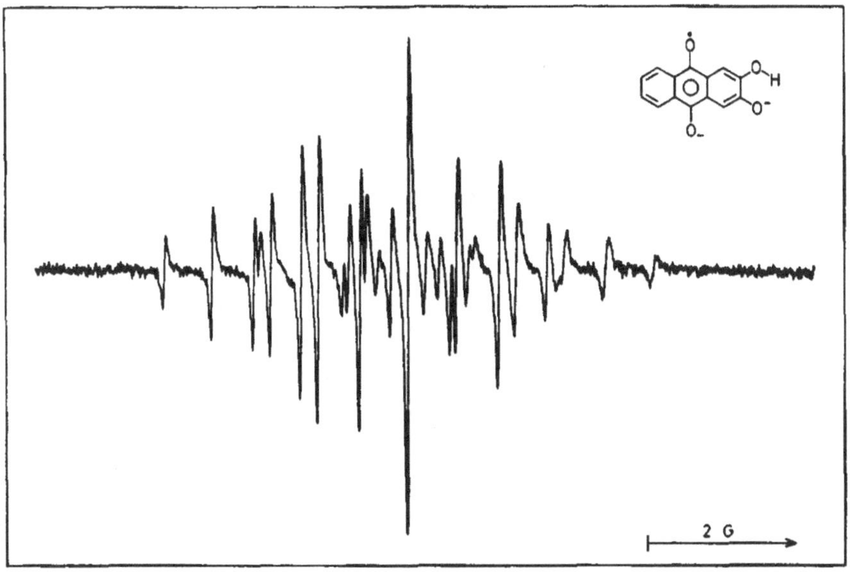

FIGURE 52.

Radical:	2-Hydroxy-3-oxidoanthrasemiquinone (no. 727)
Solvent:	w/EtOH
Method:	Re DTN/alk (pH ~12.2)
a_H/G:	0.602(2H), 1.133(2H), 1.350(2H)
Lw/G:	0.044
Instr. sett.:	10 G/2 min, 0.03 sec, 0.03 mW, 4 acc

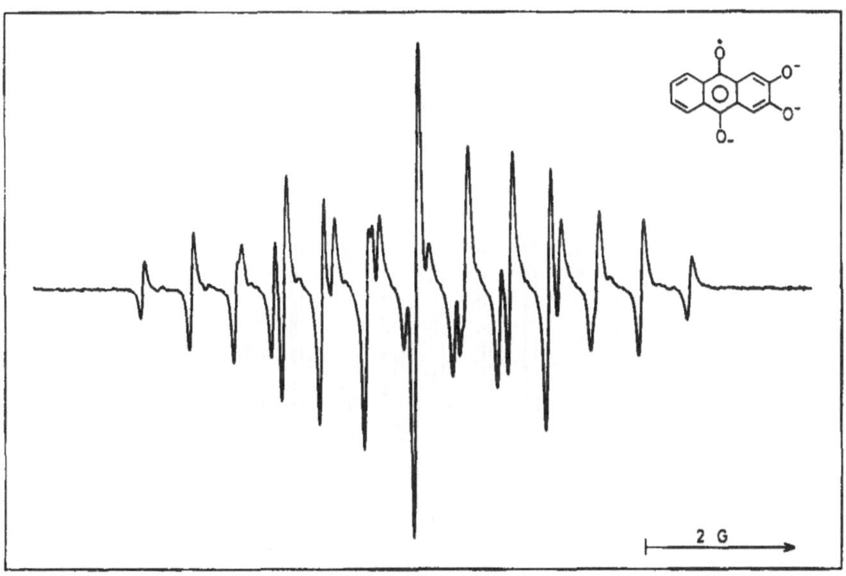

FIGURE 53.

Radical:	2,3-Dioxidoanthrasemiquinone (no. 728)
Solvent:	w/EtOH
Method:	Re DTN/alk (pH ~13.7)
a_H/G:	0.618(2H), 1.178(2H), 1.661(2H)
Lw/G:	0.059
Instr. sett.:	10 G/2 min, 0.03 sec, 0.03 mW

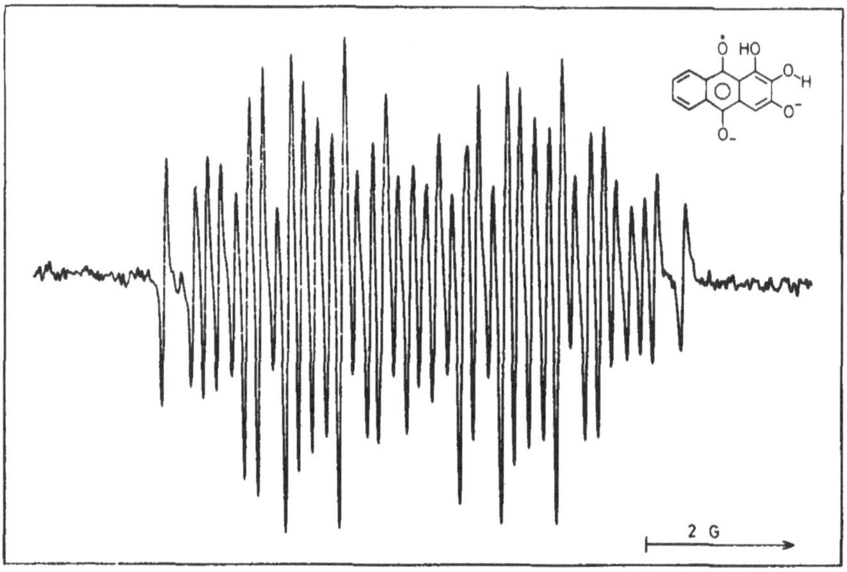

FIGURE 54.

Radical:	1,2-Dihydroxy-3-oxidoanthrasemiquinone (no. 731)
Solvent:	w/EtOH
Method:	Re alk (pH ~12.2)
a_H/G:	0.367(1H), 0.518(1H), 0.685(1H), 1.036(1H), 1.218(1H), 2.736(1H)
Lw/G:	0.049
Instr. sett.:	10 G/2 min, 0.03 sec, 0.3 mW, 12 acc

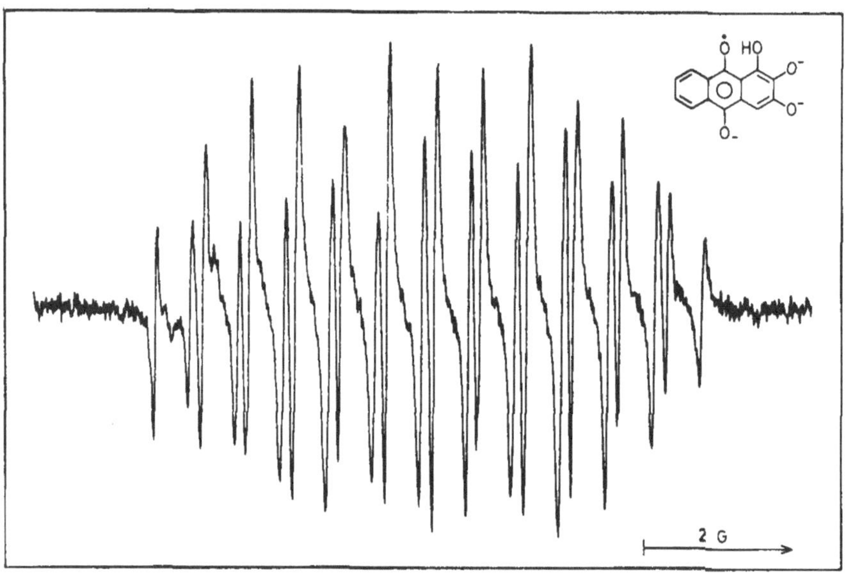

FIGURE 55.

Radical:	2,3-Dioxido-1-hydroxyanthrasemiquinone (no. 732)
Solvent:	w/EtOH
Method:	Re alk (pH ~13.7)
a_H/G:	0.437(1H), 0.583(1H), 0.613(1H), 1.147(1H), 1.185(1H), 2.940(1H)
Lw/G:	0.055
Instr. sett.:	10 G/5 min, 0.1 sec, 0.3 mW

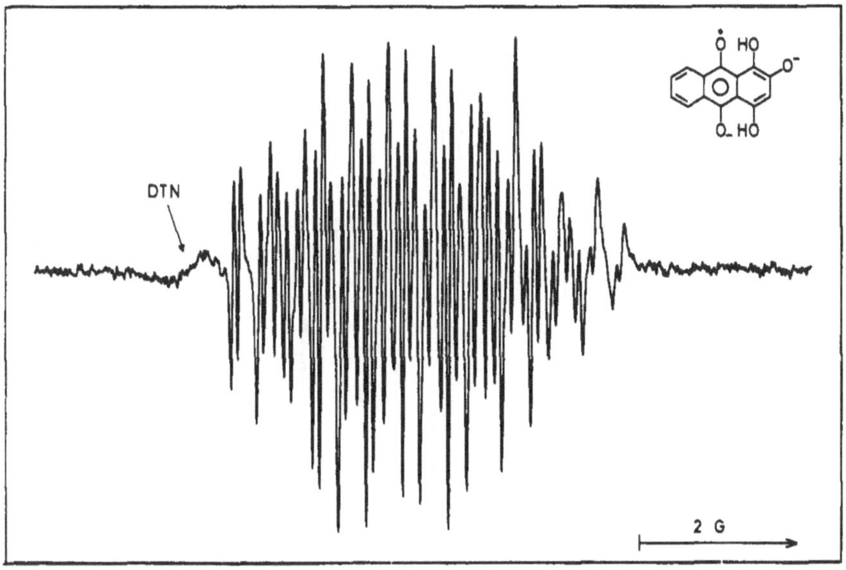

FIGURE 56.

Radical:	1,4-Dihydroxy-2-oxidoanthrasemiquinone (no. 733)
Solvent:	w/EtOH
Method:	Re DTN/alk
a_H/G:	0.09(1H), 0.33(1H), 0.46(1H), 0.59(1H), 0.81(1H), 1.04(1H), 1.62(1H)
Lw/G:	0.050
Instr. sett.:	10 G/2 min, 0.03 sec, 0.3 mW, 9 acc

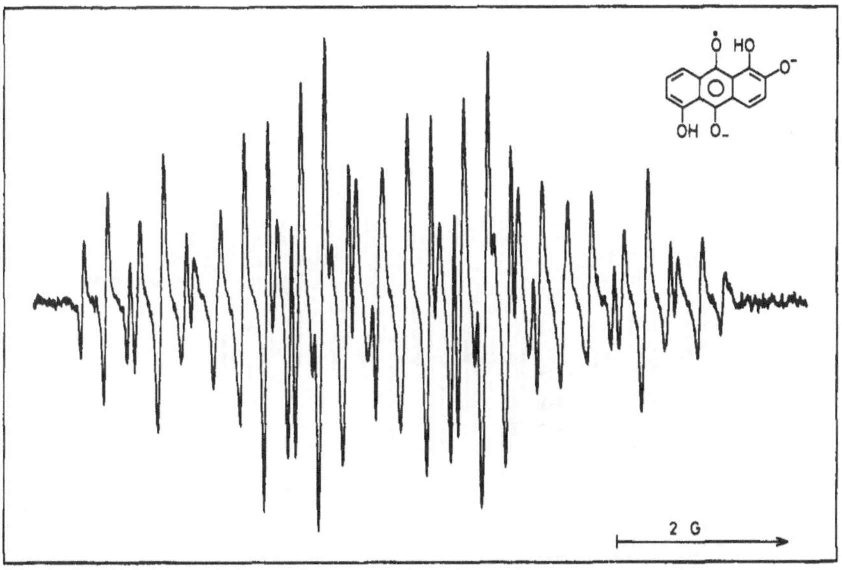

FIGURE 57.

Radical:	1,5-Dihydroxy-2-oxidoanthrasemiquinone (no. 734)
Solvent:	w/EtOH
Method:	Re DTN/alk (pH ~13.2)
a_H/G:	0.269(1H), 0.280(1H), 0.635(1H), 0.665(1H), 1.617(1H), 1.885(1H), 2.176(1H)
Lw/G:	0.042
Instr. sett.:	10 G/5 min, 0.1 sec

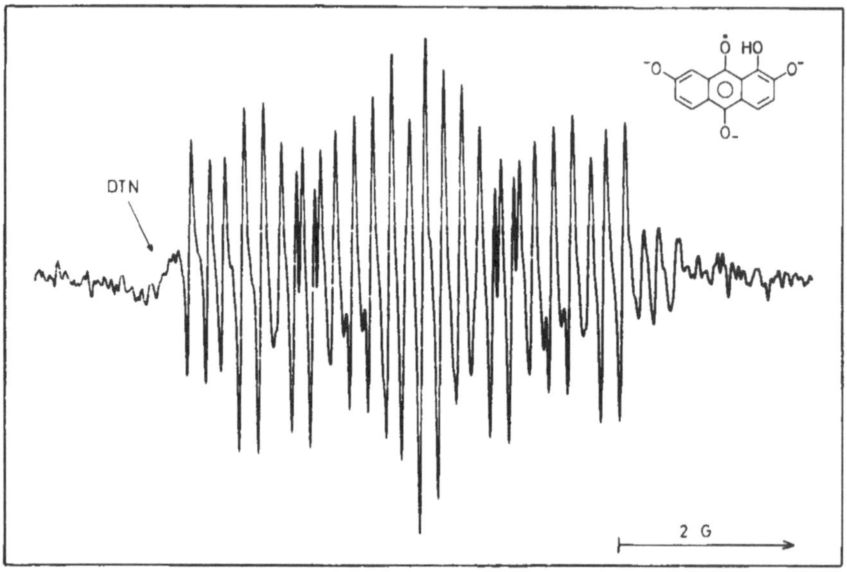

FIGURE 58.

Radical:	2,7-Dioxido-1-hydroxyanthrasemiquinone (no. 736)
Solvent:	w/EtOH
Method:	Re DTN/alk
a_H/G:	0.209(1H), 0.380(1H), 0.592(1H), 0.804(1H), 1.250(1H), 2.264(1H)
Lw/G:	0.043
Instr. sett.:	10 G/2 min, 0.1 sec, 10 acc

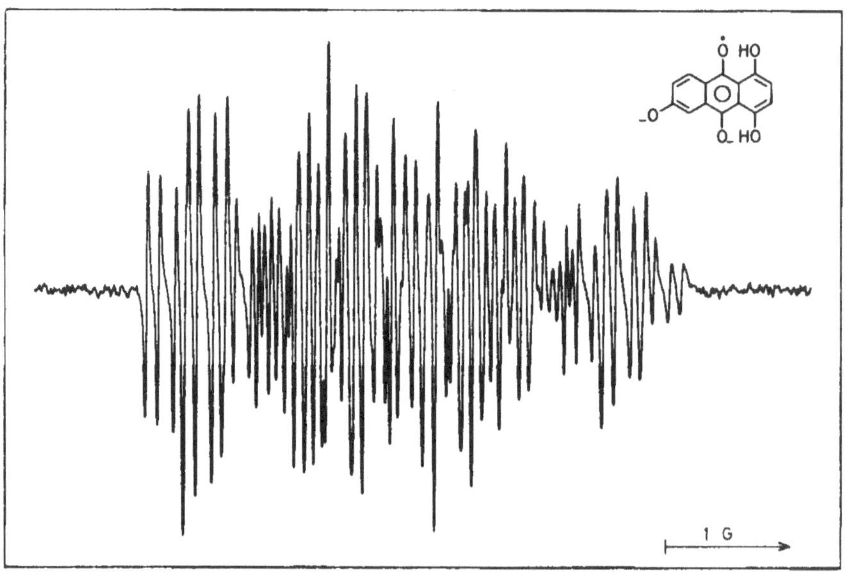

FIGURE 59.

Radical:	1,4-Dihydroxy-6-oxidoanthrasemiquinone (No. 741)
Solvent:	w/EtOH
Method:	Re DTN/alk
a_H /G:	0.187(1H), 0.429(1H), 0.565(1H), 0.582(1H), 1.679(1H), 1.979(1H), 2.730(1H)
Lw/G:	0.043
Instr. sett.:	10 G/2 min, 0.1 sec, 10 acc

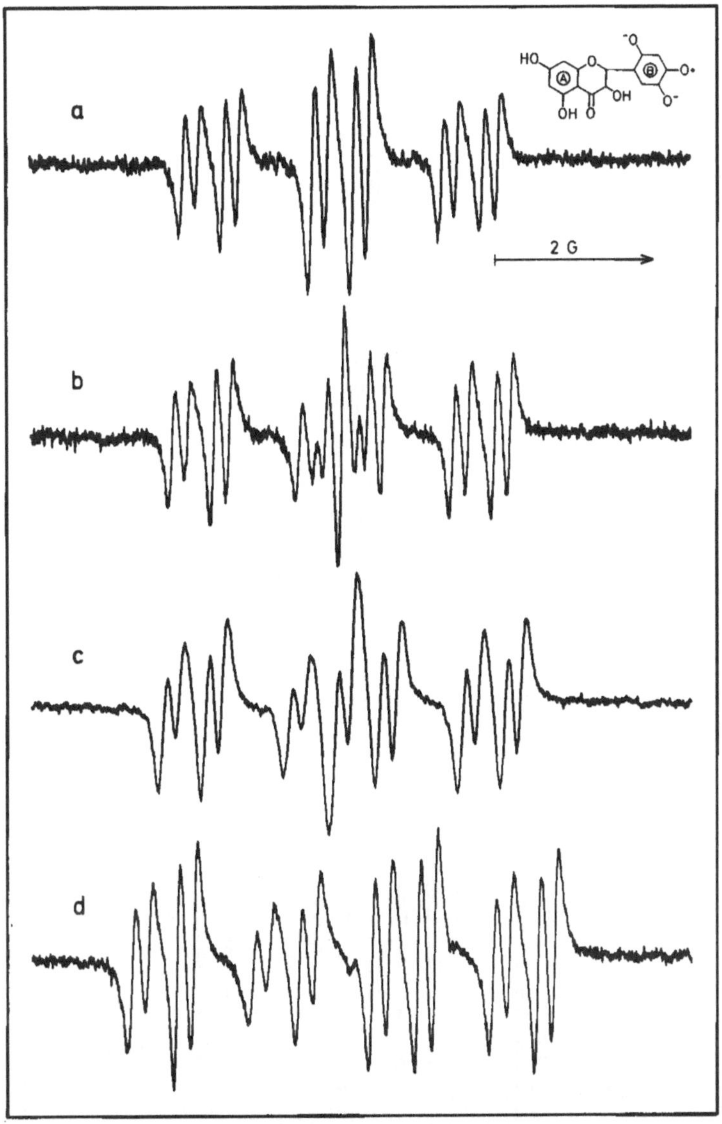

FIGURE 60.
Radical: 3′,4′,5,6′,7-Pentaoxido-2,3-dihydroflavonol anion radical (No. 876)
Method: Ox alk
Instr. sett.: 10 G/1 min, 0.03 sec, 0.1 mW

Spectrum	Solvent DMSO:w (v/v)	a_H/G				Lw/G
a	2:8	0.196	0.514	1.604	1.638	0.086
b	2:3	0.199	0.524	1.594	1.919	0.092
c	2:2	0.209	0.533	1.574	2.172	0.124
d	2:1	0.214	0.571	1.523	3.001	0.102

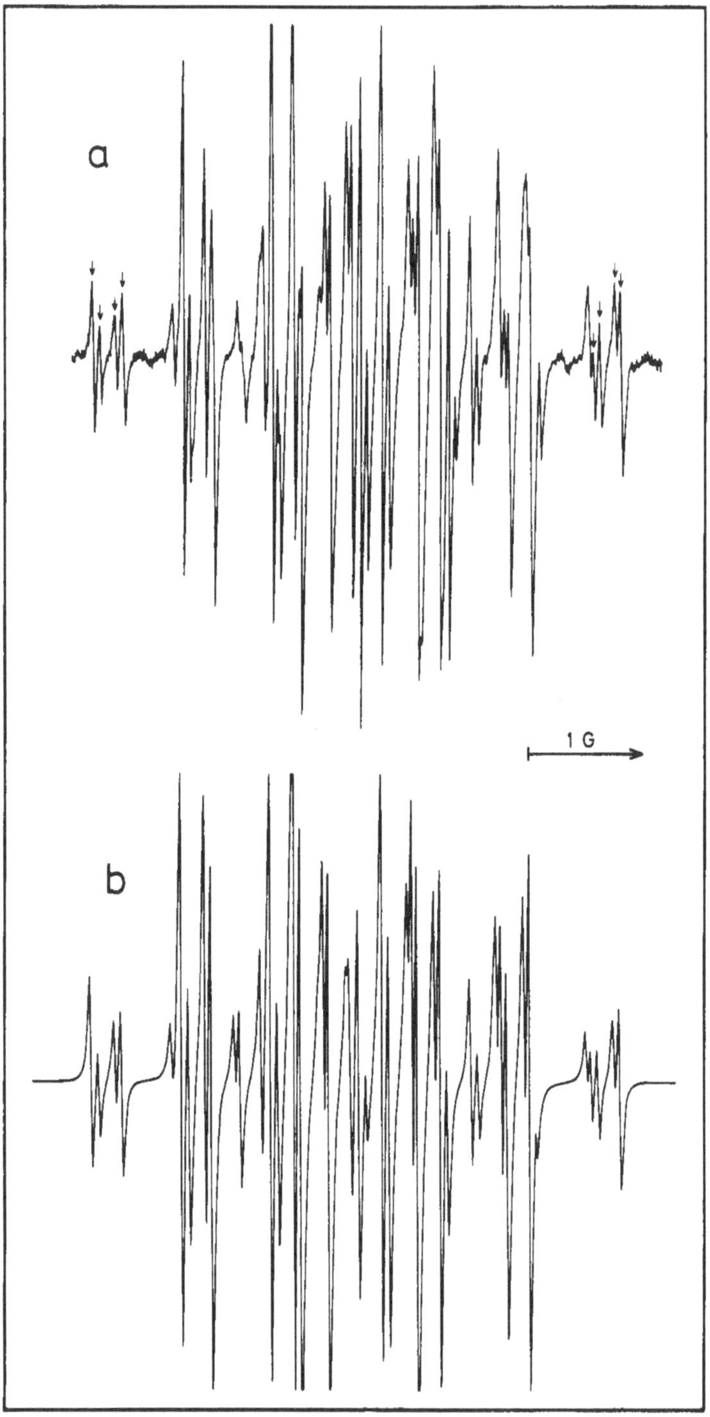

FIGURE 61. (a) Simultaneous observation of four 16-line spectra from four conformers obtained from oxidation of 4,4′-dimethoxydiquinone. The arrows indicate pairwise the first and the last line for the four spectra. For structural details and EPR data see nos. 816 to 819, Table I.20. (b) Simulation of the spectra in (a) obtained by a trial-and-error procedure. For data used see Table I.20.

EXPLANATION TO TABLES

Location of Radicals

The radicals are located in the tables according to the characteristics of the *radical nucleus*, defined as the aromatic ring system including the oxygen atoms bound directly to the rings. These oxygen atoms are denoted as "key oxygens". All other functional groups attached to the rings are termed *substituents*.

A radical is indexed under the headings **Anions** (Table I), **Neutral Species** (Table II), or **Cations** (Table III), according to the total formal charge on the key oxygens, but irrespective of any formal charges on the substituents. Semiquinone anions, for which complexes with cations have been observed, are marked [C] in Table I. The **EPR** results for the complexes are listed separately in Table IV, **Semiquinone-Cation Complexes.** Splitting constants arising from ^{13}C and ^{17}O nuclei are listed in Table V, **Isotopes.** All entries in Table V appear also in one of the Tables I to IV, where the radicals are marked [I]. Radicals for which a spectrum is shown in the handbook are marked ''.

In each table the entries are ordered with increasing number of rings in the radical nucleus and with increasing number of key oxygens. The detailed order is shown in the Table of Contents.

In Tables I to III the structure of each radical nucleus is shown with the numbering of carbon atoms adopted. In Table IV the basic structures of the semiquinone-cation complexes are similarly shown. *Substituents* are entered in the order H, C, N, P, S, F, Cl, Br, and I. In general, the names given are in accordance with IUPAC conventions. In the molecular formulas the abbreviations Me, Et, and Ph, are used for methyl, ethyl, and phenyl (C_6H_5), respectively. Asterisks or α, β in the formulas serve to link particular protons to the splitting constants similarly marked.

Splitting Constants

Unless otherwise stated, the splitting constants listed in the column a_i refer to the aromatic proton at position (i) in the radical nucleus. Splitting constants arising from substituents at position (i) are listed also in the column a_i. Constants referring to protons in aliphatic substituents have numerical superscripts according to the scheme (SQ = semiquinone):

```
    1 2 3
SQ–C–C–C
    | |
    C C
    2 3
```

For example, 1.63^1 designates a splitting from either 3 protons from $-CH_3$, 2 protons from $-CH_2R$, or 1 proton from $-CH(R)R'$ or $-CH = C(R)R'$; and 0.22^2 designates a splitting from either 9 protons from $-C(CH_3)_3$, 6 protons from $-C(CH_3)_2R$, or 4 protons from $-C(CH_2R)_2R'$, etc.

The number of equivalent protons giving rise to a particular splitting must be derived from the name or the substituent formula. In case of nonequivalent methylene protons both observed splittings are shown with a superscript 1 (see, for example, Nos. 40 to 44 in Table I.1). Splittings from nuclei other than protons are labeled explicitly, e.g., F, Si, and 2N, for fluor, silicone, and two equivalent nitrogens. Where no splitting has been attributed to the nucleus/substituent, a *u* appears.

The organization of the tables requires that all splitting constants listed are assigned. It should be noted, however, that values deviating little from each other may be interchanged. In other cases where the assignments are tentative, the values of some constants are bracketed to indicate that they may be interchanged.

Since splitting constants are often reported without explicit indication of the accuracy, limits of errors are not given in the tables. They may be estimated as ± 3 at the last decimal place.

In Tables I to IV absolute values of the splitting constants are given only, since the sign is not directly accessible in the EPR experiment. Aromatic proton splitting constants are negative with few exceptions, corresponding to positive spin densities at the carbon atoms. In Table V, **Isotopes,** the signs of the splitting constants are given.

Methods of radical generation are given in the tables in terms of the following abbreviations. For chemical methods:

1. Ox alk — autoxidation in an alkaline medium (NaOH) of quinols, phenols, etc.
2. Ox KOH — same as the above apart from the use of KOH instead of NaOH.
3. Ox agent(s) — oxidation with specified agent, e.g., I_2, KO_2, Ag_2O, PbO_2, etc.
4. Ox $AlCl_3$ — oxidation involving the combined action of $AlCl_3$ and a solvent, e.g., $MeNO_2$, CH_2Cl_2, PhMe. Method solely used to generate radical cations.
5. Re alk — reduction in an alkaline medium of quinones; reducing agent not specified.
6. Re agent(s) — reduction with specified agent, e.g., Zn, KO_2, DTN, etc.

For other methods:

1. Ox el — *in situ* electrolytic oxidation.
2. Re el — *in situ* electrolytic reduction.
3. Radiolysis — *in situ* radiolysis in an aqueous solution, saturated with N_2O to convert e^- into ·OH.
4. Phot — *in situ* photolysis with ultraviolet or visible irradiation.

Spectra are recorded at room temperature (300 K) unless otherwise stated. If reported, composition of mixed solvents are expressed in vol%. Results from the editor's laboratory[8] are with 32% w/EtOH solvents; w/EtOH is stated in the tables only. The data are obtained from the Reference number stated first, unless otherwise indicated.

The radicals are numbered consecutively in the tables. For semiquinone-cation complexes the numbers begin with a capital letter C. A capital letter D after a number indicates a deuterated compound.

I. ANIONS

BENZOSEMIQUINONE ANIONS

Table I.1.
1,2-BENZOSEMIQUINONES

Anions

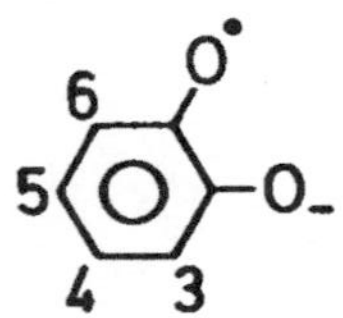

No.	Radical	Substituent(s)	Splitting constants/Gauss				g-Factor	Solvent	Method	Ref.
			a_3	a_4	a_5	a_6				
1	[C.I] 1,2-Benzosemiquinone$^{\cdot-}$		0.60	3.82	3.82	0.60		MeOH	Phot neut[a]	129
			0.74	3.66	3.66	0.74		w	Ox alk	1, 130
			0.75	3.69	3.69	0.75		w	Ox $KOBu^t$	4, 115
			0.87	3.75	3.75	0.87		75% w/MeOH	Ox KOH	2
			0.95	3.65	3.65	0.95	2.00441	w/EtOH (?)	Ox alk	3
			0.93	3.59	3.59	0.93		EtOH	Ox $KOBu^t$	4
			1.16	3.55	3.55	1.16		10% w/DMF	Ox alk	5
			1.28	3.48	3.48	1.28		DMF	Ox $KOBu^t$	4
			1.20	3.44	3.44	1.20		MeTHF	Re Na/K?	6
			1.39	3.45	3.45	1.39		DMSO	Ox $KOBu^t$	4
			1.50	3.30	3.30	1.50		MeCN	Re el	2
			1.56	3.34	3.34	1.56		HMPA	Ox $KOBu^t$	4
			1.40	3.60	3.60	1.40	2.0050	THF/DCCw	Ox KO_2	7
2	[C] -,3-methyl-	–Me	0.63[t]	2.85	4.09	0.28		w	Ox alk	1, 130
			0.722[t]	2.800	4.039	0.380	2.00462	w/EtOH	Ox alk	8
			0.80[t]	2.80	4.07	0.44		EtOH	Ox $KOBu^t$	4

Table I.1. (continued) 1,2-BENZOSEMIQUINONES

Anions

No.	Radical	Substituent(s)	Splitting constants/Gauss a_3	a_4	a_5	a_6	g-Factor	Solvent	Method	Ref.
			0.8[1]	2.9	3.7	0.8		DMF/10% w	Ox alk	9
			1.06[1]	2.90	3.66	0.98		DMF	Ox $KOBu^t$	4
			1.00[1]	2.80	3.70	1.00		DMSO	Ox $KOBu^t$	4
			1.06[1]	2.83	3.54	1.06		HMPA	Ox $KOBu^t$	4
3	[C] -,3-isopropyl-	$-CHMe_2$	0.7[1]	2.95	4.5	u		w	Ox alk	9
			0.7[1]	2.7	3.95	0.7		DMF/10% w	Ox alk	9
4	[C] -,4-methyl-	–Me	0.10	4.80[1]	3.85	0.95		w	Ox alk	1, 9
			0.25	4.80[1]	3.85	1.05		75% w/MeOH	Ox KOH	2
			0.33	4.47[1]	3.70	1.10		EtOH	Ox $KOBu^t$	4
			0.7	4.4[1]	3.7	1.4		10% w/DMF	Ox alk	9
			0.80	4.10[1]	3.72	1.54		MeCN	Re el	2
			0.80	4.08[1]	3.62	1.51		DMSO	Ox $KOBu^t$	4
			0.89	4.01[1]	3.65	1.60		DMF	Ox $KOBu^t$	4
			1.00	3.85[1]	3.55	1.69		HMPA	Ox $KOBu^t$	4
5	-,4-ethyl-	–Et	0.30	3.70[1]	3.70	1.00		75% w/MeOH	Ox KOH	2
				0.20[2]						
6	-,4-propyl-	$-CH_2CH_2Me$	0.32	3.30[1]	3.80	1.00		75% w/MeOH	Ox KOH	2
7	-,4-*s*-butyl-	$-CH(Me)CH_2Me$	0.42	1.67[1]	3.90	1.00		75% w/MeOH	Ox KOH	2
				0.32[2b]						
8	-,4-isopropyl-	$-CHMe_2$	0.39	2.21[1]	3.90	0.93		75% w/MeOH	Ox KOH	2
				0.32[2]						
9	[C] -,4-*t*-butyl-	$-CMe_3$	0.42	0.32[2]	3.96	0.83		75% w/MeOH	Ox KOH	10
			0.3	0.32[2]	3.80	1.17		50% w/MeOH	Ox alk	11
			1.16	0.30[2]	3.90	1.40		MeCN	Re el	12

10	-,5-*t*-butyl-3-methyl-	–CMe_3,–Me	0.3[1]	3.07	0.29[2]	0.3		50% w/MeOH	Ox alk	11
11	-,3-*t*-butyl-5-methyl-	–CMe_3,–Me	u	2.65	5.4[1]	0.3		w	Ox alk	9
			<0.03	2.50	5.35[1]	0.35		50% w/MeOH	Ox alk	11
			u	3.00	4.35[1]	u		10% w/DMF	Ox alk	9
12	[C]-,3,5-di-*t*-butyl-	–CMe_3	<0.03	2.67	0.31[2]	0.3		50% w/MeOH	Ox alk	11
			u	2.720	0.31[2]	0.116	2.00459	w/EtOH	Ox alk	8
			u	3.20	0.32[2]	0.60		MeCN	Re el	12
13	-,3,6-dipropyl-	–CH_2CH_2Me	u	3.26	3.26	u		Ethers	Re Na	13
14	-,3,6-diisopropyl-	–$CHMe_2$	u	3.4	3.4	u		w	Ox alk	9
			0.45[1]	3.2	3.2	0.45[1]		10% w/DMF	Ox alk	9
15	[C]-,3,6-di-*t*-butyl-	–CMe_3	u	3.3	3.3	u		w/EtOH	Ox alk	131
16	-,3,6-diallyl-	–$CH_2CH{=}CH_2$	u	3.32	3.32	u		Ethers	Re Na	13
17	-,4,5-dimethyl-	–Me	0.5	4.2[1]	4.2[1]	0.5		Ethers	Re Na	13
18	-,3,4,6-trimethyl-	–Me	u	2.7[1]	2.7	u		Ethers	Re Na	13
19	-,4,6-di-*t*-butyl-3-phenyl-	–CMe_3,–Ph	u	u	3.47	u		Ethers	Re Na	13
20	-,4-(2-carboxylatoethyl)-	–CH_2CH_2COO	0.22	3.56[1]	3.73	0.88	2.0045	w/pH=11	Phot alk	14
			0.25	3.55[1]	3.75	0.87	2.00455	w/EtOH	Ox alk	8
	-,4-(2-R-oxycarbonylethyl)-	–CH_2CH_2COOR								
	R=	R=								
21	methyl	–Me	0.35	3.42[1]	3.70	0.90		w/EtOH	Ox alk	8
22	ethyl	–Et	0.35	3.40[1]	3.70	0.91	2.00454	w/EtOH	Ox alk	8
23	butyl	–$(CH_2)_3Me$	0.36	3.30[1]	3.70	0.95		w/EtOH	Ox alk	8
24	pentyl	–$(CH_2)_4Me$	0.45	3.24[1]	3.61	1.02		w/EtOH	Ox alk	8
25	isopropyl	–$CHMe_2$	0.40	3.35[1]	3.63	0.95		w/EtOH	Ox alk	8
26	1-methyl-2-(6-pyrone)ethyl	Me | –$CHCH_2$– 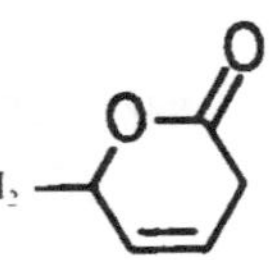	0.5	3.3[1]	3.7	1.0	2.00456	w/EtOH	Ox alk	15
27	Unknown	Unknown	0.46	3.30[1]	3.70	1.04	2.00456	30% w EtOH	Ox alk	15

Table I.1. (continued)
1,2-BENZOSEMIQUINONES

Anions

No.	Radical	Substituent(s)	Splitting constants/Gauss				g-Factor	Solvent	Method	Ref.
			a_3	a_4	a_5	a_6				
28	-,4-[2-(3,4-dioxidocinnamoyl-oxy)-2-carboxylatoethyl]-	OOCCH=CH \ $-CH_2CHCOO^-$ 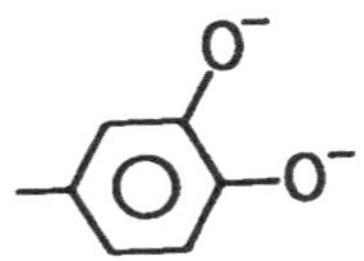	0.49	3.37^{1} 1.98^{1} 0.11^{2}	3.65	0.97	2.00456	w/EtOH	Ox alk	16
29	-,4-[4-(3,4-dioxidophenyl)-2,3-dimethylbutyl]-	Me | $-CH_2CHCHCH_2$ | Me 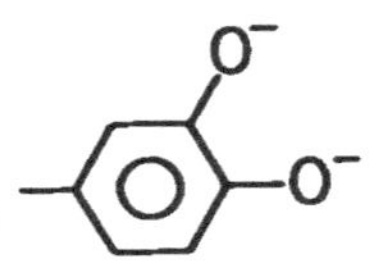	0.90	3.45^{1} 1.95^{1}	3.45	1.45		60% DMF DMSO (80°C)	Ox KoBu^t	8
30	-,4-[2-(3,5-dioxidophenylvinyl)]-	−CH=CH 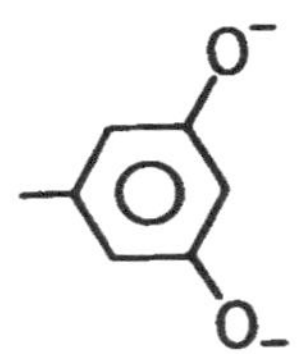	0.32	2.05^{1} 1.42^{2} 0.65^{4} 2.05^{6}	2.19	1.42		w/EtOH	Ox alk	8
		(Alternative assignment):	0.32	1.42^{1} 0.65^{2} 2.05^{4} 0.65^{6}	2.19	1.42				8

31	-,4-(2-carboxylatovinyl)-	–CH=CHCOO	0.15	2.45^{1} 1.20^{2}	2.75	1.35		w	Ox alk	26
			0.23	2.36^{1} 1.30^{2}	2.82	1.19		w/EtOH^{c}	Ox alk	17
			0.30	2.40^{1} 1.29^{2}	2.90	1.29	2.00456	30% w/EtOH	Ox alk	8, 130
	-,4-(2-R-oxycarbonylvinyl)-	–CH=CHCOOR								
	R=	R=								
32	methyl	–CH_3^*	0.70	2.35^{1} 1.19^{2} 0.17*	2.57	1.19		w/EtOH	Ox alk	8
33	ethyl	–CH_2^*Me	0.70	2.31^{1} 1.19^{2} 0.17*	2.60	1.19		w/EtOH	Ox alk	8
34	propyl	–$CH_2^*CH_2$Me	0.69	2.31^{1} 1.17^{2} 0.17*	2.61	1.17	2.00465	w/EtOH	Ox alk	8
35	isopropyl	–CH*Me_2	0.70	2.31^{1} 1.19^{2} 0.11*	2.59	1.19		w/EtOH	Ox alk	8
36	quinic acid	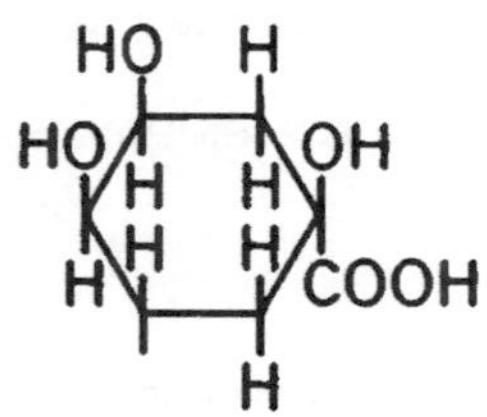	0.63	2.40^{1} 1.20^{2}	2.60	1.20		w/EtOH^{c}	Ox alk	17
			0.78	2.30^{1} 1.20^{1}	2.61	1.20	2.00463	30%/EtOH	Ox alk	15
37	[C]-,4-(2-aminoethyl)- -,4-amino-5-(2-aminoethyl)-	–$CH_2CH_2NH_2$ (see Radical no. 77)	0.28	3.11^{1}	3.70	0.91	2.0045	w/pH=11	Phot alk	14
38	[C]-,4-(2-ammonioethyl)-	–$CH_2CH_2NH_3^+$	0.46	2.98^{1}	3.58	0.94	2.0045	w/pH=7	Phot ac	14, 128
39	-,4-(2-acetylaminoethyl)-	–CH_2CH_2NHCOMe	(Spectrum not obtained)					w/EtOH	Ox alk	8
40	[C]-,4-(2-amino-2-carboxylato-ethyl)-	–$CH_2CH(NH_2)COO^-$	0.34	3.16^{1} 2.26^{1}	3.66	0.88		w	Ox alk	130

Table I.1. (continued)
1,2-BENZOSEMIQUINONES

Anions

No.	Radical	Substituent(s)	Splitting constants/Gauss a_3	a_4	a_5	a_6	g-Factor	Solvent	Method	Ref.
			0.35	3.23[i] 2.26[i]	3.70	0.88	2.0045	w/pH=11	Phot alk	14
41	[C]-,4-(2-ammonio-2-carboxylatoethyl)-	$-CH_2CH(NH_3^+)COO^-$	0.51	3.34[i] 1.94[i]	3.58	0.94	2.0045	w/pH=7	Phot ac	14
42	-,4-(2-amino-2-carboxylato-2-methylethyl)-	$-CH_2C(Me)(NH_2)$ (with COO on C)	0.38	2.75[i]	3.68	0.84		w	Ox alk	130
			0.39	2.80[i] 1.90[i]	3.70	0.89	2.0045	w/pH=11	Phot alk	14
43	-,4-(2-ammonio-2-carboxylato-2-methylethyl)-	$-CH_2C(Me)(NH_3^+)$ (with COO^- on C)	0.59	2.70[i]	3.55	0.94	2.0045	w/pH=7	Phot ac	14
44	[C]-,4-(2-amino-2-methoxycarbonylethyl)-	$-CH_2CH(NH_2)$ (with COOMe on C)	0.42	2.66[i]	3.65	0.93	2.0045	w/pH=11	Phot alk	14
45	-,4-(2-acetyl-2-aminovinyl)-	–CH=CHNHCOMe	(Spectrum not obtained)					w/EtOH	Ox alk	8
46	[C]-,4-carboxylatomethyl-	$-CH_2COO^-$	~0.25	3.0[i]	3.9	0.9		w	Ox alk	9
			0.36	3.15[i]	3.72	0.97	2.00456	w/EtOH	Ox alk	8, 130
			0.6	3.0[i]	4.1	1.15		10% w/DMF	Ox alk	9
47	-,4-ethoxycarbonylmethyl-	$-CH_2COOCH_2Me$	0.60	2.80[i]	3.52	1.01	2.00462	w/EtOH	Ox alk	8
48	-,4-propoxycarbonylmethyl-	$-CH_2COO(CH_2)_2Me$	0.61	2.77[i]	3.54	1.02		w/EtOH	Ox alk	8
49	-,4-pentoxycarbonylmethyl-	$-CH_2COO(CH_2)_4Me$	0.61	2.75[i]	3.53	1.02		w/EtOH	Ox alk	8
50	-,4-aminomethyl-	$-CH_2NH_2$	0.34	4.05[i] 0.81[d]	3.72	0.81	2.0045	w/pH=11	Phot alk	14

51	-,4-ammoniomethyl-	$-CH_2NH_3^+$	0.82	2.75^{1} 1.38^{d}	3.48	0.93	2.0045	w/pH=7	Phot ac	14
52	-,4-hydroxymethyl-	$-CH_2OH$	0.56	4.03^{1}	3.62	0.95	2.00458	w/EtOH	Ox alk	8
53	-,4-(2-amino-1-hydroxyethyl)-	$-CH(OH^*)CH_2NH_2$	0.88	2.72^{1}	3.58	0.88	2.0047	w/HCl	Phot ac	128
			0.58	2.55^{1} 0.15^{2} 0.20*	3.60	1.02		w/EtOH	Ox alk	8
54	-,4-(1-hydroxy-2-methylaminoethyl)-	$-CH(OH)CH_2NHMe$	0.88	2.72^{1}	3.58	0.88	2.0047	w/HCl	Phot ac	128
55	-,4-(1,2-dihydroxyethyl)-	$-CH(OH)CH_2OH$	~0.65	2.60^{1} ~0.0^{2}	3.60	~0.95	2.00459	w/EtOH	Ox alk	8
56	-,4-hydroxycarboxylatomethyl-	$-CH(OH)COO^-$	0.57	2.25^{1}	3.51	0.90		50% w/Dioxan	Ox alk	8
			0.577	2.295^{1}	3.541	0.908		w/EtOH	Ox alk	8
57	[C]-,4-formyl-	$-CHO$	2.35	0.45^{1}	2.35	0.90	2.00504	w/EtOH	Ox alk	8
58	[C]-,4-acetyl-	$-COMe$	2.13	~0.15^{2}	2.65	0.80		w/EtOH	Ox alk	8
59	-,3-acetyl-4-methyl-	$-COMe, -Me$	u	5.31^{1}	3.53	1.40	2.00458	w/EtOH	Ox alk	8
60	-,4-aminoacetyl	$-COCH_2NH_2$	(Spectrum not obtained)					w/EtOH	Ox alk	8
61	-,4-acetylaminoacetyl	$-COCH_2NHCOMe$	(Spectrum not obtained)					w/EtOH	Ox alk	8
62	-,4-hydroxyacetyl-	$-COCH_2OH$	(Spectrum not obtained)					w/EtOH	Ox alk	8
63	-,4-carbamoyl-	$-CONH_2$	(Spectrum not obtained)					w/EtOH	Ox alk	8
64	-,3-carboxylato-	$-COO$	—	3.92	3.45	0.67	2.00457	w	Radiolysis^{c}	18, 426
65	-,3-carboxy-	$-COOH$	—	5.15	2.9	1.7		w/0.5 *M* H_2SO_4	Ox CeIV	19
66	[C]-,3-ethoxycarbonyl-	$-COOEt$	u	(4.80	2.95)	2.00	2.00480	w/EtOH	Ox alk	8
67	[C]-,4-carboxylato-	$-COO^-$	1.25	—	3.26	0.73	2.00471	w	Radiolysis^{c}	18, 115
			1.24	—	3.23	0.72	2.0047	w/pH=11	Phot alk	9, 14^{c}
			1.35	—	3.30	0.80	2.00478	30% w/EtOH	Ox alk	15
			1.4	—	3.3	0.9		10% w/DMF	Ox alk	9^{c}
68	-,4-carboxy-	$-COOH$	1.25	—	3.25	0.75		w/0.5 *M* H_2SO_4	Ox CeIV	19
69	[C]-,4-methoxycarbonyl-	$-COOCH_3^*$	2.00	0.25*	2.75	0.82	2.00490	w/EtOH	Ox alk	8
70	-,4-ethoxycarbonyl-	$-COOCH_2^*Me$	2.09	0.25*	2.76	0.90	2.00489	w/EtOH	Ox alk	8
71	-,4-propoxycarbonyl-	$-COOCH_2^*CH_2Me$	2.07	0.27*	2.81	0.87	2.00489	w/EtOH	Ox alk	8
72	-,4-butoxycarbonyl-	$-COOCH_2^*(CH_2)_2Me$	2.07	0.27*	2.81	0.87	2.00489	w/EtOH	Ox alk	8
73	[C]-,4-isopropoxycarbonyl-	$-COOCH^*Me_2$	2.00	0.15*	2.76	0.86	2.00489	w/EtOH	Ox alk	8
74	-,3,5-dicarboxylato-	$-COO^-$	—	3.00	—	1.22	2.00471	w	Radiolysis^{c}	18

Table I.1. (continued)
1,2-BENZOSEMIQUINONES

Anions

No.	Radical	Substituent(s)	Splitting constants/Gauss a_3	a_4	a_5	a_6	g-Factor	Solvent	Method	Ref.
75	-,3-amino-	$-NH_2$	1.65[d]	0.00	5.10	0.85		EtOH/w	Ox KOH	80
			0.70[f]							
76	-,4-amino-	$-NH_2$	1.90	2.10[d]	3.10	0.85		EtOH/w	Ox KOH	80
				3.95[f]						
77	-,4-amino-5-(2-aminoethyl)-	$-NH_2$,$-CH_2CH_2NH_2$	1.60	2.75[d]	2.75[t]	0.77		w	Ox alk	181
				1.34[f]						
78	-,3,4-diamino-	$-NH_2$	1.15[d]	1.60[d]	3.20	0.89		EtOH/w	Ox KOH	80
			0.00[f]	0.95[f]						
79	-,3,5-diamino-	$-NH_2$	1.15[d]	1.50	2.50[d]	0.20		EtOH/w	Ox KOH	80
			0.20[f]		1.15[f]					
80	-,4,5-diamino-	$-NH_2$	0.20	2.25[d]	2.25[d]	0.20		EtOH/w	Ox KOH	80
				0.90[f]	0.90[f]					
	-,4,5-imidazo-	(see Nos. 92 and 93)								
81	-,4-bromo-	–Br	0.55	—	3.95	1.50	2.00456	w	Ox alk	8
82	-,4,5-dichloro-	–Cl	u	—	—	u		Ethers	Re Na	13
83	-,4-bromo-5-chloro-	–Br, –Cl	(Spectrum not obtained)					w/EtOH	Ox alk	8
84	-,4,5-dibromo-	–Br	0.5	—	—	0.5		w	Ox alk	9
			1.4	—	—	1.4		10% w/DMF	Ox alk	9
85	-,3-chloro-4,5-dimethyl-	–Cl, –Me	—	4.35[t]	4.35[t]	u		Ethers	Re Na	13
86	[C]-,3-chloro-4,6-di-t-butyl-	–Cl, $-CMe_3$	—	u	3.06	u		Ethers	Re Na	13
87	-,4-chloro-3,6-dipropyl-	–Cl, $-(CH_2)_2Me$	u	—	3.53	u		Ethers	Re Na	13
88	-,5-chloro-3,4,6-trimethyl-	–Cl, –Me	u	3.73[t]	—	u		Ethers	Re Na	13

89	[C.I]-,tetrachloro-	(see Tables IV.1.3 and IV.1.6)							
90	[C]-,tetrabromo-	(see Tables IV.1.3 and IV.1.6)							
91	-,3,5-disulfonato-	$-SO_3^-$	—	3.6	—	1.95	w	Ox alk	9
			—	3.9	—	1.8	10% w/DMF	Ox alk	9

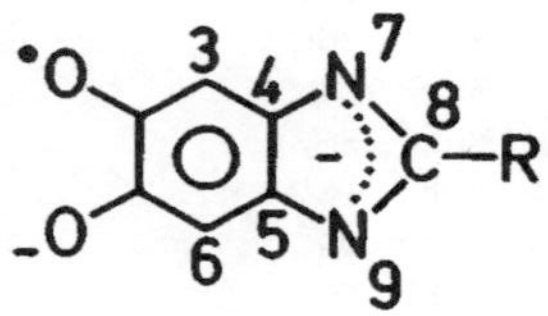

			$a_3=a_6$	$a_7^N=a_9^N$	a_8			
92	4,5-Imidazo-1,2-benzosemiquinone	(R = H)	0.53	1.06	5.04[H]	DMSO	Ox KOBu^t	220
93	-,8-methyl-	(R = Me)	0.33	1.06	5.02[Me]	DMSO	Ox KOBu^t	220

[a] Radical expected to be deprotonated in the neutral liquid (MeOH); see Reference 129.
[b] Splitting from five protons.
[c] w >50%; see Reference 17.
[d] Nitrogen splitting.
[e] *In situ* radiolysis; aqueous solutions saturated wih N_2O to convert e_{aq}^- into ·OH; see Reference 18.

Table I.2
1,3-BENZOSEMIQUINONES

Anions

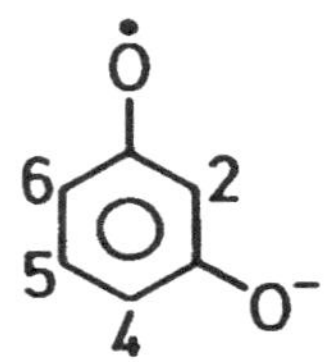

No.	Radical	Substituent(s)	Splitting constants/Gauss a_2	a_4	a_5	a_6	g-Factor	Solvent	Method	Ref.
94	[I]1,3-Benzosemiquinone		0.7	11.2	2.8	11.2		w/alk	Ox FeIII[a]	19, 83
			0.7	11.25	2.4	11.25		w/alk	Ox FeIII	27
			0.63	11.30	2.40	11.30		w/alk	Ox FeIII	1
			0.7	11.5	2.4	11.5	2.00392	w/alk	Ox CeIV	85
			0.68	11.44	2.43	11.44	2.00383	w/pH=11	Radiolysis[b]	132
95	-,2-methyl-	–Me	0.98[1]	10.98	2.25	10.98		w/alk	Ox FeIII	1
96	-,4-methyl-	–Me	(Data not available)							
97	-,4-ethyl-	–Et	u	10.55[1]	2.3	10.55		w/alk	Ox FeIII	27
98	-,4-hexyl-	–$(CH_2)_5$Me	u	10.0[1]	2.3	10.0		w/alk	Ox FeIII	27
99	-,4-benzyl-	–CH_2Ph	0.7	10.4[1]	2.3	10.4		w/alk	Ox FeIII	27
100	-,5-methyl-	–Me	0.75	11.20	3.37[1]	11.20		w/alk	Ox FeIII	1, 27
101	-,4-(2-carboxylatovinyl)-	–CH=CHCOO	2.0	4.75[1] 3.25[2]	3.5	9.0		w/alk	Ox FeIII	26'
102	[I]-,2-carboxylato-	–COO^-	—	11.25	2.35	11.25		w/alk	Ox FeIII	27
			—	11.34	2.41	11.34	2.00386	w/pH=11	Radiolysis	132
103	[I]-,4-carboxylato-	–COO	0.7	—	2.3	11.0		w/alk	Ox FeIII	27
			0.68	—	2.31	10.99	2.00389	w/pH=11	Radiolysis	132

104	[I]-,5-carboxylato-	$-COO$	0.7	11.3	—	11.3		w/alk	Ox FeIII	27
			0.68	11.41	—	11.41	2.00385	w/pH=11	Radiolysis	132
105	-,2-nitro-	$-NO_2$	u	11.5	2.6	11.5		w/alk	Ox FeIII	27
106	-,4-chloro-	$-Cl$	0.7	—	2.0	9.8		w/alk	Ox FeIII	27

[a] Steady-state generation of radicals by flow mixing (FeIII).
[b] *In situ* radiolysis, aqueous solutions saturated with N_2O.

Table I.3
1,4-BENZOSEMIQUINONES

Anions

No.	Radical	Substituent(s)	Splitting constants/Gauss a_2	a_3	a_5	a_6	g-Factor	Solvent	Method	Ref.
	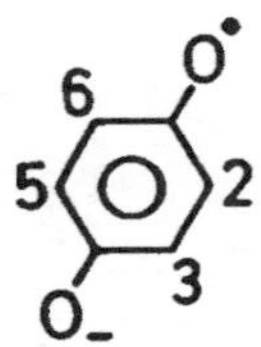									
107	[C,I] 1,4-Benzosemiquinone[c]		2.3	2.3	2.3	2.3	2.004659[a]	w	Ox alk	1, 64
			2.35	2.35	2.35	2.35	2.00457	w(pH = 5—11)	Radiolysis	115
			2.36	2.36	2.36	2.36		w	Re el	31
			2.35	2.35	2.35	2.35		50% w/MeOH	Ox alk	32
			2.37	2.37	2.37	2.37		50% w/EtOH	Ox alk	31, 33
			2.368	2.368	2.368	2.368	2.0051	DMSO/w	Ox alk[b]	34
			2.40	2.40	2.40	2.40		50% w/DMF	Ox alk	32
			2.40	2.40	2.40	2.40		DMF	Ox alk	35
			2.40	2.40	2.40	2.40		DMF	Re el	31
			2.32	2.32	2.32	2.32		MeCN	Re el	12
			2.42	2.42	2.42	2.42		MeCN	Re el	31

Table I.3 (continued)
1,4-BENZOSEMIQUINONES

Anions

No.	Radical	Substituent(s)	Splitting constants/Gauss a_2	a_3	a_5	a_6	g-Factor	Solvent	Method	Ref.
			2.395	2.395	2.395	2.395		DME	Re el	31, 39
			2.419	2.419	2.419	2.419	2.00541[c]	DMSO	Re el	36, 41
			2.42	2.42	2.42	2.42		HMPA	Re K	75
			2.45	2.45	2.45	2.45		HMPA	Ox KOBu[t]	4, 262
			2.26	2.26	2.26	2.26		$(CH_2OH)_2$	Re phot[d]	38
			2.28	2.28	2.28	2.28	2.0060	THF/DCC	Re KO_2	7
			u	u	u	u	2.004679[c]	BuOH	Ox KOH	40
			2.32	2.32	2.32	2.32	2.0050	C_6H_6	Re$(Bu_3Sn)_2$[f]	116
108	[C]-,2-methyl-	–Me	2.10[1]	1.70	2.55	2.37		w	Ox alk	1, 4
			2.12[1]	1.76	2.61	2.41		50% w/EtOH	Ox alk	33
			2.045[1]	1.764	2.537	2.462		~10% w/EtOH	Ox alk	42
			1.88[1]	1.92	2.60	2.48		DMF	Ox alk	35
			1.90[1]	1.98	2.72	2.56		MeCN	Re el	12
			1.83[1]	1.952	2.702	2.440		DMSO	Re el	36
			1.77[1]	2.05	2.73	2.43		HMPA	Ox KOBu[t]	4, 262
			1.75[1]	2.05	2.39	2.14		HMPA	Re K	37
109	-,2-ethyl-	–Et	2.02[1]	1.74	2.53	2.53		50% w/EtOH	Ox alk	33
110	-,2-butyl-	$-(CH_2)_3Me$	1.67[1]	1.67	2.50	2.50		w or w/MeOH	Ox KOH	43
111	-,2-allyl-	$-CH_2CH{=}CH_2$	1.98[1] 0.10[2]	1.98	2.52	2.47		MeCN	Ox alk	44
112	-,2-benzyl-	$-CH_2Ph$	1.57[1]	1.88	2.43	2.40		W/EtOH	Ox alk	79
113	-,2-isopropyl-	$-CHMe_2$	1.32[1]	1.75	2.60	2.42		w	Ox alk	89
			1.35[1]	1.78	2.54	2.54		w/MeOH	Ox KOH	43
			1.24[1]	1.79	2.51	2.45		C_5H_5N	Re DTN	45
			1.24[1]	2.00	2.76	2.76		MeCN	Re el	46
114	-,2-s-butyl-	$-CH(Me)CH_2Me$	0.93[1]	1.76	2.50	2.50		w or w/MeOH	Ox KOH	43

115	[I]-,2-*t*-butyl-	–CMe$_3$	0.08[z]	1.62	2.13	2.86			w	Ox alk	35
			0.06[z]	1.41	1.91	2.82			88% w/MeOH	Ox alk	11
			u	1.70	2.15	3.05			75% w/MeOH	Ox KOH	10
			0.06[z]	1.55	2.07	2.82			50% w/MeOH	Ox alk	11, 32
			u	1.60	2.05	2.90			50% w/EtOH	Ox el/pH = 9	47
			0.06[z]	1.85	2.29	2.82			50% w/DMF	Ox alk	32
			u	2.04	2.48	2.84			MeCN	Re el	12, 46
			0.06[z]	2.00	2.42	2.65			DMF	Ox alk	35
116	-,2-phenyl-[c]	–Ph	0.27[o]	1.99	2.56	2.12			50% w/EtOH	Ox alk	48, 49
			0.16[m]								
			0.27[p]								
			0.29[o]	2.24	2.5	2.38			DMSO	Re el	125
			0.16[m]								
			0.29[p]								
	(For further aryl-1,4-benzosemiquinones see Tables 1.3.1 through 1.3.4)										
156	[C]-,2,3-dimethyl-	–Me	1.74[i]	1.74[i]	2.62	2.62		2.00457[g]	50% w/EtOH	Ox alk	8, 33
			1.714[i]	1.714[i]	2.595	2.595			EtOH/w	Ox alk	42
			1.69[i]	1.69[i]	2.54	2.54			50% w/DMF	Ox alk	1
			1.68[i]	1.68[i]	2.60	2.60			BzOH	Re el	31
			1.71[i]	1.71[i]	2.74	2.74			DME/ – 86°C	Re K	50
			1.72[i]	1.72[i]	2.76	2.76			DME/ – 85°C	Re Cs	50
157	[C]-,2,5-dimethyl-	–Me	2.36[i]	1.70	2.36[i]	1.70			w	Ox alk	1
			2.295[i]	1.795	2.295[i]	1.795		2.00450[g]	50% w/EtOH	Ox alk	47, 33
			2.32[i]	1.84	2.32[i]	1.84			50% w/DMF	Ox alk	51
			2.248[i]	1.863	2.248[i]	1.863			EtOH/w	Ox alk	42
			2.24[i]	1.80	2.24[i]	1.80			MeOH	Re el	52, 53
			2.25[i]	1.86	2.25[i]	1.86			BzOH	Re el	31
			2.15[i]	1.95	2.15[i]	1.95			DMSO	Re el	52, 53
			2.24[i]	2.00	2.24[i]	2.00			MeCN	Re el	12
			2.14[i]	1.94	2.14[i]	1.94			MeCN	Re el	31
			2.12[i]	1.97	2.12[i]	1.97			DMF	Ox alk	35
			2.09[i]	1.97	2.09[i]	1.97			DMF	Re el	31
			2.13[i]	2.13	2.13[i]	2.13			DME/ – 98°C	Re K	50
158	-,5-isopropyl-2-methyl-	–CHMe$_2$, –Me	2.18[i]	1.92	1.46[i]	1.76[h]			50% w/EtOH	Ox alk	33
			2.12[i]	2.12	1.32[i]	2.12			MeCN	Re el	12

Table I.3 (continued)
1,4-BENZOSEMIQUINONES

Anions

No.	Radical	Substituent(s)	Splitting constants/Gauss a_2	a_3	a_5	a_6	g-Factor	Solvent	Method	Ref.
159	-,5-*t*-butyl-2-methyl-	–CMe$_3$, –Me	1.78[1]	2.24	u	1.78		w	Ox alk	35
			1.89[1]	2.20	u	2.00		DMF	Ox alk	35
160	-,2,5-diallyl-	–CH$_2$CH=CH$_2$	1.99[1]	2.11	1.99[1]	2.11		MeCN	Ox alk	44
			0.12[2]		0.12[2]					
161	[C]-,2,5-di-*t*-butyl-	–CMe$_3$	0.06[2]	2.04	0.06[2]	2.04		w	Ox alk	35
			0.06[2]	2.07	0.06[2]	2.07		50% w/MeOH	Ox alk	11, 32
			u	2.13	u	2.13	2.00464	w/EtOH (?)	Ox alk	3
			u	2.13	u	2.13		Bu^{t}OH/w	LiOBu^t	54
			u	2.14	u	2.14		DMSO	KOBu^t	90
			u	2.20	u	2.20		DMSO	Re KO$_2$	55
			0.06[2]	2.25	0.06[2]	2.25		50% w/DMF	Ox alk	32
			0.06[2]	2.23	0.06[2]	2.23		DMF	Ox alk	35
			u	2.32	u	2.32		MeCN	Re el	12
			u	2.33	u	2.33		HMPA	Ox KOBu^t	4
162	-,2,5-di-*t*-pentyl-	Me –C–CH$_2$Me Me	u	2.30	u	2.30		MeCN	Re el	12
163	[I]-,2,5-diphenyl-	–Ph	0.25[o]	2.20	0.25[o]	2.20		50% w/EtOH	Ox alk	48
			0.14[m]		0.14[m]					
			0.25[p]		0.25[p]					
164	-,2,5-di(*p*-chlorophenyl)-	–C$_6$H$_4$Cl(*p*)	0.31[o]	1.78	0.31[o]	1.78		50% w/EtOH	Ox alk	48
			0.16[m]		0.16[m]					
165	[C]-,2,6-dimethyl-	–Me	2.13[1]	1.88	1.88	2.13[1]		w	Ox alk	35
			2.12[1]	1.93	1.93	2.12[1]		w	Re el	31
			2.09[1]	1.93	1.93	2.09[1]	2.00453	50% w/EtOH	Ox alk	8, 33

			2.12[1]	1.87	1.87	2.12[1]		50% w/DMF	Ox alk	1
			2.125[1]	1.892	1.892	2.125[1]		EtOH/w	Ox alk	42
			2.10[1]	1.92	1.92	2.10[1]		MeOH	Re Li	50
			2.14[1]	1.86	1.86	2.14[1]		MeOH	Re el	31
			2.0[1]	2.0	2.0	2.0[1]	2.0045	w/EtOH	Ox alk	56
			1.95[1]	2.16	2.16	1.95[1]		MeCN	Re el	46
			1.86[1]	2.18	2.18	1.86[1]		DMF	Ox alk	35
			1.86[1]	2.21	2.21	1.86[1]		DMF	Re el	31
			1.85[1]	2.22	2.22	1.85[1]		HMPA	Ox KOBu[t]	4
			1.86[1]	2.30	2.30	1.86[1]		HMPA	Re el	31
			1.71[1]	2.35	2.35	1.71[1]		HMPA	Re K	37
			1.75[1]	2.34	2.34	1.75[1]		DME/ − 85°C	Re Cs	50
			1.75[1]	2.43	2.43	1.75[1]		DME/ − 85°C	Re K	50
166	-,6-ethyl-2-methyl-	–Et, –Me	2.23[1]	1.85	1.85	2.07[1]	2.00452	w/EtOH	Ox alk	8
167	-,2-methyl-6-tetraprenyl-	–Me, –R[i]	2.184[1]	1.818	1.902	2.889[1]		w/EtOH	Ox alk	8
168	-,6-isopropyl-2-methyl-	–CHMe$_2$, –Me	2.21[1]	1.76	1.76	1.35[1]		w/EtOH	Ox alk	8
169	-,6-*t*-butyl-2-methyl-	–CMe$_3$, –Me	2.56[1]	1.52	1.66	u	2.0045	w/EtOH	Ox alk	56
			2.54[1]	1.52	1.76	u		w	Ox alk	35
			2.23[1]	2.04	2.23	u		MeCN	Re el	46
			2.16[1]	1.87	2.16	u		DMF	Ox alk	35
170	-,2,6-diisopropyl-	–CHMe$_2$	1.24[1]	2.20	2.20	1.24[1]		MeCN	Re el	46
171	-,6-*t*-butyl-2-isopropyl-	–CMe$_3$, –CHMe$_2$	1.28[1]	2.14	1.94	u		MeCN	Re el	46
172	[C]-,2,6-di-*t*-butyl-	–CMe$_3$	u	1.31	1.31	u		w	Ox KOBu[t]	4
			0.09[2]	1.23	1.23	0.09[2]		w	Ox alk	35
			0.06[2]	1.17	1.17	0.06[2]		50% w/MeOH	Ox alk	11, 32
			u	1.31	1.31	u		EtOH	Re el	31
			~0.06[2]	1.67	1.67	~0.06[2]		50% w/DMF	Ox alk	32
			u	1.70	1.70	u		DMSO	Re el	31
			u	1.89	1.89	u		MeCN	Re el	31
			u	2.04	2.04	u		MeCN	Re el	46
			u	1.93	1.93	u		DMF	Re el	31
			0.06[2]	2.00	2.00	0.06[2]		DMF	Ox alk	35
			u	2.00	2.00	u	2.0052	C_6H_6	Ox Ag_2O	57
			u	2.25	2.25	u		HMPA	Ox KOBu[t]	4, 262

Table I.3 (continued)
1,4-BENZOSEMIQUINONES

Anions

No.	Radical	Substituent(s)	Splitting constants/Gauss a_2	a_3	a_5	a_6	g-Factor	Solvent	Method	Ref.
173	-,2,3,5-trimethyl-	–Me	2.33[1]	1.91[1]	1.76[1]	1.94		w	Ox alk	1
			2.33[1]	1.86[1]	1.83[1]	1.85	2.00445	50% w EtOH	Ox alk	47
			2.155[1]	1.941[1]	1.749[1]	2.112	2.00489	DMSO w	Ox alk	34
174	-,2,3-dimethyl-5-(3-methyl-2-butenyl)-	–Me, $-CH_2CH{=}CHMe_2$	1.85[1]	1.85[1]	2.50[1]	1.90	2.00450	w/EtOH	Ox alk	8
175	-,2,3-dimethyl-5-solanesyl-	–Me, $-(CH_2CH{=}C(Me)CH_2)_8CH_2CH{=}CMe_2$	1.87[1]	1.78[1]	2.37[1]	2.01		w/EtOH	Re KOH	58
			1.80[1]	1.80[1]	2.40[1]	2.00	2.00452	w/EtOH	Ox alk	8, 16
176	[C,I]-,tetramethyl-	–Me	1.96[1]	1.96[1]	1.96[1]	1.96[1]		w	Re el	31
			1.92[1]	1.92[1]	1.92[1]	1.92[1]	2.004915[k]	DME	Re el	40
			1.90[1]	1.90[1]	1.90[1]	1.90[1]		DMF, MeCN or HMPA	Re el	31
177	-,2,3,5-trimethyl-6-R-[l]	–Me, –R[l]	1.905[1]	1.905[1]	1.905[1]	0.91[1]		EtOH[m]	Re K	76
			2.10[1]	2.10[1]	1.68[1]	0.85[1]		Me_2CO	Re KOH	216
			2.21[1]	2.21[1]	1.60[1]	0.81[1]		DMF	Re el	77
178	-,2-(4-carboxylatobutyl)-	$-(CH_2)_4COO^-$	1.74[1]	1.74	2.52	2.52	2.00460	w/EtOH	Ox alk	8
179	-,2-carboxylatomethyl-	$-CH_2COO$	1.71[1]	1.87	2.40	2.56	2.00462	w/EtOH	Ox alk	8, 78
180	-,2,5-dicarboxylatomethyl-	$-CH_2COO$	1.55[1]	2.01	1.55[1]	2.01	2.00462	w/EtOH	Ox alk	8, 78
181	-,2-(2-carboxylatovinyl)-	$-CH{=}CHCOO$	1.70[1] 0.85[2]	2.13	2.13	2.40		50% w/EtOH	Ox alk	17
182	-,2-(2-methyl-2-carboxylatovinyl)-	$-CH{=}C(Me)COO^-$	1.53[1] 1.00[3]	1.80	2.30	2.36		50% w/EtOH	Ox alk	17
183	-,2-(2-phenyl-2-carboxylatovinyl)-	$-CH{=}C(Ph)COO^-$	1.36[1] 0.70[n]	1.91	2.14	2.50		50% w/EtOH	Ox alk	17
184	-,2-(1-methyl-2-carboxylatovinyl)-	$-C(CH_3^{**}){=}CH^*COO$	0.0** 0.14*	1.84	2.22	2.54		50% w/EtOH	Ox alk	17

185	-,2-(1-phenyl-2-carboxylatovinyl)-	–C(Ph)=CH*COO	0[H] 0.09[*]	2.04	2.16	2.58		50% w/EtOH	Ox alk	17
186	-,2-hydroxymethyl-	$-CH_2OH$	2.18[f]	2.05	2.41	2.41		w	Ox alk	8
187	-,2-ethoxymethyl-	$-CH_2OEt$	2.17[f]	2.17	2.40	2.48	2.00464	w/EtOH	Ox alk	8
188	-,2-cyano-	–CN	—	(u	1.5	3.0)		MeOH	Re el	59
189	[C.I]-,2,3-dicyano-	–CN	0.61[N]	0.61[N]	1.05	1.05		DME	Ox I_2[g]	60, 81
			0.66[N]	0.66[N]	1.09	1.09		DME	TDAP	168
190	[C.I]-,2,3-dicyano-5,6-dichloro-	–CN, –Cl	0.62[N]	0.62[N]	—	—		DMSO	Re KO_2	55, 81
			0.61[N]	0.61[N]	—	—	2.0052	C_6H_6	Re $(Bu_3Sn)_2$[i]	115
			0.59[N]	0.59[N]	—	—		C_6H_6	TDAP	168
			0.57[N]	0.57[N]	—	—	2.0038	DME/Me_2CO	Re Me_4NI	60
			u	u	—	—		MeCN		82
191	-,2-acetyl-	–COMe	0.25[j]	(3.89	1.25	2.49)	2.00452	w/EtOH	Ox alk	8
			—	(3.0	u	1.9)		MeOH	Re el	59
192	-,2-carboxylato-	$-COO^-$	—	(2.20	2.01	2.57)	2.00464	w/KOH	Radiolysis	18, 426
			—	(2.30	2.06	2.60)	2.00476	w/EtOH	Ox alk	8, 19
			—	(2.20	2.01	2.60)	2.00469	w(pH = 5 to 11)	Radiolysis	115
193	-,2,6-dicarboxylato-	$-COO^-$	—	1.86	1.86	—	2.00474	w/KOH	Radiolysis	18
194	-,2-carboxy-	–COOH	—	2.6	2.0	2.25		w/0.5 *M* H_2SO_4	Ox CeIV	19
195	-,2-methoxycarbonyl-	–COOMe	u	(3.05	u	1.85)		MeOH	Re el	59
196	-,2-ethoxycarbonyl-	–COOEt	0.11	(3.57	1.40	2.45)	2.00486	w/EtOH	Ox alk	8, 78
197	-,2,5-diethoxycarbonyl-	–COOEt	u	3.50	u	3.50	2.00473	w/EtOH	Ox alk	8, 78
198	-,2-trifluoromethyl-	$-CF_3$	2.66	3.37	1.62	2.19		MeCN	Ox alk	61
199	-,2-trimethylsilyl-	$-SiMe_3$	1.82[Si]	2.68	2.10	2.52		DMF	Re el	73
200	[I]-,2,5-bis(trimethylsilyl)-	$-SiMe_3$	1.51[Si]	2.79	1.51[Si]	2.79		DMF	Re el	73
201	-,2-amino-	$-NH_2$	1.75[N] 0.90[f]	0.20	4.90	1.52		EtOH/w	Ox KOH	80
202	-,2-amino-5-methyl-	$-NH_2$, –Me	1.85[N] 0.75[f]	0.18	4.65[j]	0.75		DMF/w		62
203	-,5-*t*-butyl-2-isopropylamino-	$-CMe_3$, $-NHCHMe_2$	2.05[N] 1.66[f] 3.71[f]	1.66	u	0.41		50% w/EtOH + 5% $CHCl_3$	Re el	47
204	-,2,3-diamino-	$-NH_2$	0.60[N] 0.00[f]	0.60[N] 0.00[f]	3.00	3.00		EtOH/w	Ox KOH	80

Table I.3 (continued)
1,4-BENZOSEMIQUINONES

Anions

No.	Radical	Substituent(s)	Splitting constants/Gauss a2	a3	a5	a6	g-Factor	Solvent	Method	Ref.
205 to 207	-,2,3-imidazo-	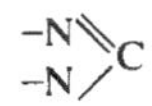	(see Table I.3.5)							
208 to 210	-,2,3-triazolo-	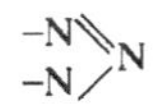	(see Table I.3.5)							
211	-,2,5-diamino-	$-NH_2$	2.6[q] 1.0[r]	0.75	2.6[q] 1.0[r]	0.75		w	Re el	52
			2.50[q] 0.90[r]	0.70	2.50[q] 0.90[r]	0.70		EtOH/w	Ox KOH	80
			2.50[q] 0.90[r]	0.70	2.50[q] 0.90[r]	0.70		DMF/w	Q NaN_3^-	62
			1.8[q] 0.2[r]	0.28	1.8[q] 0.2[r]	0.28		DMSO	Re el	52
212	-,2,5-diamino-3,6-dichloro-	$-NH_2$, –Cl	2.3[q] 1.15[r]	—	2.3[q] 1.15[r]	—		C_5H_5N/ $(CH_2OH)_2$/ 2 *M* KOH	Re DTN	63
213	-,2,5-bis(methylamino)-	–NHMe	3.00[q] 2.20[r] 3.47[t]	0.77	3.00[q] 2.20[r] 3.47[t]	0.77		50% w/EtOH +25% $CHCl_3$ (pH = 11)	Re el	47, 215
			2.03[q] 1.36[r] 2.27[t]	0.91	2.03[q] 1.36[r] 2.27[t]	0.91		MeCN or CH_2Cl_2	Re el	233
214	-,2,5-bis(ethylamino)-	–NHEt	3.02[q] 2.30[r] 3.46[t]	0.77	3.02[q] 2.30[r] 3.46[t]	0.77		50% w/EtOH +5% $CHCl_3$ (pH = 11)	Re el	47, 215

215	-,2,5-bis(propylamino)-	–NHCH$_2$CH$_2$Me	3.17[q] 2.35[r,u] 3.17[t]	0.80	3.17[q] 2.35[r,u] 3.17[t]	0.80	50% w/EtOH +5% CHCl$_3$ (pH = 11)	Re el	47, 215
			3.25[q] 2.36[r] 3.20[t] 0.1[w]	0.775	3.25[q] 2.36[r] 3.20[t] 0.1[w]	0.775	50% w/EtOH +2% CHCl$_3$ (pH = 13)	Re el	47, 215
216	-,2,5-bis(allylamino)-	–NHCH$_2$CH=CH$_2$	2.82[q] 2.10[r] 2.83[t] 0.5[s] 0.1[w]	0.79	2.82[q] 2.10[r] 2.83[t] 0.5[s] 0.1[w]	0.79	50% w/EtOH +10% CHCl$_3$ (pH = 11)	Re el	47, 215
217	-,2,5-bis(carboxylatomethylamino)-	–NHCH$_2$COO$^-$	2.95[q] 2.35[r]	0.6— 0.7	2.95[q] 2.35[r]	0.6—0.7	50% w/EtOH (pH = 13)	Re el	47
218	-,2,5-bis(isopropylamino)-	–NHCHMe$_2$	3.08[q] 2.31[r] 2.31[t]	0.77	3.08[q] 2.31[r] 2.31[t]	0.77	50% w/EtOH +10% CHCl$_3$ (pH = 11)	Re el	47, 215
219	-,2,5-bis(*t*-butylamino)-	–NHCMe$_3$	2.06[q] 1.54[r]	0.52	2.06[q] 1.54[r]	0.52	25% Bu^{t}OH/ DMSO	Ox KOBu^t	235
			2.98[q] 2.44[r] 0.05[w]	0.76	2.98[q] 2.44[r] 0.05[w]	0.76	50% w/EtOH +20% CHCl$_3$ (pH = 11)	Re el	47
			3.05[q] 2.55[r] 0.05[w]	0.777	3.05[q] 2.55[r] 0.05[w]	0.777	50% w/EtOH +2% CHCl$_3$ (pH = 13)	Re el	215
220	-,2,5-bis(phenylamino)-	–NHPh	3.1[q] 2.3[r] 2.3[x]	0.8	3.1[q] 2.3[r] 2.3[x]	0.8	50% w/EtOH +25% CHCl$_3$ (pH = 10)	Re el	47, 62
221	-,2,5-bis(1-piperidinyl)-	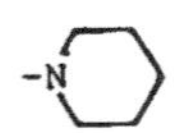	1.8[q] 1.5[t]	0.1	1.8[q] 1.5[t]	0.1	50% w/EtOH +5% CHCl$_3$ (pH = 13)	Re el	47, 62
222	-,2,5-bis(propylamino)-3,6-dimethyl-	–NHCH$_2$CH$_2$Me, –Me	1.75[q] 0.87[r] 0.87[t]	0.15	1.75[q] 0.87[r] 0.87[t]	0.15	50% w/EtOH +25% CHCl$_3$ (pH = 13)	Re el	47, 215
223	-,2,5-bis(isopropylamino)-3,6-dimethyl-	–NHCHMe$_2$, –Me	1.60[q] 0.87[r] 0.27[t]	0[l,y]	1.60[q] 0.87[r] 0.27[t]	0[l,y]	50% w/EtOH +10% CHCl$_3$ (pH = 11)	Re el	47, 215

Table I.3 (continued)
1,4-BENZOSEMIQUINONES

Anions

No.	Radical	Substituent(s)	Splitting constants/Gauss a_2	a_3	a_5	a_6	g-Factor	Solvent	Method	Ref.
224	-,2,5-bis(isopropylamino)-3,6-diphenyl-	$-NHCHMe_2$, $-Ph$	2.70[q] 1.62[r] 1.00[t]	0	2.70[q] 1.62[r] 1.00[t]	0		50% w/EtOH + 5% $CHCl_3$ (pH = 13)	Re el	47, 215
225	-,2,6-diamino-	$-NH_2$	1.00[q] 0.25[r]	1.30	1.30	1.00[q] 0.25[r]		EtOH/w	Ox KOH	80
226	-,2-tributylphosphinium-	$-P^+(CH_2CH_2CH_2Me)_3$	5.00[v]	5.15	1.70	1.70		CH_2Cl_2	—[aa]	70, 71
			4.90[v]	5.25	1.60	1.60		C_6H_6	—[aa]	70
			4.70[v]	5.55	1.55	1.75		THF	—[aa]	70
			4.70[v]	5.15	1.55	1.55		C_5H_5N	—[aa]	70
			4.80[v]	4.90	1.50	2.15		Pr^{i}OH	—[aa]	70
			4.90[v]	4.90	1.60	2.00		Bu^{t}OH	—[aa]	70
			5.05[v]	4.30	1.30	2.35		*o*-chlorophenol	—[aa]	70
227	-,2-triphenylphosphinium-	$-P^+(Ph)_3$	4.70[v]	5.80	1.55	1.70		CH_2Cl_2	—[aa]	70, 71
			4.42[v]	5.35	1.4	1.4		C_6H_6	PPh_3	168
228	[I]-,2-tri(β-cyanoethyl)phosphinium-	$-P^+(CH_2CH_2CN)_3$	4.50[v]	6.00	1.25	1.55		THF	—[aa]	70, 71
229	-,2-(phenylbisdimethylamino)-phosphinium-	$-P^+(Ph)(NMe_2)_2$	4.00[v] 0.2—0.3[q]	5.55	1.55	1.55		CH_2Cl_2	—[aa]	70
230	-,2-(phenylbisdiethylamino)-phosphinium-	$-P^+(Ph)(NEt_2)_2$	3.85[v] 0.2—0.3[q]	5.60	1.50	1.50		CH_2Cl_2	—[aa]	70
231	-,2-tris(dimethylamino)-phosphinium-	$-P^+(NMe_2)_3$	4.53[v] 0.23	5.05	1.57	1.57		C_6H_6	$P(NMe_2)_3$	168
232	-,2-tributylphosphinium-3,6-dibutyl-	$-P^+(CH_2CH_2CH_2Me)_3$, $-CH_2CH_2CH_2Me$	4.40[v]	u	1.60	u		CH_2Cl_2	—[aa]	70

233	-,2-tributylphosphinium-3,6-diphenyl-	$-P^+(CH_2CH_2CH_2Me)_3$, –Ph	4.30[c]	u	1.60	u	CH_2Cl_2	—[aa]	70
234	-,2-tris(dimethylamino)phosphinium-5,6-dicyano-	$-P^+(NMe_2)_3$, –CN	2.5[c]	3.2	0.4	0.4	CH_2Cl_2	TDAP	168
235	-,2-triphenylphosphinium-5-chloro-	$-P^+(Ph)_3$, –Cl	~4.6[c]	~5.5[bb]	—	~1.6[bb]	C_6H_6	—[aa]	72
236	-,2-benzylthio-	$-SCH_2Ph$	0.61*	1.28	2.77	2.10	EtOH/alk	Ox O_2 or Re NaOEt	74, 32
237	-,2-isopropylthio-5-methyl-	$-SCH^*Me_2$, –Me	0.6*	1.0	2.6[i]	1.6	50% w/EtOH (pH = 9)	—[cc,dd]	47
238	-,2-phenylthio-	–SPh	u	1.52	2.72	2.12	EtOH/alk	Ox O_2 or Re NaOEt	74, 32
	-,2-(*p*-R-phenylthio)-	–S–Ph–R(*p*)							
	R=	R=							
239	acetyl	COMe	u	2.24	2.33	2.24			
240	nitro	NO_2	u	2.23	2.59	2.23			
241	bromo	Br	u	1.70	2.70	2.19			
242	chloro	Cl	u	1.67	2.68	2.14	EtOH/alk	Ox O_2 or Re NaOEt	74
243	fluoro	F	u	1.53	2.78	2.15			
238	hydrogen	H	u	1.52	2.72	2.12			
244	methyl	Me	u	1.44	2.75	2.07			
245	amino	NH_2	u	1.26	2.96	2.10			
246	-,2-(α-naphthylthio)-	$-S-C_{10}H_7$	—	1.51	2.77	2.21	EtOH/alk	Ox O_2 or Re NaOEt	74
247	-,2-(β-naphthylthio)-	$-S-C_{10}H_7$	—	1.51	2.73	2.04	EtOH/alk	Ox O_2 or Re NaOEt	74
248	-,2,5-bis(methylthio)-	$-SCH_3$	0.868*	0.868	0.868*	0.868	50% w/EtOH + 20% $CHCl_3$ (pH = 11)	—[dd]	47
249	-,2,5-bis(ethylthio)-	$-SCH_2Me$	0.698*	0.958	0.698*	0.958	50% w/EtOH + 25% $CHCl_3$ (pH = 10)	—[dd]	47
250	-,2,5-bis(butylthio)-	$-SCH_2CH_2CH_2Me$	0.710*	0.915	0.710*	0.915	50% w/EtOH + 25% $CHCl_3$ (pH = 10)	Re el	47

Table I.3 (continued)
1,4-BENZOSEMIQUINONES

Anions

No.	Radical	Substituent(s)	Splitting constants/Gauss a_2	a_3	a_5	a_6	g-Factor	Solvent	Method	Ref.
251	-,2,5-bis(benzylthio)-	$-SCH_2Ph$	0.79*	0.91	0.79*	0.91		EtOH/alk	Ox O_2 or Re NaOEt	74
252	-,2,5-bis(2-carboxylatoethylthio)-	$-SCH_2CH_2COO^-$	0.697*	0.884	0.697*	0.884		50% w/EtOH (pH = 10)	—[d,k,l]	47
253	-,2,5-bis(2-dimethylaminoethylthio)-	$-SCH_2CH_2NMe_2$	u	1.77	u	1.77		50% w/EtOH (PH = 7.5)	—[d,k,l]	47
254	-,2,5-bis(isopropylthio)-	$-SCH*Me_2$	0.442*	1.137	0.442*	1.137		50% w/EtOH + 20% $CHCl_3$ (pH = 11)	—[d,k,l]	47
255	-,2,5-bis(phenylthio)-	$-SPh$	—	1.25	—	1.25		EtOH/alk	Ox O_2 or Re NaOEt	74
			—	1.298	—	1.298		50% w/EtOH + 20% $CHCl_3$ (pH = 10)	—[d,k,l]	47
256	-,2,5-bis(pentachlorophenylthio)-	$-SC_6Cl_5$	—	1.37	—	1.37		50% w/EtOH + 20% $CHCl_3$ (pH = 10)	—[d,k,l]	47
257	-,2,6-bis(benzylthio)-	$-SCH_2Ph$	0.57*	1.77	1.77	0.57*		EtOH/alk	Ox O_2 or Re NaOEt	74
258	-,2,6-bis(phenylthio)-	$-SPh$	—	1.91	1.91	—		EtOH/alk	Ox O_2 or Re NaOEt	74
259	-,2-fluoro-	$-F$[cc]	3.50[t]	1.85	3.15	1.95		MeCN	Re el	65, 66
260	-,2-chloro-	$-Cl$	—	2.147	2.459	2.241		w/EtOH	Ox alk	8, 217
			—	2.208	2.453	2.208		EtOH/w	Ox alk	42
			—	2.30	2.30	2.30		MeOH	Re el	59
261	-,2-bromo-	$-Br$	—	2.159	2.367	2.355		w/EtOH	Ox alk	8
262	-,2-bromo-5-t-butyl-	$-Br$, $-CMe_3$	—	2.9	u	1.4		w/MeOH	Ox alk	93

263	-,2-iodo-	–I		(Data not available)						
264	-,2,3-dichloro-	–Cl	—	—	2.319	2.319		EtOH/w	Ox alk	42
265	[I]-,2,5-difluoro-	–F	5.24^F	1.42	5.24^F	1.42		80% DME/ MeCN	Re el	66
			5.35^F	1.41	5.35^F	1.41		MeCN	Re el	65
266	[C,I]-,2,5-dichloro-	–Cl	—	2.030	—	2.030		EtOH/w	Ox alk	42, 54
			—	2.15	—	2.15		DMSO	Re KO_2	55
			—	2.16	—	2.16		MeCN	Re el	65
			—	1.98	—	1.98	2.0055	C_6H_6	Re $(Me_3Sn)_2$[f]	115
267	[C]-,2,6-dichloro-	–Cl	—	2.320	2.320	—		EtOH/w	Ox alk	42, 54
			—	2.35	2.35	—		DMSO	Re KO_2	55
			—	2.38	2.38	—	2.0053	C_6H_6	Re $(Me_3Sn)_2$[f]	115
268	-,2,6-diiodo-	–I	—	2.30	2.30	—	2.0061	w/BuOH	Ox O_2[d]	69
269	-,trichloro-	–Cl	—	—	—	2.163		EtOH/w	Ox alk	42
270	[C,I]-,tetrafluoro-	–F	4.14^F	4.14^F	4.14^F	4.14^F	2.0048	w/EtOH Me_2CO	Ox KOH	67
			4.08^F	4.08^F	4.08^F	4.08^F		Pr^iOH	Re phot	68
			4.08^F	4.08^F	4.08^F	4.08^F		EtOH	Re phot	68
			4.02^F	4.02^F	4.02^F	4.02^F		MeOH	Re phot	68
			4.0^F	4.0^F	4.0^F	4.0^F		DMSO	Re KO_2	55
			3.95^F	3.95^F	3.95^F	3.95^F		DME/ – 22°C	Re el	65
271	[C,I]-,tetrachloro-	–Cl	—	—	—	—	2.005819[e]	DME	Re Na	40
			—	—	—	—	2.005859	DMSO	Re el	64
			—	—	—	—	2.0061	C_6H_6	Re $(Bu_3Sn)_2$[f]	115
272	[I]-,tetrabromo-	–Br	—	—	—	—	2.008433	DMSO	Re el	64
273	-,tetraiodo-	–I	—	—	—	—	2.011399	DMSO	Re el	64
274		(See Table 1.4)								

[a] From Reference 64.
[b] DMSO plus one drop of 1% aqueous NaOH.
[c] From Reference 41.
[d] From a CIDEP experiment.
[e] g-Factor, corrected for second-order shifts, often used as a standard.

Table I.3 (continued)
1,4-BENZOSEMIQUINONES

Anions

[f] Reaction with oxidative cleavage of the Sn–Sn bond.
[g] From Reference 8.
[h] The assignment in Reference 33 changed by the editor.
[i] Water <1% added to obtain the "free" ion.

```
                       Me
                       |
R = -(CH2CH=CH-)4Me.
```

[k] Corrected for second-order shifts.

```
                  OH          Me
                  |           |
[l]  R = CH2CH2-C-(CH2CH2CH2CH)3Me.
                  |
                  Me
```

[m] Reduction performed in DME and radical transferred to EtOH.
[n] Total splitting from phenyl group protons.
[o] Splitting from cyano nitrogen.
[p] KOBu^t added to form the dianion; this then oxidized by iodine.
[q] Nitrogen splitting.
[r] NH_2 or NH proton splitting.
[s] Radical obtained by treatment of quinone with sodium azid.
[t] Splitting from NCH_i protons (i = 1, 2, or 3).
[u] Assignment confirmed by deuteration.
[v] Splitting from "β-proton(s)".
[w] Splitting from "γ-protons"; estimated from line broadening.
[x] Splitting from phenyl protons; no details given; spectrum poorly resolved.

[y] Spectral assignment for this radical is ambiguous.
[z] Splitting from phosphorus.
[aa] Radical obtained by reaction between the quinone and tertiary phosphine.
[bb] Constants estimated from spectrum in Reference 72.
[cc] Spectrum very poorly resolved.
[dd] Radical obtained by spontaneous decomposition.
[ee] Fluorine splittings assigned with an F.
[ff] Sodium phosphate buffer used.
[gg] Corrected for second-order shifts.

Table I.3.1
2′-SUBSTITUTED PHENYL-1,4-BENZOSEMIQUINONES

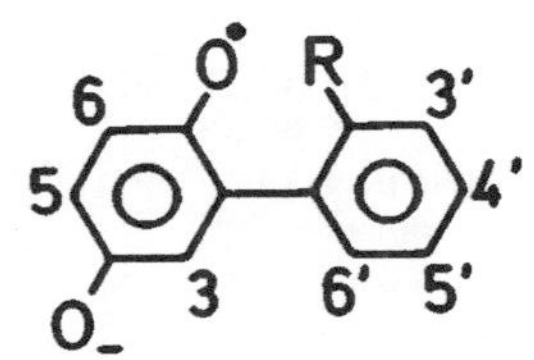

No.	Radical	R substituent	a_3	a_5	a_6	$a_{3'}$	$a_{4'}$	$a_{5'}$	$a_{6'}$	$a^R_{2'}$	Ref.
	1,4-Benzosemiquinone										
117	-,2-(2′-oxidophenyl)-	$-O^-$	1.82	2.65	2.28	0.14	0.14	0.14	0.14	—	49[a]
118	-,2-(2′-hydroxyphenyl)-	–OH	1.60	3.20	3.20	u	u	u	u	—	5[b]
119	-,2-(2′-methoxyphenyl)-	–OMe	2.05	2.63	2.17	0.15	0.15	0.15	0.15	—	49
120	-,2-(2′-methylphenyl)-	–Me	2.09	2.51	2.20	0.13	0.13	0.13	0.13	0.05[H,c]	49
121	-,2-(2′-phenylphenyl)-	–Ph	2.06	2.60	2.06	u	u	u	u	—	49
116	-,2-phenyl-	–H	1.99	2.56	2.12	0.16	0.27	0.16	0.27	0.27	48, 49
122	-,2-(2′-fluorophenyl)-	–F	2.07	2.52	2.21	0.16	0.16	0.16	0.16	0.56[F]	49
123	-,2-(2′-chlorophenyl)-	–Cl	2.12	2.52	2.22	u	u	u	u	—	49
124	-,2-(2′-bromophenyl)-	–Br	2.10	2.50	2.20	0.10	0.10	0.10	0.10	—	49
125	-,2-(2′-nitrophenyl)-	$-NO_2$	2.10	2.56	2.10	0.16	0.16	0.16	0.16	0.06[d]	49

Splitting constants/Gauss (a_3 to $a^R_{2'}$).

[a] Autoxidation of pertinent quinol in 50% w/EtOH (alk).[48,49]
[b] Reduction of pertinent quinone in anhydrous DMF.
[c] Estimated splitting.
[d] Estimated splitting from nitrogen nucleus.

Table I.3.2
3′-SUBSTITUTED PHENYL-1,4-BENZOSEMIQUINONES

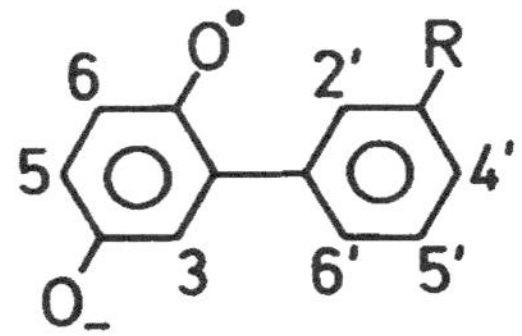

1,4-Benzosemiquinone

No.	Radical	R substituent	a_3	a_5	a_6	$a_{2'}$	$a_{4'}$	$a_{5'}$	$a_{6'}$	$a^R_{3'}$	Ref.
			Splitting constants/Gauss								
126	-,2-(3′-oxidophenyl)-	$-O^-$	1.81	2.57	2.17	0.26	0.26	0.15	0.26	—	49[a]
127	-,2-(3′-methoxyphenyl)-	–OMe	2.02	2.54	2.05	0.26	0.26	0.15	0.26	—	49
128	-,2-(3′-methylphenyl)-	–Me	1.96	2.58	2.14	0.27	0.27	0.18	0.27	0.09[b]	49
116	-,2-phenyl-	–H	1.99	2.56	2.12	0.27	0.27	0.16	0.27	0.16	48, 49
129	-,2-(3′-fluorophenyl)-	–F	2.02	2.56	2.13	0.28	0.28	0.15	0.28	0.04[b]	49
130	-,2-(3′-chlorophenyl)-	–Cl	2.00	2.52	2.13	0.27	0.27	0.15	0.27	—	48
131	-,2-(3′-nitrophenyl)-	$-NO_2$	2.26	2.44	1.94	0.27	0.27	0.17	0.27	0.04[b]	49

[a] Autoxidation of pertinent quinol in 50% w/EtOH (alk).[48,49]
[b] Estimated splitting from nitrogen nucleus.

Table I.3.3
4′-SUBSTITUTED PHENYL-1,4-BENZOSEMIQUINONES

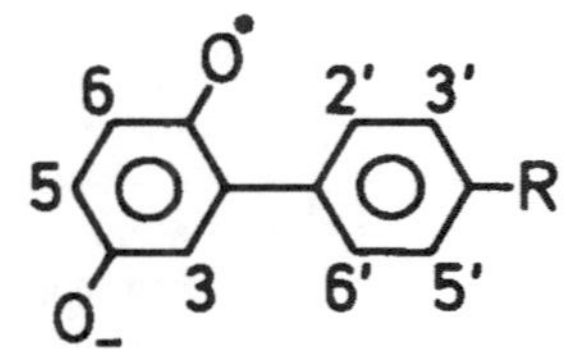

No.	Radical	R Substituent	Splitting constants/Gauss a_3	a_5	a_6	$a_{2'} = a_{6'}$	$a_{3'} = a_{5'}$	$a_{4'}^R$	Ref.
	1,4-Benzosemiquinone								
132	-,2-(4′-oxidophenyl)-	–O	1.38	2.53	2.33	0.32	0.14	—	48, 49[a]
133	-,2-(4′-methoxyphenyl)-	–OMe	1.85	2.56	2.16	0.30	0.15	—	48
134	-,2-(4′-methylphenyl)-	–Me	1.88	2.55	2.13	0.28	0.16	0.28[f]	48
			2.05	2.29	2.20	0.31	0.17	0.28[f]	47[b]
135	-,2-(4′-phenylphenyl)-	–Ph	1.98	2.58	2.10	0.28	0.20	—	48
116	-,2-phenyl-	–H	1.99	2.56[c]	2.12[c]	0.27	0.16	0.27	48, 49
136	-,2-(4′-fluorophenyl)-	–F	2.01	2.54	2.12	0.28	0.15	0.57[f]	49
137	-,2-(4′-chlorophenyl)-	–Cl	2.10	2.54[d]	2.10	0.28	0.16	—	48
138	-,2-(4′-bromophenyl)-	–BR	2.04	2.48	2.04	0.28	0.16	—	48
139	-,2-(4′-carboxylatophenyl)-	–COO	2.10	2.56	2.10	0.27	0.17	—	48
140	-,2-(4′-methoxycarbonylphenyl)-	–COOMe	2.26	2.56	1.95	0.29	0.18	—	48
141	-,2-(4′-nitrophenyl)-	$–NO_2$	2.38	2.50	1.92	0.30	0.15	0.07[e]	48, 49

[a] Autoxidation of pertinent quinol in 50% w/EtOH (alk).[48,49]
[b] Fifty percent w/EtOH, pH = 9.
[c] By additivity, a_5 is assigned the value 2.12.
[d] Assignment of the value 2.54 to a_5 agrees with additivity.
[e] Reference 49 states the splitting 0.07 from the nitrogen nucleus. Reference 48 states a^N = 0.75, which is assumed to be incorrect.

Table I.3.4
POLYSUBSTITUTED ARYL-1,4-BENZOSEMIQUINONES

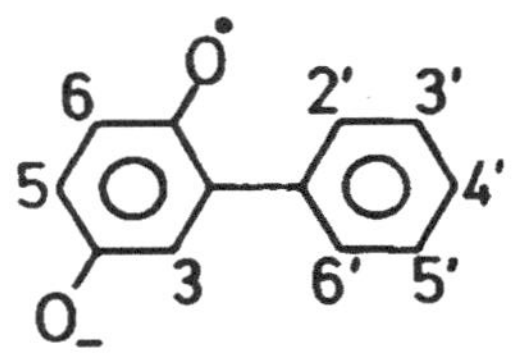

No.	Radical	Aryl substituent	Splitting constants/Gauss								Ref.
			a_3	a_5	a_6	$a_{2'}$	$a_{3'}$	$a_{4'}$	$a_{5'}$	$a_{6'}$	
	1,4-Benzosemiquinone										
142	-,2-(2',3'-dimethylphenyl)-	–Me	2.05	2.52	2.18	u[1]	u[1]	0.12	0.12	0.12	49[a]
143	-,2-(2',4'-dimethylphenyl)-	–Me	2.05	2.55	2.18	0.13[1]	0.13	u[1]	0.13	0.13	49
144	-,2-(2',5'-dimethylphenyl)-	–Me	2.05	2.55	2.17	u[1]	0.12	0.12	u[1]	0.12	49
145	-,2-(2',6'-dimethylphenyl)-	–Me	2.10	2.50	2.26	~0.05[1,b]	0.15	u	0.15	~0.05[1,b]	49
146	-,2-(2',4',6'-trimethylpenyl)-	–Me	2.11	2.54	2.26	~0.04[1,b]	0.16	u[1]	0.16	~0.04[1,b]	49
147	-,2-(3'-chloro-2'-methylphenyl)-	–Cl, –Me	2.20	2.52	2.20	u[1]	—	u	u	u	49
148	-,2-(6'-chloro-2'-methylphenyl)-	–Cl, –Me	2.14	2.50	2.18	u[1]	0.13	u	0.13	—	49
149	-,2-(4'-chloro-2',6'-dimethylphenyl)-	–Cl, –Me	2.20	2.48	2.20	~0.05[1,b]	0.15	—	0.15	~0.05[1,b]	49
150	-,2-(4'-bromo-2',6'-dimethylphenyl)-	–Br, –Me	2.20	2.49	2.20	~0.05[1,b]	0.14	—	0.14	~0.05[1,b]	49
151	-,2-(3'-bromo-2',4',6'-trimethylphenyl)-	–Br, –Me	2.16	2.50	2.16	~0.04[1,b]	—	u[1]	0.11	~0.04[1,b]	49
152	-,2-(2',6'-dichlorophenyl)-	–Cl	2.10	2.50	2.42	—	~0.10	u	~0.10	—	49
153[b]	-,2-(2',5'-dimethoxyphenyl)-	–OMe	2.11	2.48	2.11	u	0.12	0.12	u	0.12	105[c]
154[c]	-,2-(2',5'-dihydroxyphenyl)-	–OH	1.93	2.81	2.40	—	(0.52	0.21	—	0.21[d]	214)
155	-,2-(α-naphthyl)-	$-C_{10}H_7$	2.16	2.52	2.16	a_H = 0.10(2H)			a_H = 0.16(1H)		49

[a] Autoxidation of pertinent quinol in 50% w/EtOH (alk).
[b] Estimated splitting.
[c] Solvent: 5% w/DMSO; method: Re KOH.
[d] Solvent: alcohol; base: BaOH or CaOH, g = 2.00545.

Table I.3.5
IMIDAZO- AND TRIAZOLO-1,4-BENZOSEMIQUINONES

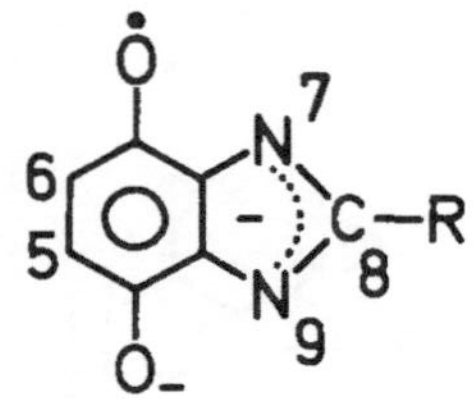

No.	Radical		Splitting constants/Gauss a_5	a_6	a_7^N	a_9^N	a_8	Solvent	Method	T [K]	Ref.
205	2,3-Imidazo-1,4-benzosemiquinone	(R = H)	3.50	3.50	0.94	0.94	0.37	DMF	Re el	295.5	218, 219
			3.54	3.54	0.96	0.96	0.40			253	219
			3.59	3.59	0.98	0.98	0.44			233	219
			3.64	3.64	0.85	0.85	0.42			213	219
			3.59	3.59	0.81	0.81	0.40	MeCN	Re el	295.5	219
			3.54	3.54	0.85	0.85	0.40			253	219
			3.54	3.54	0.86	0.86	0.45			233	219
			3.56	3.56	0.92	0.92	0.47			193	219
			3.49	3.49	0.95	0.95	0.40	$DMF:D_2O$ (100:1)	Re el	295.5	219
			3.82	3.82	0.80	0.80	0.50	DMSO	Ox $KOBu^t$	RT	220
206	-,8-methyl-	(R=Me)	3.77	3.77	0.81	0.81	0.69[a]	DMSO	Ox $KOBu^t$	RT	220

Table I.3.5 (continued)
IMIDAZO- AND TRIAZOLO-1,4-BENZOSEMIQUINONES

No.	Radical	a_5	a_6	a^N_7	a^N_9	a_8	Solvent	Method	T [K]	Ref.
	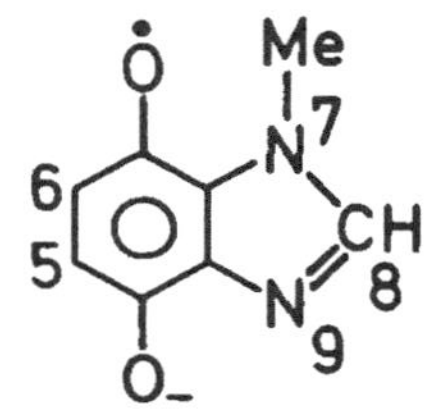									
207	7-Methyl-2,3-imidazo-1,4-benzosemiquinone	(2.96	3.67)[b]	0.35 1.09[a]	0.94	0.35	DMSO	Ox KOBu[t]	RT	220
	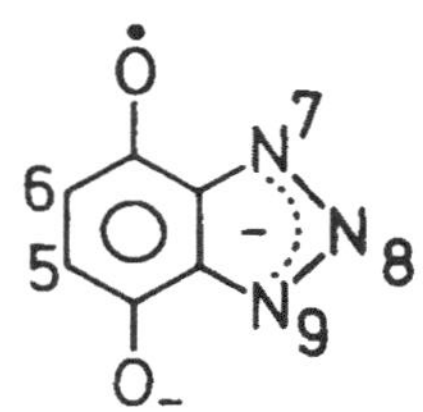									
208	2,3-Triazolo-1,4-benzosemiquinone	3.5	3.5	0.90	0.90	0.65[c]	w/DMF	Q/NaN$_3$[d]		62
209	-,5-methyl-	3.45[1]	2.55	0.90	0.90	0.70[c]	w/DMF	Q/NaN$_3$[d]		62

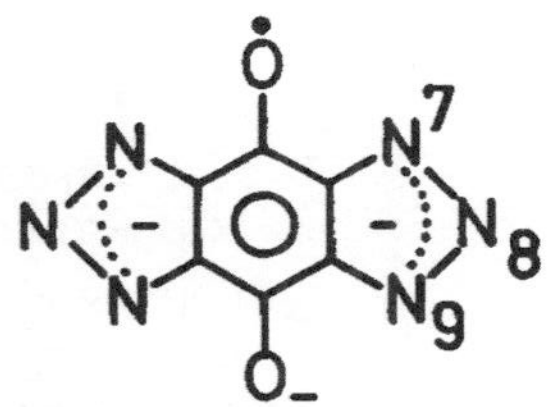

210	2,3: 5,6-Bis(triazolo)-1,4-benzo-semiquinone	—	—	1.22[e]	1.22[e]	0.88[e]	w/DMF	Q/NaN$_3$[d]	RT	62

[a] Splitting from protons of Me-group.
[b] Assignment unknown for the values within the bracket.
[c] Splitting from nitrogen nucleus.
[d] Quinone (0.1 g) in water or DMF (5 cm^3) treated with saturated aqueous sodium azide (1 to 2 cm^3).
[e] Splitting from two equivalent positioned nitrogen nuclei.

Table I.4
PYROGALLOL SEMIQUINONES

Dianions

No.	Radical	Substituent(s)	Splitting constant/Gauss a_4	a_5	a_6	g-Factor	Solvent	Method	Ref.
	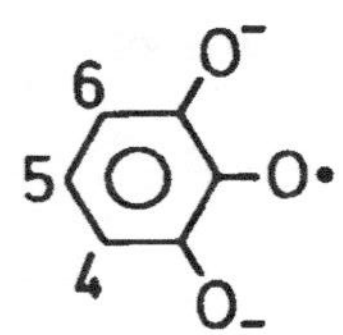								
274	[C]-Pyrogallol semiquinone anion		0.99	5.66	0.99		w/pH ~14	Ox alk	86
			0.95	5.5	0.95		w	Ox alk	83, 19

Table I.4 (continued)
PYROGALLOL SEMIQUINONES

Dianions

No.	Radical	Substituent(s)	Splitting constant/Gauss a_4	a_5	a_6	g-Factor	Solvent	Method	Ref.
			0.97	5.51	0.97	2.00471	w/EtOH	Ox alk	8
			0.96	5.41	0.96	2.00468	w/others[a]	Ox alk	3
			0.85	5.13	0.85	2.00463	20% w/DMSO	Ox alk	178
275	-,4,6-di-*t*-butyl-	$-CMe_3$	u	5.45	u		w	Ox alk	83
276	-,4-(1-methyl-2-carboxylatovinyl)-	$-C(Me)=CHCOO^-$	0.06[b]	5.40	0.83		50% w/EtOH	Ox alk	17
277	-,4-(2-methyl-2-carboxylatovinyl)-	$-CH=C(Me)COO^-$	u	5.4	0.8		w	Ox alk	26
278	-,4-formyl-	−CHO	u	4.4	0.80[c]		w	Ox alk	9
279	-,5-formyl-	−CHO	0.97	1.07[l]	0.97	2.00520	w/EtOH	Ox alk	8
280	[C]-,4-acetyl-	−COMe	u	4.8	0.82[c]		w	Ox alk	9
			0.16[2]	4.70	0.8	2.00481	w/ETOH	Ox alk	8
281	-,4-acetyl-6-methylw/EtOH-	−COMe, −Me	0.16[2]	4.77	0.9[l]	2.00481	w/EtOH	Ox alk	8
282	-,4-propionyl-	−COEt	u	4.8	0.80[c]		w	Ox alk	9
283	-,4-carboxylato-	$-COO^-$	—	5.1	0.9		w	Ox alk	83
			—	5.095	0.781	2.00475	w/EtOH	Ox alk	8
			—	4.95	0.84	2.00459	20% w/DMSO	Ox alk	178
284	[C]-,4-methoxycarbonyl-	$-COOCH_3^*$	0.058*	4.860	0.859		w/EtOH	Ox alk	8
285	-,4-ethoxycarbonyl-	$-COOCH_2^*Me$	0.048*	4.865	0.847		w/EtOH	Ox alk	8
286	-,4-butoxycarbonyl-	$-COOCH_2^*(CH_2)_2Me$	0.048*	4.845	0.840		w/EtOH	Ox alk	8
287	-,4-pentyloxycarbonyl-	$-COOCH_2^*(CH_2)_3Me$	0.051*	4.888	0.846		w/EtOH	Ox alk	8
288	-,5-carboxylato-	$-COO^-$	1.0	—	1.0		w	Ox alk	83
			1.03	—	1.03	2.00473	20% w/DMSO	Ox alk	178
			1.088	—	1.088		w/EtOH	Ox alk	8
			1.07	—	1.07	2.00489	w/others[a]	Ox alk	3

289	-,5-methoxycarbonyl-	$-COOCH_3^*$	1.11	0.38*	1.11		w	Ox alk	103
			1.14	0.38*	1.14		50% w/EtOH	Ox KOH	87
			1.111	0.376*	1.111		w/EtOH	Ox alk	8
290	-,5-ethoxycarbonyl-	$-COOCH_2^*Me$	1.12	0.36*	1.12		50% w/EtOH	Ox KOH	87
			1.104	0.366*	1.104		w/EtOH	Ox alk	8
291	-,5-propoxycarbonyl-	$-COOCH_2^*CH_2Me$	1.13	0.38*	1.13	2.00490	w/others[a]	Ox alk	3
			1.105	0.379*	1.105		w/EtOH	Ox alk	8
292	-,5-butoxycarbonyl-	$-COOCH_2^*(CH_2)_2Me$	1.111	0.384*	1.111		w/EtOH	Ox alk	8
293	-,5-pentyloxycarbonyl-	$-COOCH_2^*(CH_2)_3Me$	1.107	0.381*	1.107		w/EtOH	Ox alk	8
294	-,5-hexyloxycarbonyl-	$-COOCH_2^*(CH_2)_4Me$	1.108	0.381*	1.108		w/EtOH	Ox alk	8
295	-,5-octyloxycarbonyl-	$-COOCH_2^*(CH_2)_6Me$	1.101	0.381*	1.101		w/EtOH	Ox alk	8
296	-,5-nonyloxycarbonyl-	$-COOCH_2^*(CH_2)_7Me$	1.107	0.387*	1.107		w/EtOH	Ox alk	8
297	-,5-dodecyloxycarbonyl-	$-COOCH_2^*(CH_2)_8Me$	1.114	0.384*	1.114		w/EtOH	Ox alk	8
298	-,5-isopropoxycarbonyl-□	$-COOCH^*Me_2$	1.13	0.24*	1.13		50% w/EtOH	Ox KOH	87
			1.104	0.220*	1.104		w/EtOH	Ox alk	8
299	-,4,5-dibromo-	$-Br$	—	—	0.70		w	Ox alk	9
300	-,4,6-disulfonato-	$-SO_3^-$	—	4.6	—		w	Ox alk	9

[a] Ethanol, acetone, or mixtures thereof.
[b] Splitting from methyl-group protons.
[c] The assignment in Reference 9 is incorrect and has been corrected (cf. compounds Nos. 280 and 281).

Table I.5
PYROGALLOL SEMIQUINONES

Monoanions

No.	Radical	Substituent	Splitting constants/Gauss a_3	a_4	a_5	a_6	g-Factor	Solvent	Method	Ref.
11	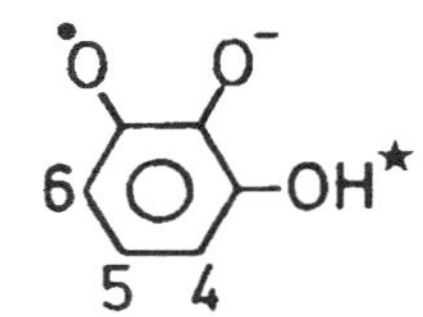									
301	[C] Pyrogallol semiquinone		0.36*	1.90	3.50	0.33		HMPA	Ox KOBu^t	262
			0.33*	1.24	4.89	0.57		w/H_2SO_4	Ox H_2O_2/TiIII	86[a]
			u	0.63	5.05	0.23	2.00466	20% w/DMSO	Ox alk	178
302	-,4-carboxylato-	–COO$^-$	0.65*	—	4.30	0.65	2.00470	20% w/DMSO	Ox alk	178
303	[C]-,5-carboxylato-	–COO	u	1.25	—	0.30	2.00474	20% w/DMSO	Ox alk	178
	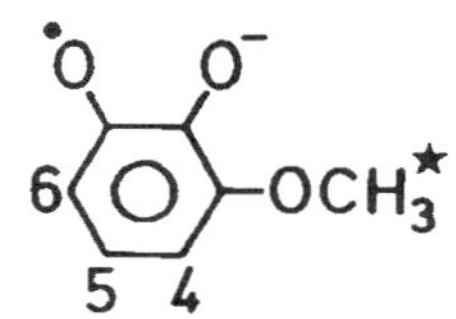									
304	[C] 3-Methoxy-1,2-benzosemiquinone		0.68*	1.25	4.83	0.59	2.00464	w/NaOH (pH = 5—11)	Radiolysis	115
			0.65*	1.25	4.76	0.59		w	Ox KOBu^t	4
			0.65*	1.3	4.75	0.55		w/0.5 *M* H_2SO_4	Ox CeIV	19
			0.47*	1.90	3.78	0.46		HMPA	Ox KOBu^t	4, 262

305	-,5-propanoyl-	–COEt	0.55*	1.75	0.28	u		10% w/DMF	Ox alk	103
306	-,5-carboxylato-	–COO⁻	0.60*	1.0	—	0.20		w	Ox alk	103
			0.63*	0.99	—	0.24	2.00479	w/NaOH (pH = 5—11)	Radio-lysis	115
307	-,6-carboxylato-	–COO⁻	0.73*	1.37	4.75	—	2.00465	w/EtOH (pH ~13.2)	Ox alk	8
			0.71*	1.41	4.65	—	2.00472	w/EtOH (pH ~13.7)	Ox alk	8

[a] Radical generated by continuous flow-mixing; pH ~6.5 to 8.0

Table I.6
4-OXIDO-1,2-BENZOSEMIQUINONES

Dianions

No.	Radical	Substituent(s)	Splitting constants/ Gauss a_3	a_5	a_6	g-Factor	Solvent	Method	Ref.
	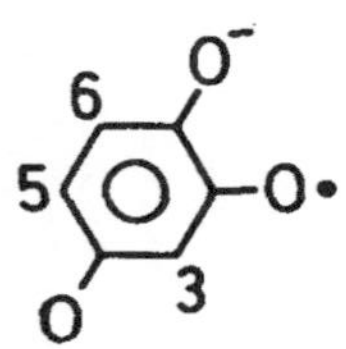								
308	[C]4-Oxido-1,2-benzosemiquinone[i]		0.60	4.81	1.33		w	Ox alk	1, 9, 33
			0.60	4.90	1.37	2.00449	w/EtOH	Ox alk	8
			0.62	4.82	1.46		MeOH/Et_3N	Phot alk[a]	100, 99
			0.69	3.7	1.7		10% w/DMF	Ox alk	9
309	-,3-methyl-	–Me	0.95[1]	4.90	1.35		w	Ox alk	88
310	-,5-methyl-	–Me	0.58	4.91[1]	0.66		w	Ox alk	1
			0.62	5.06[1]	0.67		w	Ox alk	88
			0.54	5.01[1]	0.71	2.00434	w/EtOH	Ox alk	8

Table I.6 (continued)
4-OXIDO-1,2-BENZOSEMIQUINONES

Dianions

No.	Radical	Substituent(s)	Splitting constants/ Gauss a_3	a_5	a_6	g-Factor	Solvent	Method	Ref.
			0.58	5.12[1]	0.69		50% w/EtOH	Ox alk	33, 51
			0.50	5.2	0.80		10% w/DMF	Ox alk	33
311	-,5-ethyl-	–Et	0.54	4.56[1]	0.71		50% w/EtOH	Ox alk	33
312	-,5-hexyl-[a]	–(CH$_2$)$_5$Me	0.50	4.00[1]	0.77	2.00439	w/EtOH	Ox alk	8, 130
				0.11[2]					
313	-,5-isopropyl-	–CHMe$_2$	0.70	2.8[1]	0.70		w	Ox alk	9
			0.64	2.68[1]	0.64		w	Ox alk	89
				0.08[2]					
314	-,5-*t*-butyl-	–CMe$_3$	0.85	0.16[2]	0.43		w	Ox alk	10, 32, 33
315	-,5-phenyl-	(see Tables I.6.1 through I.6.4 for 5-phenyl-4-oxido-1,2-benzosemiquinones, radical nos. 315 to 349)							
350	-,5-(α-naphthyl)-	–C$_{10}$H$_7$	0.53	0.34[b]	1.16		w	Ox alk	49
				0.23[c]					
351	-,6-methyl-	–Me	0.55	4.15	0.95[1]		w	Ox alk	33, 51
			0.60	4.20	1.00[1]		w	Ox KOH	43
			0.54	4.17	0.95[1]	2.00446	w/EtOH	Ox alk	8
352	-,6-ethyl-	–Et	0.56	4.14	1.08[1]		w	Ox alk	33
353	-,6-butyl-	–(CH$_2$)$_3$Me	0.60	4.15	0.90[1]		w	Ox KOH	43
354	-,6-pentyl-	–(CH$_2$)$_4$Me	0.53	4.06	0.90		w	Ox alk[d]	8
355	-,6-isopropyl-	–CHMe$_2$	0.60	4.20	0.75[1]		w	Ox KOH	43
			0.54	4.15	0.71[1]		50% w/DMF	Ox alk	8
356	-,6-*s*-butyl-	–CH(Me)CH$_2$Me	0.55[e]	4.15	0.55[1,e]		w	Ox KOH	43
357	-,6-*t*-butyl-	–CMe$_3$	0.66	4.02	0.06[2,f]	2.00453	w	Ox alk	33, 32
			0.66	4.10	u		w	Ox KOH	10, 43
			0.66	3.92	0.07[2]		w/EtOH	Ox alk	8
								Re alk[g]	
			0.50	4.75	u		DMSO	Ox alk[h]	90
358	-,6-phenyl-	(see Tables I.6.5 through I.6.8 for 6-phenyl-4-oxido-1,2-benzosemiquinones, radical nos. 358 to 392)							

393	-,6-(α-naphthyl)-	$-C_{10}H_7$	0.64	4.60[1]	u		w	Ox alk	49
394	-,3,5-dimethyl-	–Me	0.95[1]	5.02[1]	0.70		w	Ox alk	88
			0.93[1]	5.08[1]	0.76		50% w/DMF	Ox alk	1, 33, 51
395	-,5-ethyl-3-methyl-	–Et, –Me	0.81[1]	3.28[1]	0.81	2.00456	w/EtOH	Ox alk	8
396	-,3,5-di-*t*-butyl-	$-CMe_3$	u	0.16[2]	1.05		DMSO	Re KOBu[t]	90
397	-,3,6-dimethyl-	–Me	0.95[1]	4.13	1.00[1]		w	Ox alk	88, 1
			0.88[1]	4.20	0.98[1]		50% w/DMF	Ox alk	33, 51
398	-,6-isopropyl-3-methyl-	$-CHMe_2$, –Me	0.92[1]	4.20	0.75[1]		w	Ox alk	33
399	-,5,6-dimethyl-	–Me	0.55	4.70[1]	0.55[1]		w	Ox alk	33, 88
			0.47	4.60[1]	0.62[1]		50% w/DMF	Ox alk	1
400	-,3,5,6-trimethyl-	–Me	0.84[1]	4.56[1]	0.62[1]	2.00440	w	Ox alk	1, 88
401	-,5-(4-carboxylatobutyl)-	$-CH_2CH_2CH_2CH_2COO$	0.54	3.98[1] 0.12[2]	0.78		w	Ox alk	130
402	-,5-(2-carboxylatoethyl)-	$-CH_2CH_2COO$	0.55	4.22[1] 0.11[2]	0.81	2.00437	w/EtOH	Ox alk	15
			0.55	3.97[1] 0.10[2]	0.78		w	Ox alk	130
	-,5-(2-R-oxycarbonylethyl)-	$-CH_2CH_2COOR$							
	R=	R=							
403	butyl	$-(CH_2)_3Me$	0.52	3.72[1] 0.1[2]	0.95		w/EtOH	Ox alk	8
404	unknown	unknown	0.55	3.76[1] 0.075[2]	0.95	2.00437	w/EtOH	Ox alk	15
405	1-methyl-2-(6-pyrone)ethyl	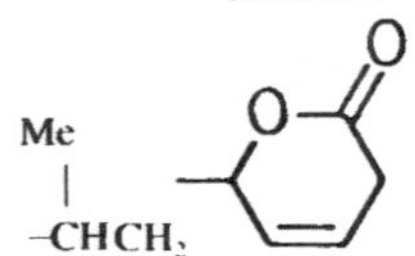	(Data approximately equal to those of no. 404; see Reference 15)						
406	-,5-(2-carboxylatovinyl)-	–CH=CHCOO	0.45	2.9[1] 1.9[2]	1.35		w	Ox alk	26
							w/EtOH	Ox alk	15
			0.40	2.92[1] 1.80[2]	1.37	2.00438	50% w/EtOH[1]	Ox alk	17, 91

Table I.6 (continued)
4-OXIDO-1,2-BENZOSEMIQUINONES

Dianions

No.	Radical	Substituent(s)	Splitting constants/Gauss a_3	a_5	a_6	g-Factor	Solvent	Method	Ref.
			0.44	2.83[1] 1.75[2]	1.32				
407	-,5-(2-methyl-2-carboxylatovinyl)-	–CH=C(Me)COO⁻	0.47	2.94[1] 1.77[j]	0.69		50% w/EtOH[i]	Ox alk	17, 91
408	-,5-(2-phenyl-2-carboxylatovinyl)-	–CH=C(Ph)COO⁻	0.44	2.94[1] 1.77[k]	0.95		50% w/EtOH[i]	Ox alk	17
409	-,5-(2,2-dicarboxylatovinyl)-	–CH=C(COO⁻)$_2$	0.44	2.50[1]	1.25		50% w/EtOH[i]	Ox alk	17, 91
410	-,5-(1-methyl-2-carboxylatovinyl)-	–C(Me)=CHCOO⁻	0.47	0.01[1,c] 0.24[2]	0.96		50% w/EtOH[i]	Ox alk	17, 91
411	-,5-(1-phenyl-2-carboxylatovinyl)-	–C(Ph)=CHCOO	0.45	0.0[k] 0.19[2]	1.13		50% w/EtOH[i]	Ox alk	17
412	-,5-(1,2-dimethyl-2-carboxylatovinyl)-	–C(Me)=C(Me)COO	0.50	0.0[2] 0.58[3]	0.96		50% w/EtOH[i]	Ox alk	17
413	-,6-(2,2-dicarboxylatovinyl)-	–CH=C(COO⁻)$_2$	0.67	5.01	0.67[1]		50% w/EtOH[i]	Ox alk	17
414	-,6-(1,2-dimethyl-2-carboxylatovinyl)-	–C(Me)=C(Me)COO	0.62	4.22	0.0[2] 0.10[3]		50% w/EtOH[i]	Ox alk	17
415	-,5-(2-aminoethyl)-	–CH$_2$CH$_2$NH$_2$	0.57	3.36[1]	0.85		w	Ox alk	92
			0.58	3.24[1]	0.87	2.0043	w	Phot alk	14
			0.542	3.332[1] 0.054[2]	0.896		w/EtOH	Ox alk	8
416	-,5-(2-acetylaminoethyl)-	–CH$_2$CH$_2$NHCOMe	0.58	3.28[1]	0.95		50% w/EtOH	Ox alk	8
417	-,5-(2-amino-2-carboxylatoethyl)-	–CH$_2$CH(NH$_2$)COO	0.55	3.03[1] 0.25[2]	0.95		w	Ox alk	8
418	-,5-(2-amino-2-carboxylato-2-methylethyl)-	–CH$_2$C(Me)(NH$_2$)COO	0.50	2.98[1]	0.95		w	Ox alk	8
419	-,5-carboxylatomethyl-	–CH$_2$COO	0.52	3.99[1]	0.87		w	Ox alk	130
420	-,5-aminomethyl-	–CH$_2$NH$_2$	0.60	4.15[1] 0.65[m]	0.95		w/EtOH	Ox alk	8

421	-,5-hydroxymethyl-	$-CH_2OH$	0.57	4.38[f]	1.12		w	Ox alk	130
			0.56	4.60[f]	1.13	2.00444	w/EtOH	Ox alk	8
422	-,5-ethoxymethyl-	$-CH_2OCH_2Me$	0.51	3.60[f] 0.05*	1.26		w	Ox alk	130
			0.50	4.64[f]	1.16		60% w/DMF	Ox alk	130
423	-,6-hydroxymethyl-	$-CH_2OH$	0.61	4.47	0.92[f]		w	Ox alk	130
424	-,6-ethoxymethyl-	$-CH_2OEt$	0.64	4.64	0.72[f]		w	Ox alk	130
425	-,6-phenoxymethyl-	$-CH_2OPh$	0.60	4.80	0.60[f]		w	Ox alk	130
426	-,5-(2-amino-1-hydroxyethyl)-	$-CH(OH)CH_2NH_2$	0.55	2.90[f]	1.20		w/EtOH	Ox alk	8
427	-,5-(1,2-dihydroxyethyl)-	$-CH(OH)CH_2OH$	0.59	2.94[f] 0.08[ℓ]	1.30	2.00442	w/EtOH	Ox alk	8
428	-,3-carboxylato-	$-COO$	—	5.16	1.16		w	Ox alk	130
429	[C]-,5-carboxylato-	$-COO$	0.44	—	1.55	2.00459	w/EtOH	Ox alk	8
430	-,6-carboxylato-	$-COO$	0.69	4.76	—		w	Ox alk	130
431	-,5-amino-	$-NH_2$	0.15	1.30[m] 0.63[n]	1.50		w/DMF[o]	$ReNaN_3^p$	62
432	(See Table 1.7)								

[a] Radical obtained when fluorescein is photolyzed in MeOH in the presence of Et_3N and NaOH (0.5 M). Radical assigned by the editor.
[b] Splitting from the α-naphthyl nucleus (1H).
[c] Splitting from the α-naphthyl nucleus (2H).
[d] Spectrum obtained from primin by demethylation.
[e] Accuracy ±0.10 G due to insufficient resolution of the spectrum.
[f] From Reference 32.
[g] Radical obtained from 3-*t*-butyl-5-methoxy-1,2-benzoquinone by reduction.
[h] Radical obtained from 3,5-di-*t*-butylguaiacol or 2,3,5-triacetoxybutylbenzene by autoxidation.
[i] Reference 49 states 1.60 G, a value which must be a misprint if the data correspond to radical no. 393.
[j] Contains less than 50% EtOH. See Reference 17 for details.
[k] Total splitting from phenyl group.
[ℓ] Splitting from methyl group protons.
[m] Splitting from amino nitrogen.
[n] Splitting from two amino protons.
[o] Exact solvent composition not given.
[p] Spectrum obtained by autoxidation of the products from the reaction between 1,4-benzoquinone and aqueous sodium azide.

Table I.6.1
2′-SUBSTITUTED 5-PHENYL-4-OXIDO-1,2-BENZOSEMIQUINONES[a]

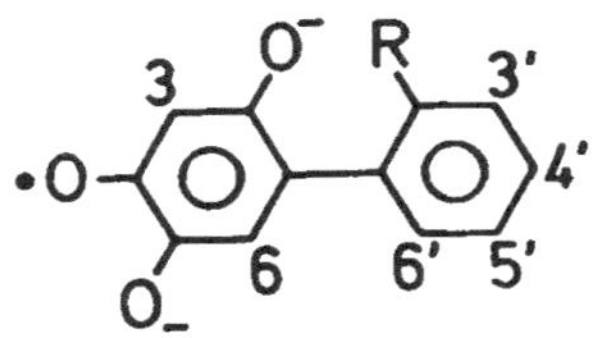

No.	Radical	R substituent	a_3	a_6	$a_{3'}$	$a_{4'}$	$a_{5'}$	$a_{6'}$	$a_{2'}^R$	Ref.
	4-Oxido-1,2-benzosemiquinone									
316	-,5-(2′-oxidophenyl)-	–O	0.49	0.82	0.24	0.30	0.24	0.24	—	49
317	-,5-(2′-methoxyphenyl)-	–OMe	0.52	1.12	0.26	0.26	0.26	0.26	—	49
318	-,5-(2′-methylphenyl)-	–Me	0.51	0.96	0.28	0.28	0.28	0.28	0.11[b]	49
319	-,5-(2′-phenylphenyl)-	–Ph	0.48	1.04	0.30	0.20	0.30	0.30	0.10[b]	49
315	-,5-phenyl-	–H	0.50	1.08	0.31	0.54	0.31	0.54	0.54	48, 49
320	-,5-(2′-fluorophenyl)-	–F	(Spectrum not analyzed)							49
321	-,5-(2′-chlorophenyl)-	–Cl	0.50	1.27	0.25	0.20	0.25	0.25	—	49
322	-,5-(2′-bromophenyl)-	–Br	0.48	1.23	0.24	0.17	0.24	0.24	—	49
323	-,5-(2′-nitrophenyl)-	$–NO_2$	(Spectrum not analyzed)						0.13[c]	49

[a] Radicals obtained as secondary product by adding an aqueous solution of the corresponding phenyl-1,4-hydroquinone to an equal volume of 15% NaOH solution.

[b] Splittings from 2H of phenyl group.

[c] Splitting from nitrogen nucleus.

Table I.6.2
3′-SUBSTITUTED 5-PHENYL-4-OXIDO-1,2-BENZOSEMIQUINONES[a]

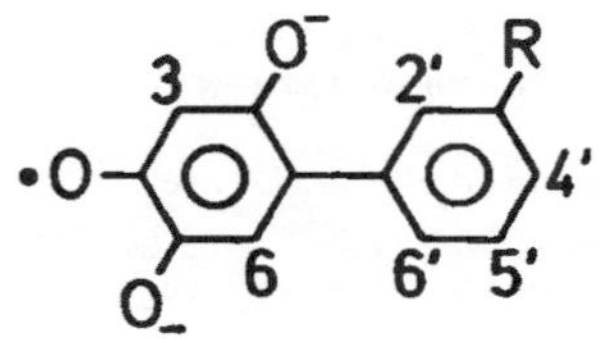

No.	Radical	R substituent	a_3	a_6	$a_{2'}$	$a_{4'}$	$a_{5'}$	$a_{6'}$	$a^R_{3'}$	Ref.
			Splitting constants/Gauss							
	4-Oxido-1,2-benzosemiquinone									
324	-,5-(3′-oxidophenyl)-	–O	0.51	0.96	0.51	0.51	0.27	0.51	—	49
325	-,5-(3′-methoxyphenyl)-	–OMe	0.54	1.08	0.54	0.54	0.29	0.54	—	49
326	-,5-(3′-methylphenyl)-	–Me	0.52	1.06	0.52	0.52	0.32	0.52	0.16[l]	49
315	-,5-phenyl-	–H	0.50	1.08	0.54	0.54	0.31	0.54	0.31	48, 49
327	-,5-(3′-fluorophenyl)-	–F	0.53	1.15	0.53	0.53	0.28	0.53	0.08[l]	49
328	-,5-(3′-chlorophenyl)-	–Cl	0.52	1.19	0.54	0.54	0.29	0.54	—	48
329	-,5-(3′-nitrophenyl)-	$–NO_2$	0.43	1.42	0.60	0.60	0.30	0.60	0.09[b]	49

[a] Radicals obtained as secondary product by adding an aqueous solution of the corresponding phenyl-1,4-hydroquinone to an equal volume of 15% NaOH solution.

[b] Splitting from nitrogen nucleus.

Table I.6.3
4′-SUBSTITUTED 5-PHENYL-4-OXIDO-1,2-BENZOSEMIQUINONES[a]

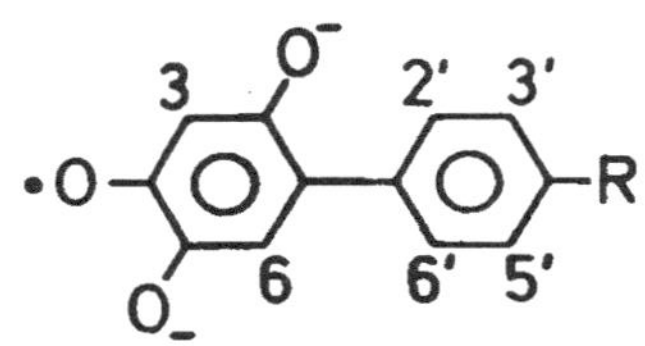

No.	Radical	R substituent	a_3	a_6	$a_{2'}=a_{6'}$	$a_{3'}=a_{5'}$	$a_{4'}^R$	Ref.
			Splitting constants/Gauss					
	4-Oxido-1,2-benzosemiquinone							
330	-,5-(4′-oxidophenyl)-	$-O^-$	0.53	0.53	0.65	0.27	—	48, 49
			(0.53	0.65	0.53	0.27)[b]		
331	-,5-(4′-methoxyphenyl)-	–OMe	0.53	0.90	0.59	0.29	u	48
332	-,5-(4′-methylphenyl)-	–Me	0.50	0.90	0.58	0.31	0.58[d]	48
333	-,5-(4′-phenylphenyl)-	–Ph	0.50	1.12	0.58	0.31	u	48
315	-,5-phenyl-	–H	0.50	1.08	0.54	0.31	0.54	48, 49
334	-,5-(4′-fluorophenyl)-	–F	0.51	1.09	0.58	0.30	1.09[e]	49
335	-,5-(4′-chlorophenyl)-	–Cl	0.48	1.16	0.59	0.31	—	48
336	-,5-(4′-bromophenyl)-	–Br	0.49	1.13	0.59	0.31	—	49
337	-,5-(4′-carboxylatophenyl)-	$-COO^-$	0.48	1.17	0.58	0.32	—	48
338	-,5-(4′-nitrophenyl)-	$-NO_2$	0.45	1.48	0.60	0.30	0.15[c]	48, 49

[a] Radicals obtained as secondary product by adding an aqueous solution of the corresponding phenyl-1,4-hydroquinone to an equal volume of 15% NaOH solution.
[b] Assignment suggested by the editor.
[c] Splitting from nitrogen nucleus.

Table I.6.4
POLYSUBSTITUTED 5-PHENYL-4-OXIDO-1,2-BENZOSEMIQUINONES[a]

No.	Radical	Phenyl substituent	Splitting constants/Gauss							Ref.
			a_3	a_6	$a_{2'}$	$a_{3'}$	$a_{4'}$	$a_{5'}$	$a_{6'}$	
	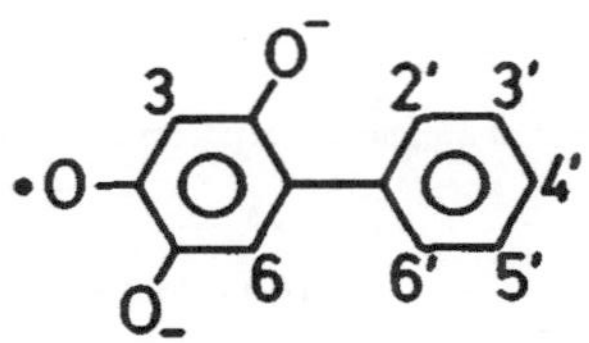 4-Oxido-1,2-benzosemiquinone									
339	-,5-(2′,3′-dimethylphenyl)-	–Me	0.48	1.10	u	u	0.24	0.24	0.24	49
340	-,5-(2′,4′-dimethylphenyl)-	–Me	(Spectrum not resolved and analyzed)							49
341	-,5-(2′,5′-dimethylphenyl)-	–Me	(Spectrum not resolved and analyzed)							49
342	-,5-(2′,6′-dimethylphenyl)-	–Me	0.51	1.08	0.09[t]	0.31	u	0.31	0.09[t]	49
343	-,5-(2′,4′,6′-trimethylphenyl)-	–Me	0.49	1.00	0.08[t]	0.32	~0.00[t]	0.32	0.08[t]	49
344	-,5-(2′-methyl-3′-chlorophenyl)-	–Me, –Cl	(Spectrum not resolved and analyzed)							49
345	-,5-(2′6′-dimethyl-4′-chlorophenyl)-	–Me, –Cl	0.51	1.08	0.09[t]	0.30	—	0.30	0.09[t]	49
346	-,5-(2′,6′-dimethyl-4′-bromophenyl)-	–Me, –Br	0.50	1.09	0.09[t]	0.30	—	0.30	0.09[t]	49
347	-,5-(2′-methyl-6′-chlorophenyl)-	–Me, –Cl	0.54	1.18	u	0.30	u	0.30	—	49
348	-,5-(2′,6′-dichlorophenyl)-	–Cl	0.54	1.38	—	0.27	u	0.27	—	49
349	-,5-(2′,4′,6′-trimethyl-3-bromophenyl)-	–Me, –Br	(Spectrum not resolved and analyzed)							49

[a] Radicals obtained as secondary product by adding an aqueous solution of the corresponding phenyl-1,4-hydroquinone to an equal volume of 15% NaOH solution.

Table 1.6.5
2′-SUBSTITUTED 6-PHENYL-4-OXIDO-1,2-BENZOSEMIQUINONES[a]

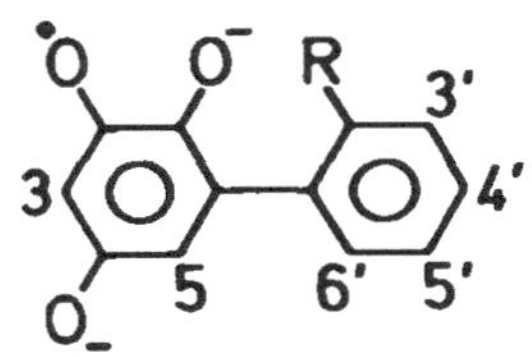

No.	Radical	R substituent	Splitting constants/Gauss							Ref.
			a_3	a_6	$a_{3'}$	$a_{4'}$	$a_{5'}$	$a_{6'}$	$a_{2'}^{R}$	
	4-Oxido-1,2-benzosemiquinone									
359	-,6-(2′-oxidophenyl)-	$-O^-$	0.64	4.16	u	u	u	u	u	49
360	-,6-(2′-methoxyphenyl)-	–OMe	0.67	4.65	u	u	u	u	u	49
361	-,6-(2′-methylphenyl)-	–Me	0.64	4.56	u	u	u	u	u	49
362	-,6-(2′-phenylphenyl)-	–Ph	0.64	4.44	u	u	u	u	u	49
358	-,6-phenyl-	–H	0.67	4.52	0.12	0.12	0.12	0.12	0.12	48, 49
363	-,6-(2′-fluorophenyl)-	–F	0.68	4.72	u	u	u	u	0.40[f]	49
364	-,6-(2′-chlorophenyl)-	–Cl	0.65	4.72	u	u	u	u	u	49
365	-,6-(2′-bromophenyl)-	–Br	0.64	4.74	u	u	u	u	u	49
366	-,6-(2′-nitrophenyl)-	$-NO_2$	(Spectrum not obtained)							49

[a] Radicals obtained as secondary product by adding an aqueous solution of the corresponding phenyl-1,4- hydroquinone to an equal volume of 15% NaOH solution.

Table I.6.6
3′-SUBSTITUTED 6-PHENYL-4-OXIDO-1,2-BENZOSEMIQUINONES[a]

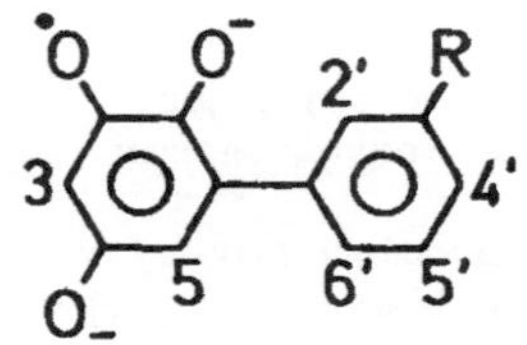

No.	Radical	R substituent	a_3	a_5	$a_{2'}$	$a_{4'}$	$a_{5'}$	$a_{6'}$	$a^R_{3'}$	Ref.
			Splitting constants/Gauss							
	4-Oxido-1,2-benzosemiquinone									
367	-,6-(3′-oxidophenyl)-	–O	0.66	4.36	0.12	0.12	0.12	0.12	—	49
368	-,6-(3′-methoxyphenyl)-	–OMe	0.68	4.56	0.12	0.12	0.12	0.12	—	49
369	-,6-(3′-methylphenyl)-	–Me	0.65	4.45	0.12	0.12	0.12	0.12	u	49
358	-,6-phenyl-	–H	0.67	4.52	0.12	0.12	0.12	0.12	0.12	48, 49
370	-,6-(3′-fluorophenyl)-	–F	0.66	4.64	0.12	0.12	0.12	0.12	—	49
371	-,6-(3′-chlorophenyl)-	–Cl	0.67	4.62	0.12	0.12	0.12	0.12	—	48
372	-,6-(3′-nitrophenyl)-	$-NO_2$	(Spectrum not obtained)							49

[a] Radicals obtained as secondary product by adding an aqueous solution of the corresponding phenyl-1,4-hydroquinone to an equal volume of 15% NaOH solution.

Table I.6.7
4′-SUBSTITUTED 6-PHENYL-4-OXIDO-1,2-BENZOSEMIQUINONES[a]

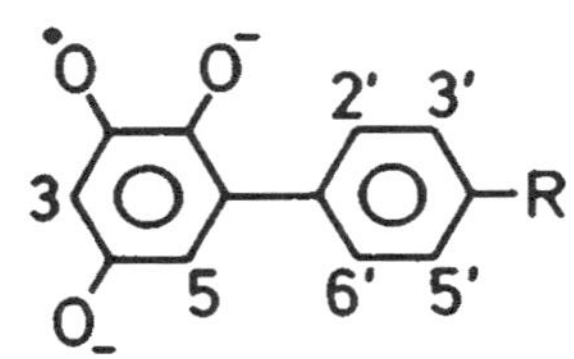

No.	Radical	R substituent	a_3	a_5	$a_{2'}=a_{6'}$	$a_{3'}=a_{5'}$	$a_{4'}^R$	Ref.
	4-Oxido-1,2-benzosemiquinone							
373	-,6-(4′-oxidophenyl)-	–O	0.63	4.00	0.14	0.10	—	48, 49
374	-,6-(4′-methoxyphenyl)-	–OMe	0.65	4.40	0.12	0.12	u	48
375	-,6-(4′-methylphenyl)-	–Me	0.67	4.42	0.13	0.13	0.13′	48
376	-,6-(4′-phenylphenyl)-	–Ph	0.66	4.52	0.12	0.12	u	48
358	-,6-phenyl-	–H	0.66	4.52	0.12	0.12	0.12	48, 49
377	-,6-(4′-fluorophenyl)-	–F	0.66	4.48	0.12	0.12	0.24′	49
378	-,6-(4′-chlorophenyl)-	–Cl	0.68	4.58	0.12	0.12	—	48
379	-,6-(4′-bromophenyl)-	–Br	0.67	4.56	0.12	0.12	—	49
380	-,6-(4′-carboxylatophenyl)-	–COO	0.67	4.66	0.12	0.12	—	48
381	-,6-(4′-nitrophenyl)-	$–NO_2$	0.70	4.86	0.12	0.12	u	48, 49

[a] Radicals obtained as secondary product by adding an aqueous solution of the corresponding phenyl-1,4-hydroquinone to an equal volume of 15% NaOH solution.

Table I.6.8
POLYSUBSTITUTED 6-PHENYL-4-OXIDO-1,2-BENZOSEMIQUINONES[a]

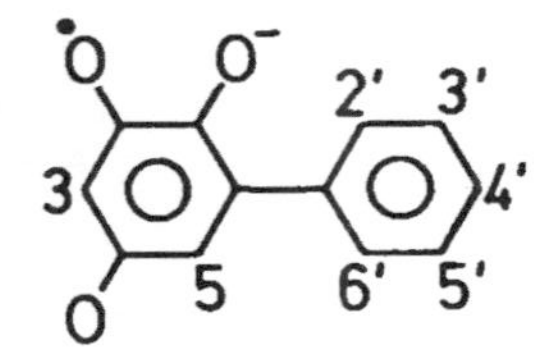

No.	Radical	Phenyl substituents	Splittingconstants/Gauss a_3	a_5	$a_{3'}$	$a_{5'}$	Ref.
	4-Oxido-1,2-benzosemiquinone						
382	-,6-(2',3'-dimethylphenyl)-	–Me	0.64	4.48	u	u	49
383	-,6-(2',4'-dimethylphenyl)-	–Me	0.64	4.52	u	u	49
384	-,6-(2',5'-dimethylphenyl)-	–Me	0.64	4.52	u	u	49
385	-,6-(2',6'-dimethylphenyl)-	–Me	0.63	4.52	0.08[b]	0.08[b]	49
386	-,6-(2',4',6'-trimethylphenyl)-	–Me	0.65	4.52	0.09[b]	0.09[b]	49
387	-,6-(2',6'-dimethyl-4'-chlorophenyl)-	–Me, –Cl	0.63	4.62	u	u	49
388	-,6-(2',6'-dimethyl-4'-bromophenyl)-	–Me, –Br	0.64	4.62	u	u	49
389	-,6-(2',4',6'-trimethyl-3'-bromophenyl)-	–Me, –Br	0.64	4.60	u	u	49
390	-,6-(2'-methyl-3'-chlorophenyl)-	–Me, –Cl	0.64	4.64	u	u	49
391	-,6-(2'-methyl-6'-chlorophenyl)-	–Me, –Cl	0.56	5.86	u	u	49
392	-,6-(2',6'-dichlorophenyl)-	–Cl	1.20[c]	5.16	u	u	49

[a] Radicals obtained as secondary product by adding an aqueous solution of the corresponding phenyl-1,4-hydroquinone to an equal volume of 15% NaOH solution.

[b] *Only splittings from 3' and 5' positions of the phenyl group have been reported for the compounds in this* table.

[c] This large splitting does not make sense (editor's remark).

Table I.7
4-HYDROXY- AND 4-ALKOXY/ARYLOXY-1,2-BENZOSEMIQUINONES

Anions

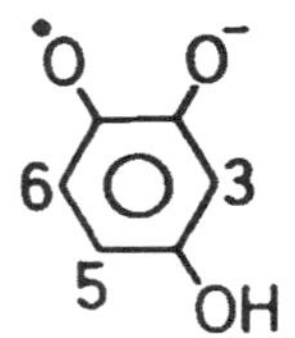

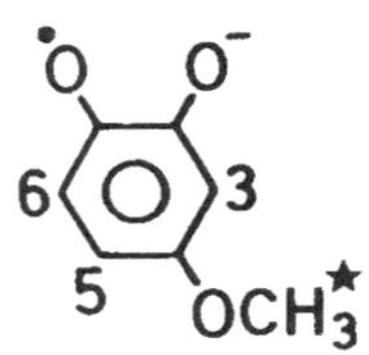

No.	Radical	Substituent(s)	Splitting constants/Gauss a_3	a_4	a_5	a_6	g-Factor	Solvent	Method	Ref.
432	4-Hydroxy-1,2-benzosemiquinone		(Data not available)							
433	4-Methoxy-1,2-benzosemiquinone		0.0	1.1*	3.7	0.75		w H_2SO_4	Ox CeIV	19
434	-,5-*t*-butyl-	–CMe₃	0.63	1.14*	0.16'	0.92		75% w MeOH	Ox KOH	10
435	-,6-*t*-butyl-	–CMe₃	0.6	1.1*	3.2	u		10% w DMF	Ox alk	9
			0.8	1.1*	3.1	u		w	Ox alk	9
			0.78	1.02*	2.88	u		MeOH	Re alk	93
			0.55	1.00*	3.02	u		50% MeOH DMSO	Re alk	93
436	-,3-6-di-*t*-butyl-	–CMe₃	u	0.4*	3.8	u		w	Ox alk	131
437	-,6-carboxylato-	–COO	0.67	1.17*	3.93	—	2.00459	w NaOH (pH=5—11)	Radio-lysis'	115

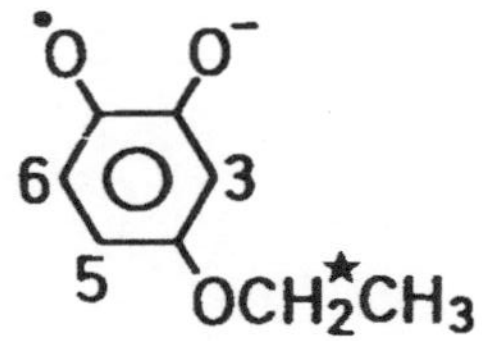

438	4-Ethoxy-1,2-benzosemiquinone		(Data not available)						
439	-,6-*t*-butyl-	–CMe₃	0.78	1.29*	2.95	u	EtOH	Re alk	93
			0.55	1.18*	3.08	u	50% EtOH/ DMSO	Re alk	93
440	-,3,6-di-*t*-butyl-	–CMe₃	u	0.35*	3.9	u	w	Ox alk	131

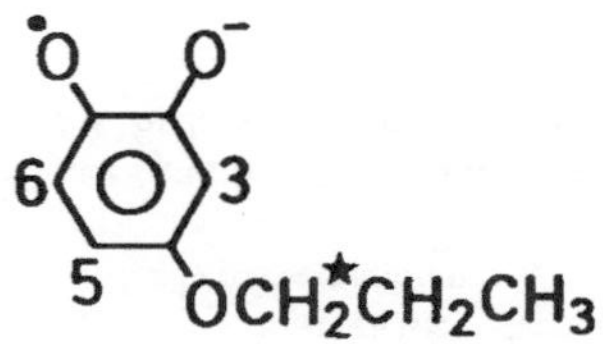

441	4-Propoxy-1,2-benzosemiquinone		(Data not available)						
442	-,6-*t*-butyl-	–CMe₃	0.78	1.30*	2.96	u	PrOH	Re alk	93
			0.58	1.25*	3.10	u	50% PrOH/ DMSO	Re alk	93

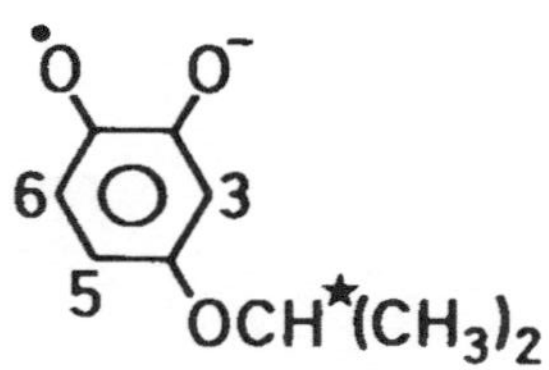

443	4-Isopropoxy-1,2-benzosemiquinone		(Data not available)						
444	-,6-*t*-butyl-	–CMe₃	0.73	0.73*	2.95	u	PrⁱOH	Re alk	93
			0.66	0.66*	3.16	u	50% PrⁱOH/ DMSO	Re alk	93

Table I.7 (continued)
4-HYDROXY- AND 4-ALKOXY/ARYLOXY-1,2-BENZOSEMIQUINONES

Anions

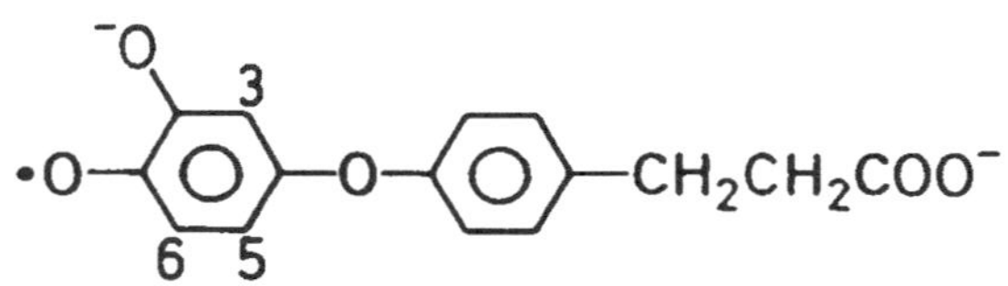

No.	Radical	Substituent(s)	Splitting constants/Gauss				g-Factor	Solvent	Method	Ref.
			a_3	a_4	a_5	a_6				
445	4-[4-(2-Carboxylatoethyl)phenoxy]-1,2-benzosemiquinone		0.43	—	3.96	1.53	2.0039	w (pH = 9.6)	Ox alk[b]	94

[a] Secondary product from 5-methoxysalicylic acid. Temperature 5°C.
[b] Autoxidation of 3′-hydroxythyropropionic acid in 0.1 *M* sodium phosphate buffer (pH = 9.6).

Table I.8
2-HYDROXY- AND 2-ALKOXY/ARYLOXY-1,4-BENZOSEMIQUINONES

Anions

No.	Radical	Substituent(s)	Splitting constants/Gauss a_2	a_3	a_5	a_6	g-Factor	Solvent	Method	Ref.
	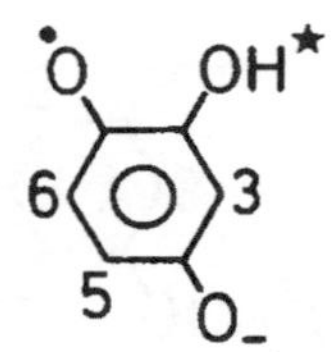									
446	2-Hydroxy-1,4-benzosemiquinone		0.31*	1.13	4.75	1.58		HMPA	Ox KOBu^t	262
	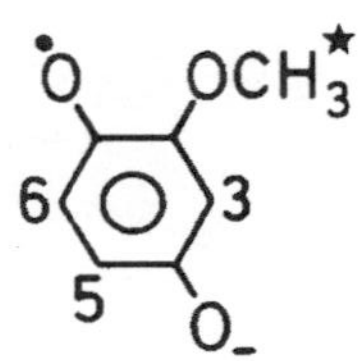									
447	2-Methoxy-1,4-benzosemiquinone		0.85*	0.54	3.65	1.95		w (pH = 9.18)	Re alk	96, 19
			0.85*	0.54	3.65	1.95		w	Ox alk (1*M*)	103
			0.82*	0.52	3.60	1.94		w	Ox KOBu^t	4
			0.796*	0.589	3.606	2.016	2.00461	w/EtOH (pH ~12.7)	Ox alk	8, 130
			0.63*	0.93	3.58	2.28		DMF	Re el	78, 95
			0.47*	1.16	3.48	2.31		HMPA	Ox KOBu^t	4, 262
448	-,3-methyl-	–Me	0.0*	1.52[i]	2.79	2.35		w	Ox alk	8

Table I.8 (continued)
2-HYDROXY- AND 2-ALKOXY/ARYLOXY-1,4-BENZOSEMIQUINONES

Anions

No.	Radical	Substituent(s)	Splitting constants/Gauss a_2	a_3	a_5	a_6	g-Factor	Solvent	Method	Ref.
449	-,3,5-dimethyl-	–Me	0.0*	1.56[1]	2.59[1]	1.64		w	Ox alk	8
			u	1.48[1]	2.59[1]	1.72		MeOH	Ox alk[a]	98
450	-,5,pentyl-	–$(CH_2)_4Me$	0.79*	0.71	2.71[1]	1.40		50% w/DMF	Ox alk	130
451	-,5-*t*-butyl-	–CMe_3	0.70*	1.10	0.09[2]	1.30		25% w/MeOH	Ox KOH	10
452	-,6-pentyl-	–$(CH_2)_4Me$	0.75*	0.75	2.59	1.37[1]		w	Ox alk	8
453	-,6-*t*-butyl-	–CMe_3	0.84*	0.44	2.72	u		25% w/MeOH	Ox KOH	10, 97
			0.78*	0.41	2.54	u		MeOH	Re alk	93
			0.66*	0.66	2.64	u		50% MeOH/DMSO	Re alk	93
			0.64*	1.00	3.12	u		MeCN	Re el	97

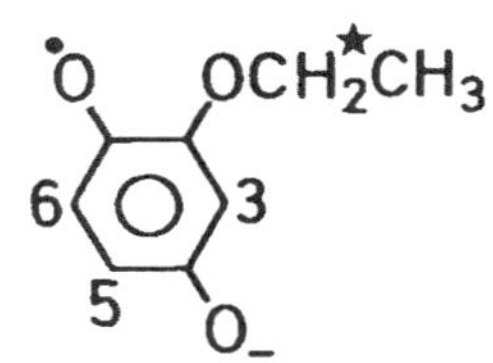

No.	Radical	Substituent(s)	a_2	a_3	a_5	a_6	g-Factor	Solvent	Method	Ref.
454	2-Ethoxy-1,4-benzosemiquinone									
455	-,5-pentyl-	–$(CH_2)_4Me$	0.95*	0.65	2.75[1]	1.35	2.00453	w/EtOH	Ox alk	8
456	-,5-*t*-butyl-	–CMe_3	0.63*	1.26	0.09[2]	1.89		EtOH	Re alk	93
457	-,6-methyl-	–Me	0.96*	0.70	2.91	1.67[1]	2.00459	w/EtOH	Ox alk	8
458	-,6-pentyl-	–$(CH_2)_4Me$	0.961*	0.604	2.790	1.409[1]		w/EtOH	Ox alk	8
459	-,6-*t*-butyl-	–CMe_3	1.00*	0.44	2.72	u		25% w/EtOH	Ox KOH	97
			0.89*	0.42	2.60	u		EtOH	Re alk	93
			0.85*	0.57	2.68	u		50% EtOH/DMSO	Re alk	93

460 -,3,5-dimethyl-	–Me	0.54*	1.41[1]	2.57[1]	1.76	EtOH	Ox alk[a]	98

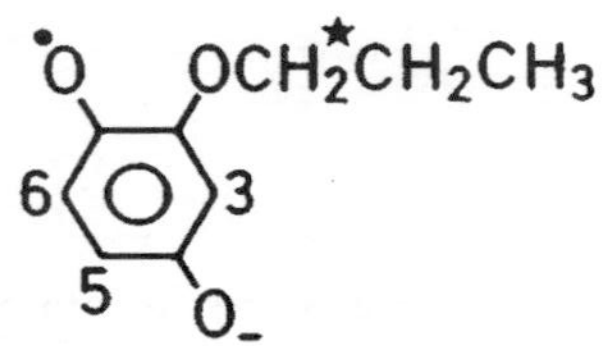

461 2-Propoxy-1,4-bensosemiquinone								
462 -,5-*t*-butyl-	–CMe_3	0.63*	1.26	0.09[2]	1.89	PrOH	Re alk	93
463 -,6-*t*-butyl-	–CMe_3	0.90*	0.45	2.35	u	PrOH	Re alk	93
		0.87*	0.58	2.65	u	50% PrOH: DMSO	Re alk	93

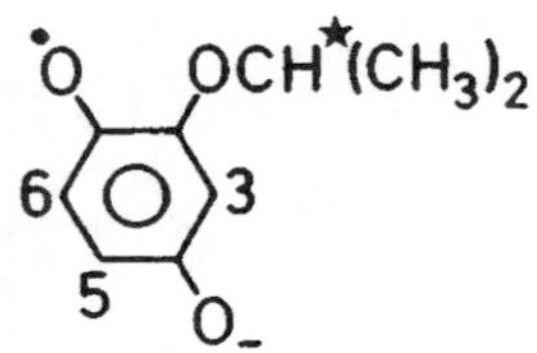

464 2-Isopropoxy-1,4-benzosemiquinone								
465 -,5-pentyl-	–$(CH_2)_4Me$	0.540*	0.687	2.739[1]	1.344	w PrOH	Ox alk	8
466 -,5-*t*-butyl-	–CMe_3	0.42*	1.21	0.10[2]	1.21	PrOH	Re alk	93
467 -,6-methyl-	–Me	(Data not available)				w PrOH	Ox alk	8
468 -,6-pentyl-	–$(CH_2)_4Me$	0.528*	0.608	2.753	1.440[1]	w/PrOH	Ox alk	8
469 -,6-*t*-butyl-	–CMe_3	0.53*	0.53	2.65	u	PrOH	Re alk	93
		0.49*	0.60	2.64	u	50% PrOH DMSO	Re alk	93
470 -,3,5-dimethyl-	–Me	u	1.30	2.54	1.81	PrOH	Ox alk[a]	98

Table I.8 (continued)
2-HYDROXY- AND 2-ALKOXY/ARYLOXY-1,4-BENZOSEMIQUINONES

Anions

No.	Radical	Substituent(s)	Splitting constants/Gauss a_2	a_3	a_5	a_6	g-Factor	Solvent	Method	Ref.
	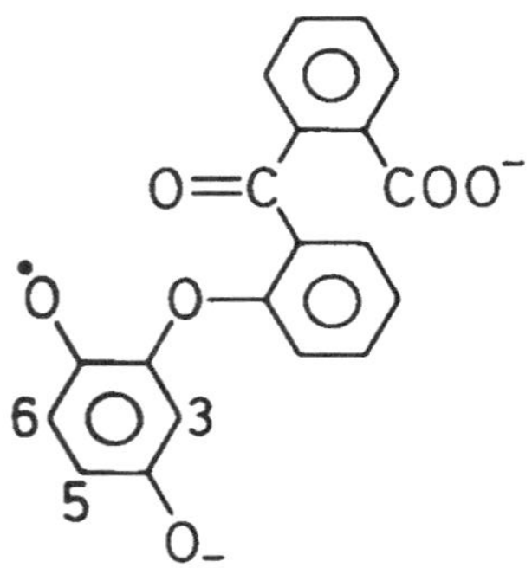									
471	2-[2-(2-Carboxylatobenzoyl)phenoxy]-1,4-benzosemiquinone		u	0.67	3.72	1.95		w/alk	Ox H_2O_2[b]	99, 100

[a] ENDOR experiment at 273 K.
[b] Radical obtained from oxidation of fluorescein.

Table I.9
BENZOSEMIQUINONES WITH FOUR KEY OXYGENS [1,2,3,4]

Anions

No.	Radical	Substituent(s)	Splitting constants/Gauss a_2	a_3	a_5	a_6	g-Factor	Solvent	Method	Ref.
	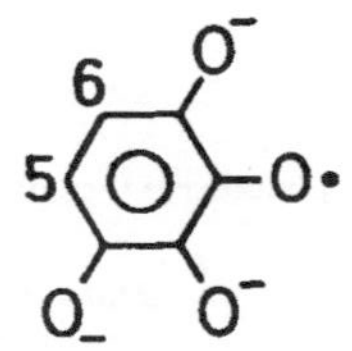									
472	2,3-Dioxido-1,4-benzosemiquinone									
473	-,5-methyl-	–Me	—	—	3.50[1]	2.30		w	Ox KOH	43
	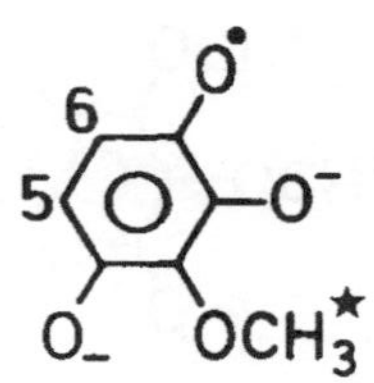									
474	3-Methoxy-2-oxido-1,4-benzosemiquinone									
475	-,5-formyl-	–CHO	—	0.60*	u	3.5		10% w/DMF	Ox alk[b]	103[a]
476	-,5-acetyl-	–COMe	—	u	0.20[2]	4.65		w/alk	[c]	103
477	-,5-propinoyl-	–COEt	—	u	0.15[2]	4.5		w/alk	[c]	103
478	-,6-methyl-	–Me	—	0.0*	4.0	1.18[1]		w	Ox alk	102

Table I.9 (continued)
BENZOSEMIQUINONES WITH FOUR KEY OXYGENS [1,2,3,4]

Anions

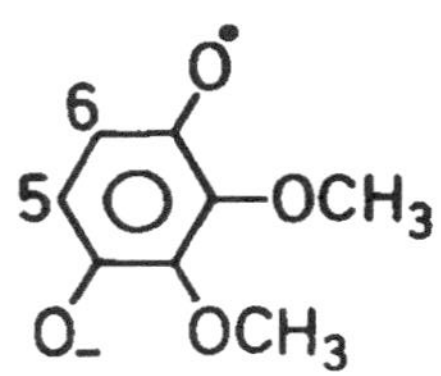

No.	Radical	Substituent(s)	Splitting constants/Gauss a_2	a_3	a_5	a_6	g-Factor	Solvent	Method	Ref.
479	2,3-Dimethoxy-1,4-benzosemiquinone		u	u	2.86	2.86		DMF	Re el	95
			0.0	0.0	2.65	2.65		w	Ox alk	102
480	-,5-methyl-	–Me	0.0	0.0	2.45[l]	1.98		w	Ox alk	102
481	-,5,6-dimethyl-	–Me	0.0	0.0	2.15[l]	2.15[l]		w	Ox alk	102
482	-,5-methyl-6-decaprenyl-	–Me,–[CH_2CH=C(Me)CH_2]$_{10}$–H	0.0	0.0	2.037[l]	1.018[l]		EtOH	Re K	76
			0.0	0.0	2.089[l]	1.044[l] 1.342[l]		DME (40°C)	Re K	76
			0.0	0.0	2.089[l]	0.736[l]		DME(–50°C)[d]	Re K	76

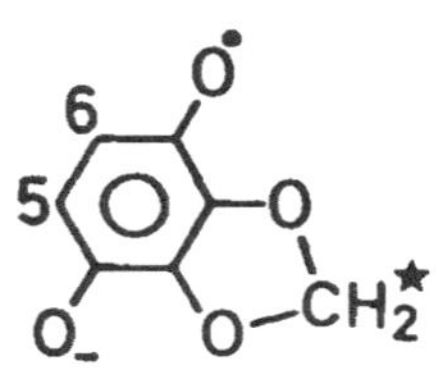

483	2,3-Methylenedioxy-1,4-benzo-semiquinone		$a^{H}_{CH_2}$ = 0.18*	2.63	2.63			w	Ox alk	102
			a_3	a_4	a_5	a_6				
484	3,4-Dimethoxy-1,2-benzosemiquinone		0.0*	0.44*	3.92	1.70		w	Ox alk	102
			0.0*	0.38*	3.91	1.67	2.0045	w/EtOH	Ox alk	8
			0.0*	0.53*	3.92	1.70		EtOH	Ox KOH(w)	80
			<0.05*	0.47*	3.82	1.73	2.00453	w/NaOH (pH = 5—11)	Radiolysis[c]	115
485	-,6-methyl-	–Me	0.0*	0.53*	2.95	1.85'		w	Ox alk	102
486	-,6-carboxylato-	–COO	0.0*	0.38*	3.95	—		w/EtOH	Ox alk	8
487	3,6-Dimethoxy-1,2-benzosemiquinone									
488	-,4-methyl-	–Me	0.0*	5.61'	1.70	0.61*		w	Ox alk	102

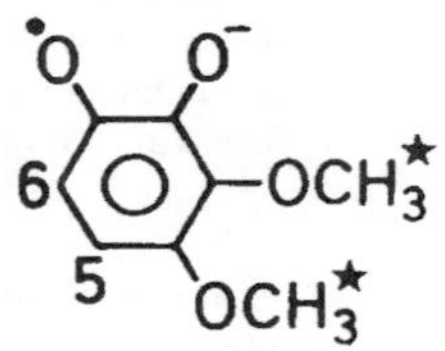

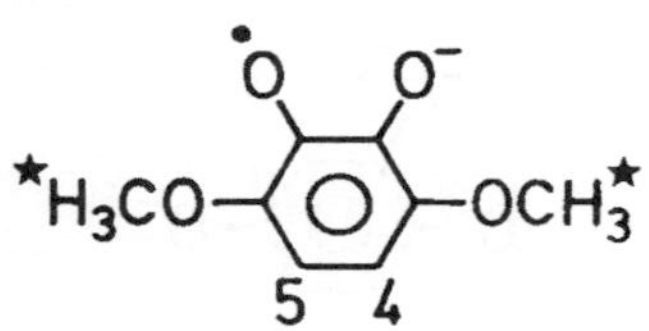

[a] Reference 103 states the radicals 475, 476, and 477 as derived from 2-hydroxy-3-methoxy-1,4-benzosemiquinone.
[b] Radical obtained from 5-hydroxyvanillin.
[c] Radical obtained by treatment of parent syringyl derivative by 1.0 *M* NaOH.
[d] The methylene proton splittings (a_6) are inequivalent at this temperature.
[e] Radical obtained as secondary product from 2,3-dimethoxyphenol at 5°C.

Table I.10
BENZOSEMIQUINONES WITH FOUR KEY OXYGENS [1,2,3,5]

Anions

No.	Radical	Substituent	Splitting constants/Gauss a_2	a_3	a_5	a_6	g-Factor	Solvent	Method	Ref.
	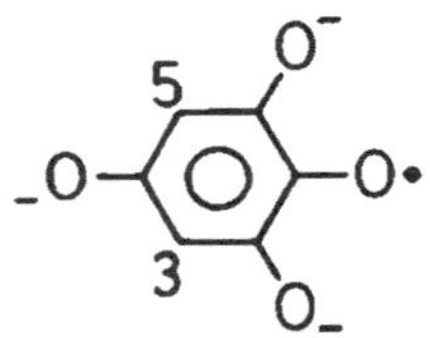									
489	2,6-Dioxido-1,4-benzosemiquinone									
490	-,3-methyl-	–Me	—	0.52[1]	0.52	—		w	Ox KOH	43
	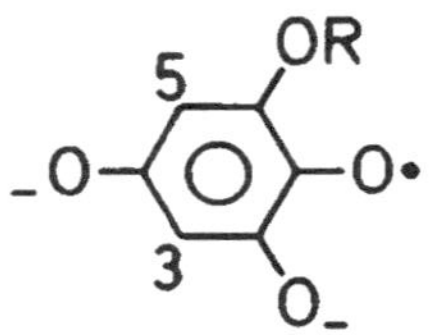									
491	**6-Methoxy-2-oxido-1,4-benzosemiquinone**	[R = CH_3]	—	0.28	2.90	0.60*		Bu'OH	0.1 *M* NaOBu'	96
			—	0.20	2.50	0.70*				
								w	1 *M* NaOH	101, 103
492	6-Ethoxy-2-oxido-1,4-benzosemiquinone	[R = CH_2Me]	—	0.6	2.80	0.68*		EtOH abs	1 *M* NaOEt	96
493	6-Isopropoxy-2-oxido-1,4-benzosemiquinone	[R = CH^*Me_2]	—	0.15	2.90	0.40*		Pr'OH/abs	1 *M* NaOPr'	96
			—	0.19	2.81	0.44*		Pr'OH	Ox alk	98*

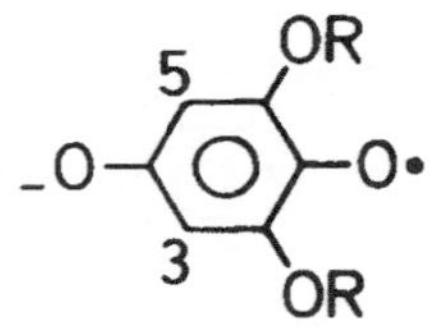

494	2,6-Dimethoxy-1,4-benzosemiquinone	[R = CH_3]	0.80*	1.47	1.47	0.80*	w	Ox KOBu[t]	4
			0.80*	1.49	1.49	0.80*	w	Ox alk	101, 102
			0.77*	1.50	1.50	0.77*	50% w/Bu[t]OH	0.1 *M* NaOH	96
			0.74*	1.48	1.48	0.74*	MeOH	Re alk	93, 96
			0.68*	1.60	1.60	0.68*	50% MeOH/DMSO	Re alk	93
			0.61*	1.83	1.83	0.61*	DMF	Re el	95
			0.50*	1.92	1.92	0.50*	HMPA	Ox KOBu[t]	4
495	-,3-ethyl-	–Et	u	1.35[i]	1.35	0.61*	DMF	Re el	95
496[h]	2,6-Diethoxy-1,4-benzosemiquinone	[R = CH_2Me]	0.97*	1.38	1.38	0.97*	90% w/EtOH	0.1 *M* NaOH	96
			0.96*	1.41	1.41	0.96*	75% w/EtOH	0.1 *M* NaOH	96
			0.94*	1.42	1.42	0.94*	50% w/EtOH	0.1 *M* NaOH	96
			0.90*	1.46	1.46	0.90*	20% w/EtOH	0.1 *M* NaOH	96
			0.87*	1.50	1.50	0.87*	EtOH/abs	0.1 *M* NaOH	96
			0.80*	1.60	1.60	0.80*	EtOH/abs	0.1 *M* NaOEt	96
			0.74*	1.70	1.70	0.74*	50% EtOH DMSO	Re alk	93
497[h]	2,6-Dipropoxy-1,4-benzosemiquinone	[R = CH_2CH_2Me]	0.79*	1.58	1.58	0.79*	PrOH	Re alk	93
			0.76*	1.70	1.70	0.76*	50% PrOH/DMSO	Re alk	93

Table I.10 (continued)
BENZOSEMIQUINONES WITH FOUR KEY OXYGENS [1,2,3,5]

Anions

No.	Radical	Substituent	Splitting constants/Gauss				g-Factor	Solvent	Method	Ref.
			a_2	a_3	a_5	a_6				
498[b]	2,6-Diisopropoxy-1,4-benzosemiquinone	[R = CH*Me$_2$]	0.50*	1.40	1.40	0.50*		50% w/Pr^iOH	0.1 *M* NaOH	96
			0.49*	1.57	1.57	0.49*		Pr^iOH	Ox alk	98[a]
			0.45*	1.65	1.65	0.45*		Pr^iOH/abs	0.1 *M* $NaOPr^i$	96
			0.36*	1.62	1.62	0.36*		50% Pr^iOH/DMSO	Re alk	93
	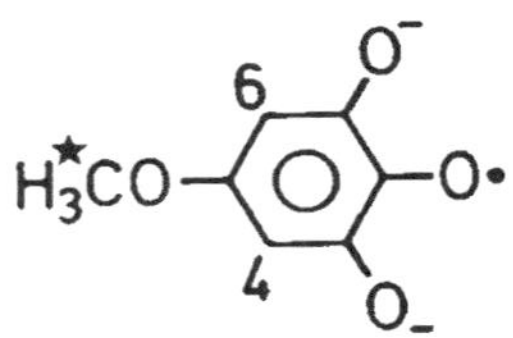									
			a_3	a_4	a_5	a_6				
499	5-Methoxy-3-oxido-1,2-benzosemiquinone		—	0.44	0.70*	0.44	2.00455	33% w/DMSO	Ox alk	8, 178
	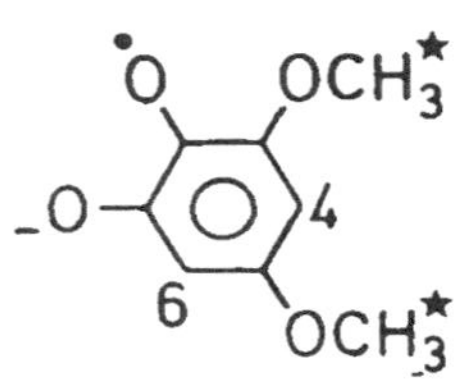									

500	3,5-Dimethoxy-1,2-benzosemiquinone		0.75[a]	1.70	1.00[a]	0.95		w	Ox alk	102, 80

[a] ENDOR experiment at 273 K.
[b] Structure (see no. 494).

Table I.11
BENZOSEMIQUINONES WITH FOUR KEY OXYGENS [1,2,4,5]

Anions

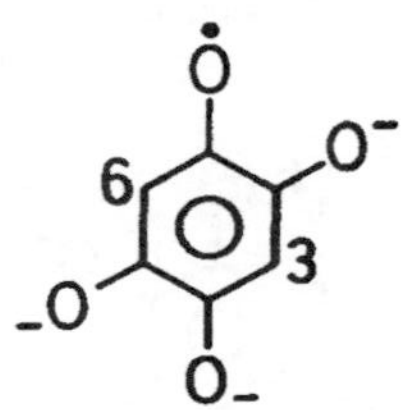

No.	Radical	Substituent(s)	Splitting constants/Gauss a_2	a_3	a_5	a_6	g-Factor	Solvent	Method	Ref.
501	[I] 2,5-Dioxido-1,4-benzosemiquinone		—	0.790	—	0.790		w	Re KOH	39
			—	0.798	—	0.798		w/EtOH	Re alk	8
502	-,3-methyl-	–Me	—	1.30'	—	0.80		w	Ox KOH	43
503	-,3-undecyl-	$-C_{11}H_{23}$	—	0.70'	—	0.79		w/EtOH	Ox alk	8
504	[I]-,3,6-difluoro-	–F	—	3.85[a]	—	3.85[a]		w	Re alk	65
			—	3.7[a]	—	3.7[a]		w	Re alk[b]	229
505	[I]-,3,6-dichloro-	–Cl	—	0.18[c]	—	0.18[c]		w	Re alk	134

Table I.11 (continued)
BENZOSEMIQUINONES WITH FOUR KEY OXYGENS [1,2,4,5]

Anions

No.	Radical	Substituent(s)	a_2	a_3	a_5	a_6	g-Factor	Solvent	Method	Ref.
			Splitting constants/Gauss							
	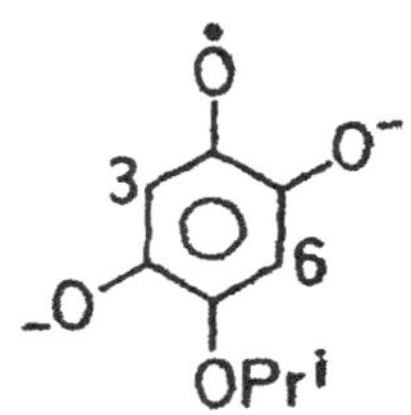									
506	5-Isopropoxy-2-oxido-1,4-benzosemiquinone	[Pr^i = CH*Me_2]	—	0.20	0.71*	0.65		Pr^iOH	Ox KOH^d	98
	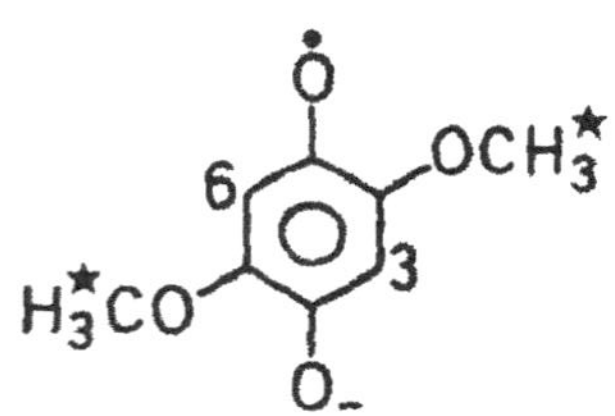									
507	2,5-Dimethoxy-1,4-benzosemiquinone		1.05*	0.30	1.05*	0.30		50% w C_5H_5N	Re KOH	87
			0.98*	0.31	0.98*	0.31		MeOH abs	Re KOH^d	104, 234
			0.91*	0.25	0.91*	0.25		MeOH	Re alk	93, 234, 440

		0.89*	0.41	0.89*	0.41		50% MeOH/ DMSO	Re alk	93
		0.75*	0.72	0.75*	0.72		DMSO	Re el	52
		0.75*	0.75	0.75*	0.75		DMF	Re el	95
508 -,3,6-dimethyl-	–Me	u	1.45[l]	u	1.45[l]		MeOH/abs	Re KOH[d]	104

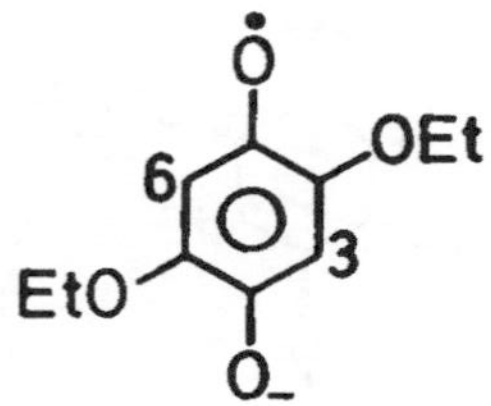

509 2,5-Diethoxy-1,4-benzosemiquinone	[Et = $CH_2^{*}CH_3^{**}$]	1.17*	u	1.17*	u		50% w/C_5H_5N	Re KOH	87
		1.19* 0.06**	0.31	1.19* 0.06**	0.31		EtOH/abs	Re KOH[c]	104
		1.08*	0.30	1.08*	0.30		EtOH	Re alk	93
		1.08*	0.41	1.08*	0.41		w/C_5H_5N	Re KOH	45
		1.00*	0.45	1.00*	0.45		50% EtOH/ DMSO	Re alk	93
		0.89*	0.76	0.89*	0.76	2.0050	MeCN or CH_2Cl_2	Re el	233
510 -,3,6-dimethyl-	–Me	0.07*	1.39[l]	0.07*	1.39[l]		EtOH/abs	Re KOH[d]	104
		<0.2*	1.47[l]	<0.2*	1.47[l]		EtOH	Re NaOEt	133

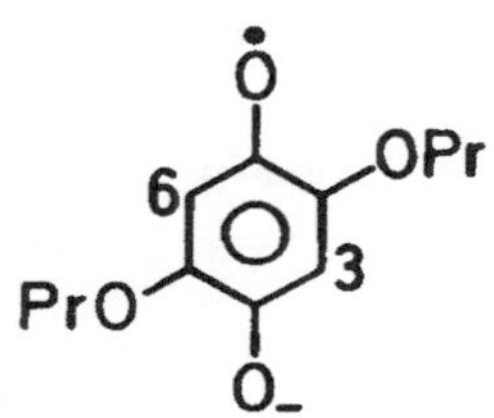

511 2,5-Dipropoxy-1,4-benzosemiquinone	[Pr = $CH_2^{*}CH_2Me$]	1.08*	0.32	1.08*	0.32		PrOH	Re alk	93
		1.00*	0.50	1.00*	0.50		50% PrOH/ DMSO	Re alk	93

Table I.11 (continued)
BENZOSEMIQUINONES WITH FOUR KEY OXYGENS [1,2,4,5]

Anions

No.	Radical	Substituent(s)	Splitting constants/Gauss a_2	a_3	a_5	a_6	g-Factor	Solvent	Method	Ref.
	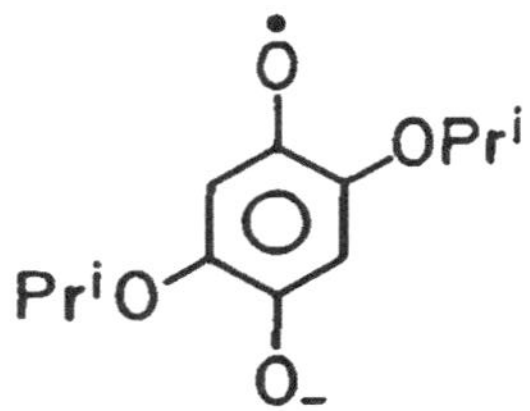									
512	2,5-Diisopropoxy-1,4-benzosemiquinone	[Pr^i = CH*Me_2]	0.62*	0.34	0.62*	0.34		Pr^iOH	Ox KOH	98
			0.55*	0.35	0.55*	0.35		Pr^iOH	Re alk	98
			0.57*	0.47	0.57*	0.47		50% Pr^iOH/DMSO	Re alk	93
	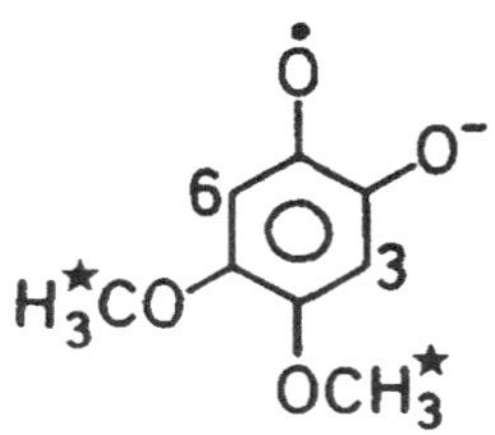									
			a^3	a^4						
513	4,5-Dimethoxy-1,2-benzosemiquinone		0.26	1.20*	1.20*	0.26		MeOH/w	Ox alk	28
			0.32	1.10*	1.10*	0.32		w	Ox alk	102

	0.32	1.05*	1.05*	0.32	w	Ox KOBu^t	4
	0.33	0.85*	0.85*	0.33	HMPA	Ox KOBu^t	4
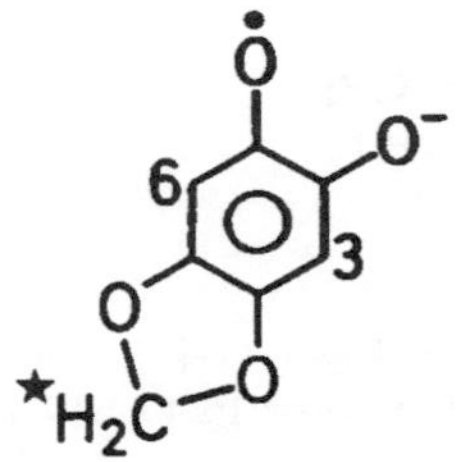							
514 4,5-Methylenedioxy-1,4-benzosemiquinone	0.33	4.22*	4.22*	0.33	w	Ox alk	102
	0.33	4.22*	4.22*	0.33	w	Ox KOBu^t	4
	0.50	2.39*	2.39*	0.50	HMPA	Ox KOBu^t	4
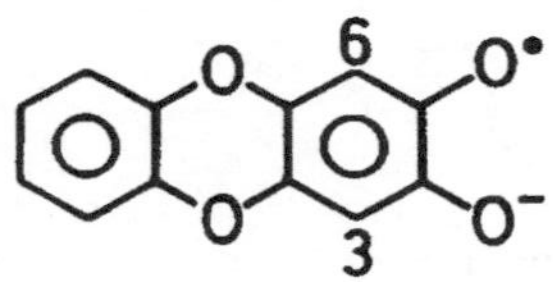							
515 Dibenzo-[1,4]-dioxin-1,2-benzosemiquinone	0.64	—	—	0.64	DMF	Re alk(w)	5

[a] Splitting from fluorine nucleus.
[b] An additional doublet splitting of 0.3 G of unknown origin was observed in the spectrum. See Reference 229.
[c] Splitting from ^{35}Cl nucleus; simulation performed as a superposition of the spectra due to ^{35}Cl and ^{37}Cl and with a linewidth of 0.155 G.
[d] ENDOR experiment at 273 K. Radical obtained from benzoquinone.
[e] ENDOR experiment at 253 K.

Table I.12
BENZOSEMIQUINONES WITH FIVE AND SIX KEY OXYGENS

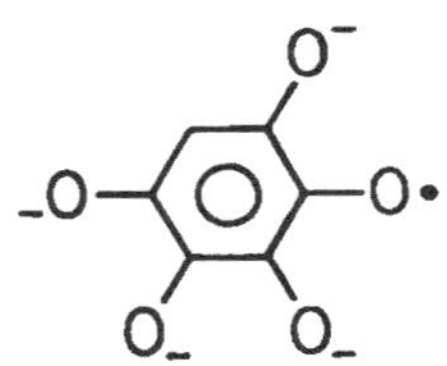

No.	Radical	Splitting constants/Gauss a_3	a_4	a_5	a_6	g-Factor	Solvent	Method	Ref.
516	3,4,5-Trioxido-1,2-benzosemiquinone[a]	—	—	—	0.70	2.00476	w/EtOH	Ox alk	8
517	4,5-Dioxido-3-(pentaoxidophenoxy)-1,2-benzosemiquinone[a]	—	—	—	0.80		w/alk	KCO[b]	236

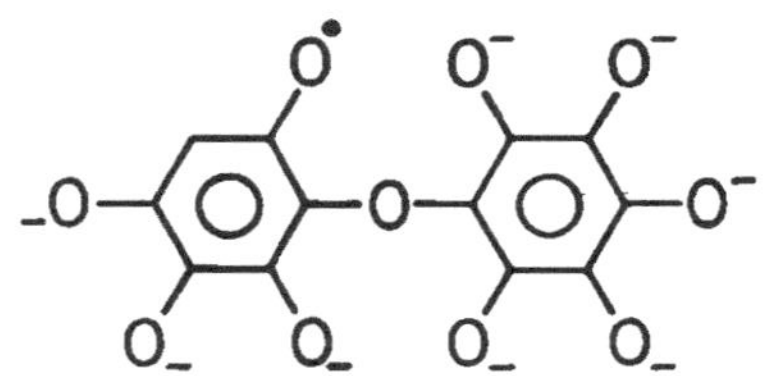

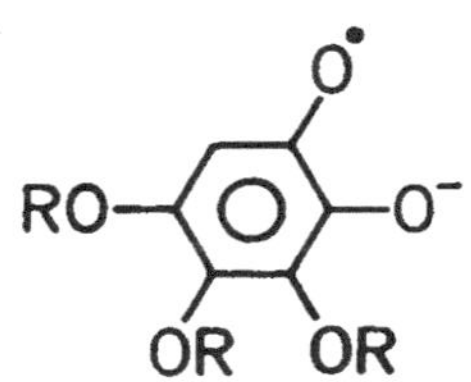

518	3,4,5-Trimethoxy-1,2-benzosemiquinone [R = CH_3]	0.0*	0.10*	1.20*	0.60		w	Ox alk	102
519	3,4,6-Trimethoxy-1,2-benzosemiquinone [R = CH_3]	0.0*	0.90*	1.10	0.68*		w	Ox alk	102
		0.0*	0.90*	1.15	0.66*		w	Ox KOBu^t	4
		0.0*	0.50*	2.13	0.25*		HMPA	Ox KOBu^t	4
520	[I] Tetraoxido-1,4-benzosemiquinone	—	—	—	—	2.00487[c]	w	Ox alk[d]	8,236

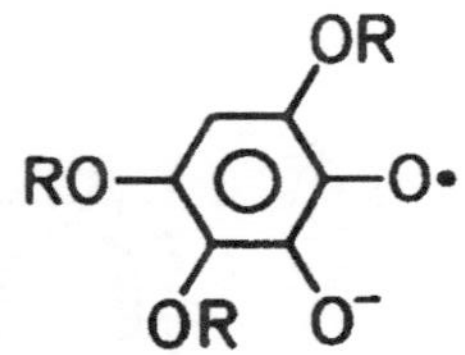

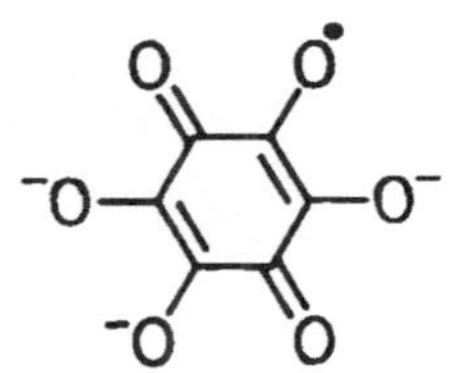

[a] Proposed radical structure.
[b] Radical obtained in the "potassium carbonyl" (KCO) reaction. See Reference 236 for details.
[c] Single line observed. Linewidth 30 mG, minimum observed.
[d] Spectrum obtained by treating tetrahydroxy-1,4-benzoquinone with aqueous alkali.

Table I.13.1
BICYCLOBENZOSEMIQUINONES

No.	Radical	Substituent(s)	Splitting constants/Gauss $a_{R'}$	$a_{R''}$	$a_{R'''}$	$a_{R''''}$	a_5	a_6	Solvent	Method	Ref.
	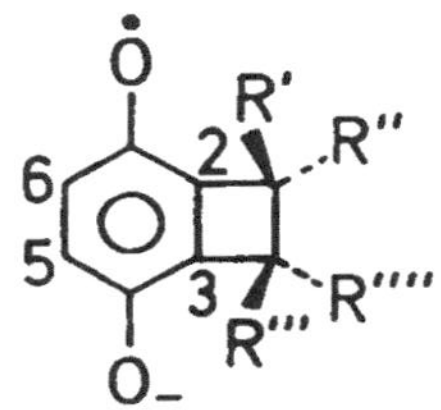										
		R′ R″ R‴ R⁗									
521	*p*-Benzocyclobutanesemiquinone	H H H H	1.93	1.93	1.93	1.93	2.46	2.46	DMSO	O_2/alk	417, 418
522		Me H Me H	u	1.77	u	1.77	2.51	2.51	DMSO	O_2/alk	417, 418
521 D		D H D H	u	1.83	u	1.83	0.31^D	0.31^D	DMSO-d6	O_2/alk	418
523		Me H H Me	u	1.70	1.70	u	2.51	2.51	DMSO	O_2/alk	417, 418
	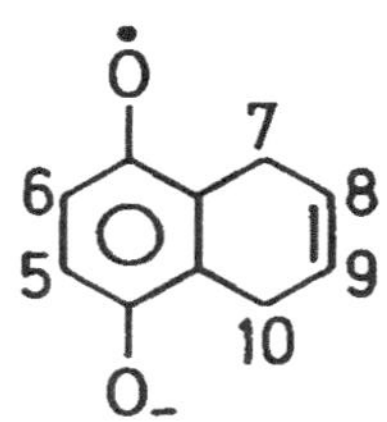										
			a_7	a_{10}							
524	7,10-Dihydro-1,4-naphthosemiquinone		2.05[a]	2.05[a]			2.80	2.80	MeCN	Ox alk	44

[a] Splitting from two equivalent methylene protons; splittings from protons at C-8 and C-9 not resolved; magnitude estimate, $a_8 = a_9 \approx 0.05$ G.

Table I.13.2
SEMIQUINONES DERIVED FROM 3′,6′-DIHYDROXYBENZOBICYCLO[2.2.1]HEPT-2-ENE

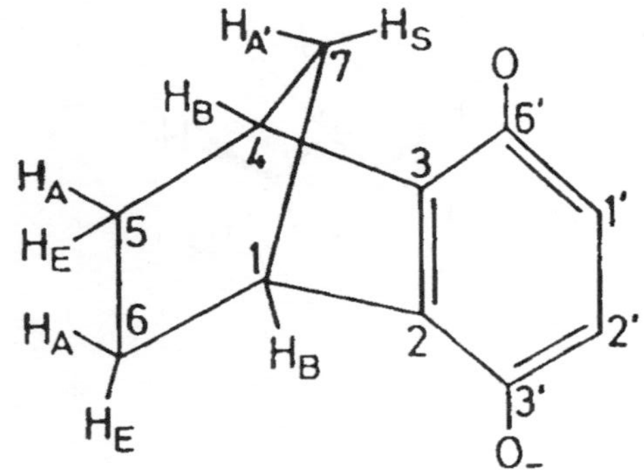

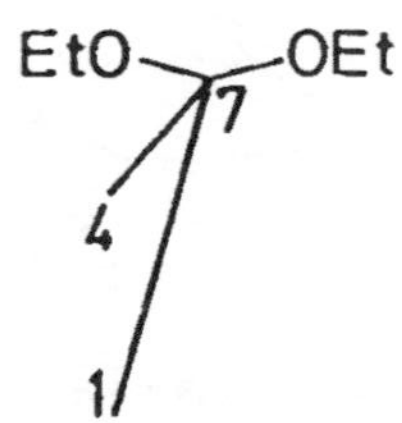

No.	Radical structure	Splitting constants/Gauss										Solvent	Method	Ref.
		$a_{1'}$	$a_{2'}$	$a_1^{H_B}$	$a_4^{H_B}$	$a_5^{H_A}$	$a_6^{H_A}$	$a_5^{H_E}$	$a_6^{H_E}$	$a_7^{H_{A'}}$	$a_7^{H_{S'}}$			
525	No. 1	2.50	2.50	u	u	0.70	0.70	u	u	0.70	u	MeCN	Ox alk	44, 107
		2.55	2.55	u	u	0.66	0.66	u	u	0.66	u	DMSO	Ox KOBu^t	108
526	No. 1	2.48	2.48	u	u	0.84	0.84	u	u	u	u	MeCN	Ox alk	44

Table I.13.2 (continued)
SEMIQUINONES DERIVED FROM 3′,6′-DIHYDROXYBENZOBICYCLO[2.2.1]HEPT-2-ENE

No.	Radical structure	$a_{1'}$	$a_{2'}$	$a_{1}^{H_B}$	$a_{4}^{H_B}$	$a_{5}^{H_A}$	$a_{6}^{H_A}$	$a_{5}^{H_E}$	$a_{6}^{H_E}$	$a_{7}^{H_A'}$	$a_{7}^{H_S'}$	Solvent	Method	Ref.
		Splitting constants/Gauss												
527 No. 1	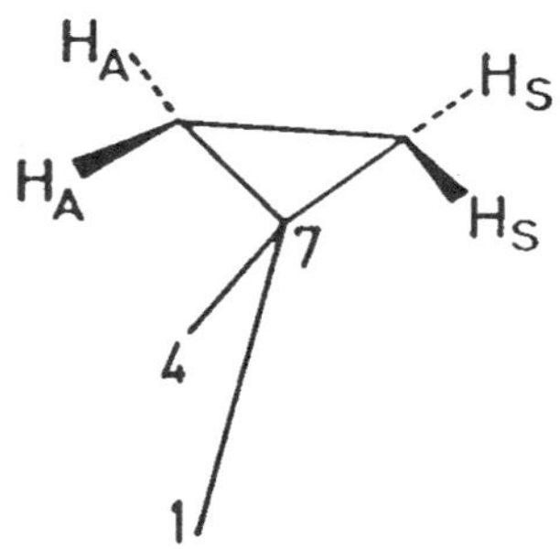	2.55	2.55	u	u	0.81	0.81	0.10	0.10	0.23[a]	0.10[a,b]	MeCN	Ox alk	107
528 No. 1	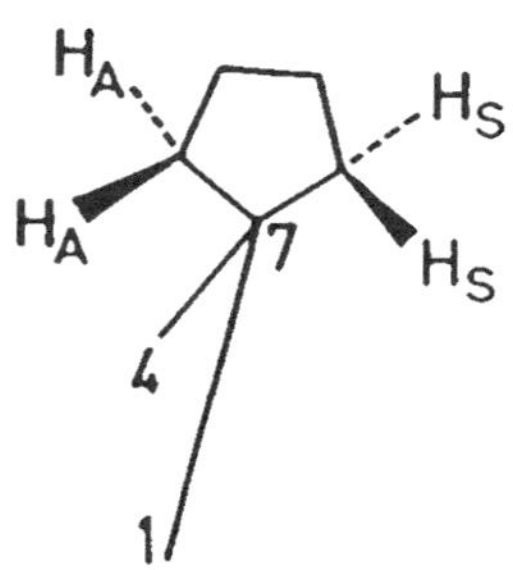	2.60	2.60	u	u	0.85	0.85	u	u	0.25[a]	u	DMSO	Ox KOBu^t	107

529	No. 1	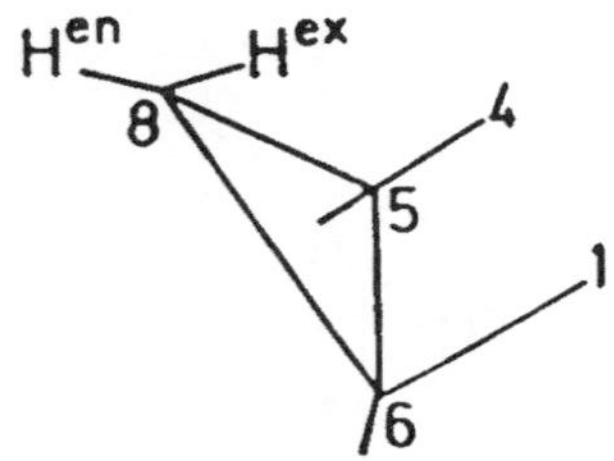	2.42	2.42	0.10	0.10	0.22[c]	u	0.22	0.22	0.65	0.22	MeCN	Ox alk	44
530	No. 1	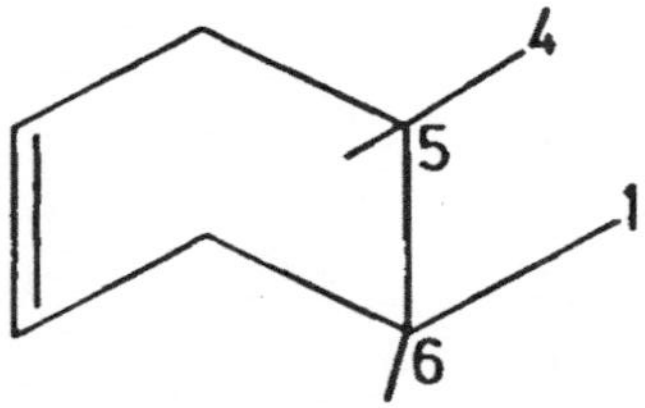	2.65	2.65	u	u	u	u	u	u	0.60	u	MeCN	Ox alk	44

[a] Splitting from two equivalent protons.

[b] This constant may alternatively be assigned the H_B protons.

[c] This constant derives from H_8^{en} or H_8^{ex}.

Table I.13.3
SEMIQUINONES DERIVED FROM 3′,6′-DIHYDROXYBENZOBICYCLO[2.2.2.]OCT-2-ENE

No.	Radical structure	$a_{1'}$	$a_{2'}$	$a_5^{H_A}$	$a_6^{H_A}$	$a_5^{H_S}$	$a_6^{H_S}$	$a_7^{H_A}$	a_8^{H}	$a_7^{H_S}$	$a_8^{H_S}$	Solvent	Method	Ref.
		Splitting constants/Gauss												
531	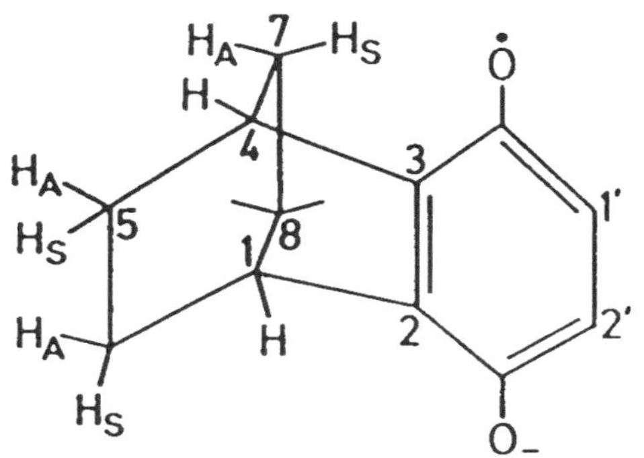 No.1	2.78	2.78	0.45	0.45	u	u	0.45	0.45	u	u	MeCN	Ox alk	44, 109
		2.67	2.67	0.40	0.40	0.09	0.09	0.40	0.40	0.09	0.09	DMSO	Ox KOBu^t	111
532	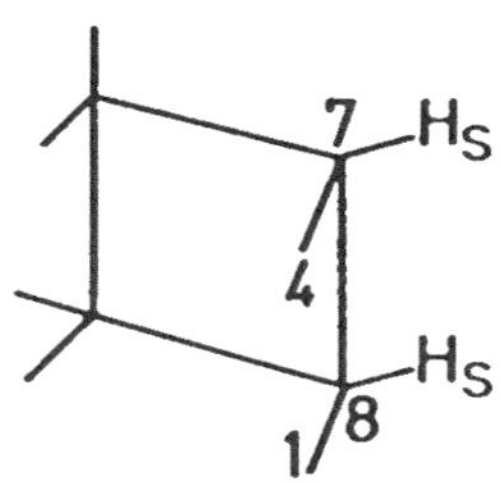 No. 1	2.70	2.70	0.49	0.49	u	u	u	u	u	u	MeCN	Ox alk	44, 109

Table I.13.4
SEMIQUINONES DERIVED FROM 3′,6′-DIHYDROXYBENZOBICYCLO[2.2.1.] HEPTA-2,5-DIENE

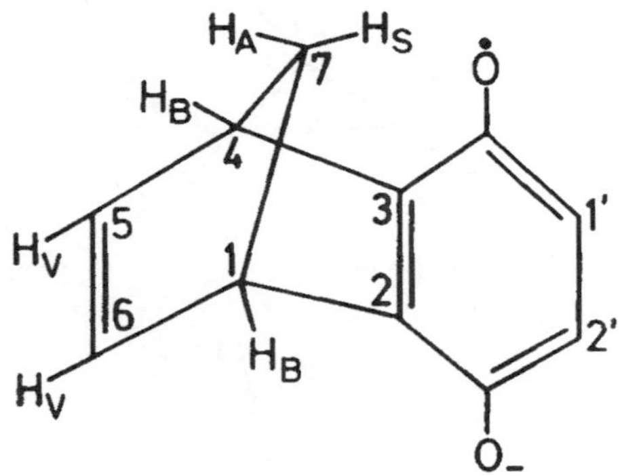

No.	Radical structure	Splitting constant/Gauss $a_{1'}$	$a_{2'}$	a_1^{HB}	a_4^{HB}	a_5^{HV}	a_6^{HV}	a_7^{HA}	a_7^{HS}	Solvent	Method	Ref.
533	No. 1	2.36	2.36	u	u	0.40	0.40	0.80	0.40	MeCN	Ox alk	44, 10, 106
		2.36	2.36	u	u	0.50	0.50	0.81	0.36	EtOH	Ox alk	44
		2.35	2.35	u	u	0.40	0.40	0.80	0.40	DMSO	Ox KOBu^t	108
534	No. 1 (4,6-dimethyl)	(2.51	2.18)[a]	u	u	0.54	0.05[b]	0.81	0.34	MeCN(?)	Ox alk	44, 106
535	No. 1 (7,7-diethoxy)	2.38	2.38	u	u	0.52	0.52	u	u	MeCN	Ox alk	44, 110
536	No. 1 (1-bromo-7,7-dimethoxy)	(2.45	1.90)[c]	u	u	0.45	0.45	u	u	MeCN	Ox alk	44, 110

Table I.13.4 (continued)
SEMIQUINONES DERIVED FROM 3′,6′-DIHYDROXYBENZOBICYCLO[2.2.1.] HEPTA-2,5-DIENE

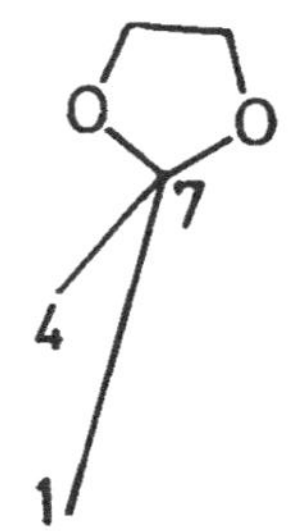

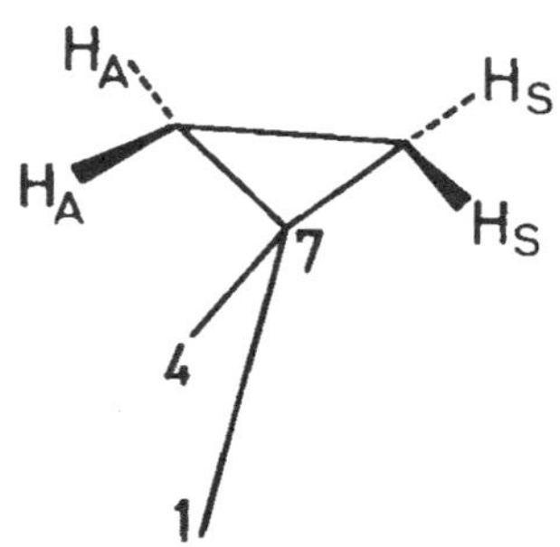

	No.	Radical structure	Splitting constant/Gauss								Solvent	Method	Ref.
			$a_{1'}$	$a_{2'}$	a_1^{HB}	a_4^{HB}	a_5^{HV}	a_6^{HV}	a_7^{HA}	a_7^{HS}			
537	No. 1	(5-*t*-butyl)	(2.48	2.28)[a]	u	u	u	0.52	u	u	MeCN	Ox alk	44, 110
538	No. 1		2.40	2.40	0.10[c]	0.10[c]	0.47	0.47	0.38[d]	0.18[c,d]	MeCN	Ox alk	107

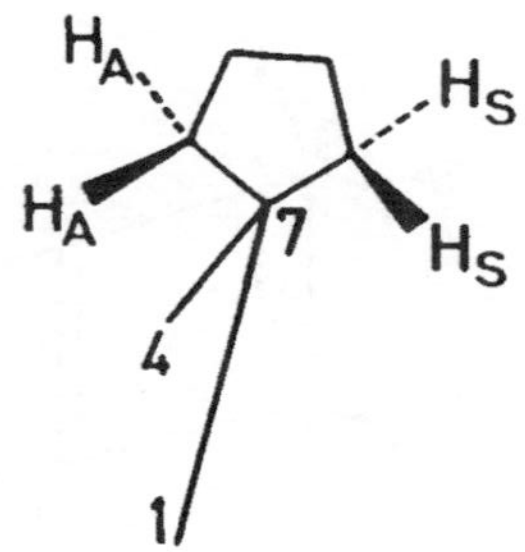

539	No. 1	2.40	2.40	u[c]	u[c]	0.40	0.40	0.40[d]	0.08[d,e]	MeCN	Ox alk	107

[a] Assignment unknown.
[b] a_{Me} <0.05 G.
[c] The constants 0.10 and 0.18 tentatively assigned.
[d] Splitting from two equivalent protons.
[e] A less likely assignment is the one where the 0.08 G splitting is assigned the H_B protons.

Table I.13.5
SEMIQUINONES DERIVED FROM 3′,6′-DIHYDROXYBENZOBICYCLO[2.2.2]OCTA-2,5-DIENE

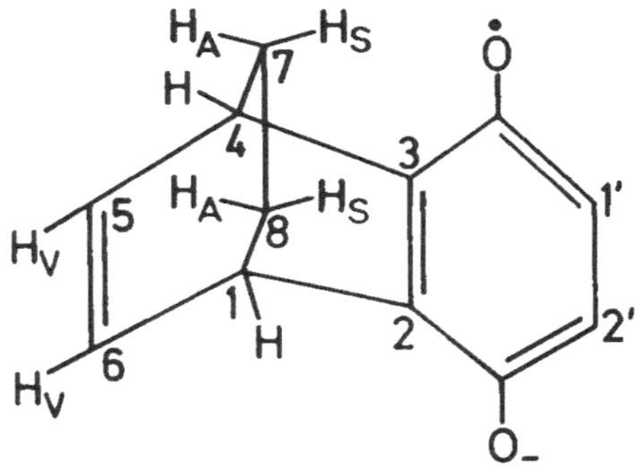

No.	Radical structure	$a_{1'}$	$a_{2'}$	$a_5^{H_V}$	$a_6^{H_V}$	$a_7^{H_A}$	$a_8^{H_A}$	$a_7^{H_S}$	$a_8^{H_S}$	Solvent	Method	Ref.
		Splitting constants/Gauss										
540	No. 1	2.72	2.72	0.54	0.54	0.54	0.54	0.13	0.13	MeCN	Ox alk	44, 106, 109
		2.62	2.62	0.49	0.49	0.49	0.49	0.18	0.18	DMSO	Ox KOBu^t	111
540D	No. 1 (H_1=H_4=D)[a]	2.62	2.62	0.49	0.49	0.49	0.49	0.18	0.18	DMSO	Ox KOBu^t	111
541	No. 1 (1-methyl)	(2.66	2.10)[b]	0.57	0.57	0.57	0.57	u	u	MeCN	Ox alk	44
542	No. 1 (7-isopropyl-5-methyl)	2.55	2.55	<0.05[c]	0.47	u	0.47	u	u	MeCN	Ox alk	44

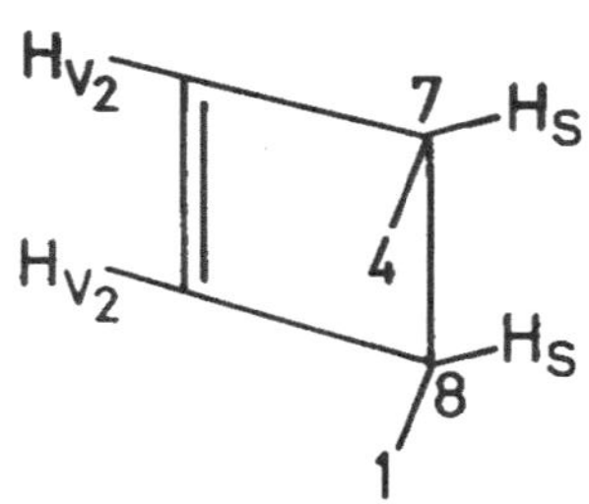

543	No. 1	2.65	2.65	0.44	0.44	0.14[d]	0.14[d]	0.14	0.14	MeCN	Ox alk	44, 109
	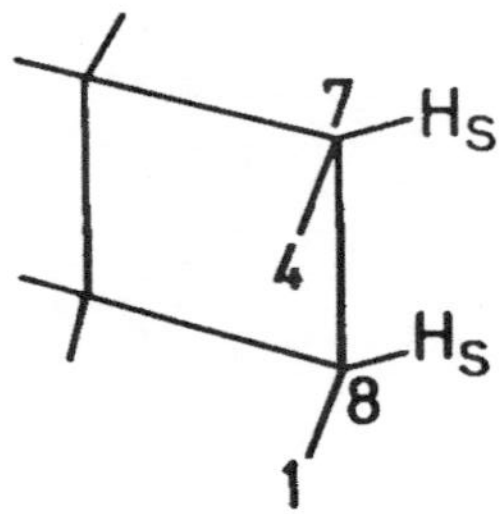											
544	No. 1	2.70	2.70	0.48	0.48	—	—	0.13	0.13	MeCN	Ox alk	44, 109
	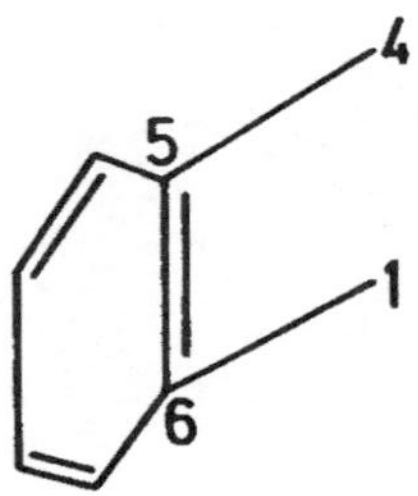											
545	No. 1	2.68	2.68	u	u	0.54	0.54	0.09	0.09	MeCN	Ox alk	44

[a] Deuterium exchange clearly demonstrates that no splitting is derived from the protons at C-1 and C-4.
[b] Assignment unknown.
[c] a_{Me} <0.05 G.
[d] Splitting from H_{v_2}.

Table I.14
SEMIQUINONE RADICALS DERIVED FROM TRIPTYCENE

No.	Radical	R′	R″	R‴	Splitting constants/Gauss a₂	a₃	Others	g-Factor	Solvent	Method	Ref.
	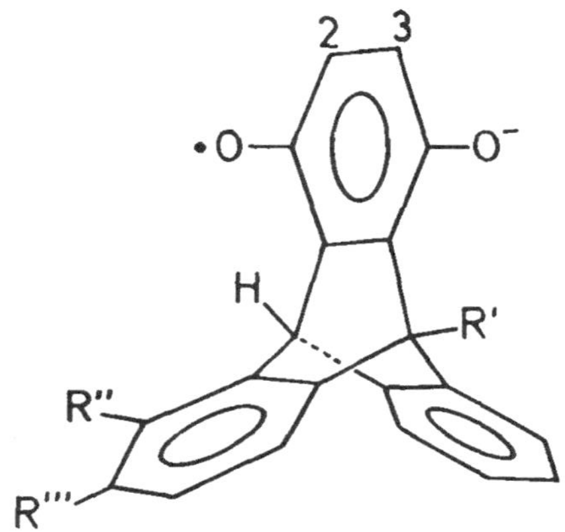										
546	Triptycene-*p*-semiquinone	H	H	H	2.31	2.31	u		MeCN	Ox alk	61
					2.40	2.40	u		DMF	Re el	153[a]
547		Me	H	H	2.62	1.98	u		MeCN	Ox alk	61
548		F	H	H	2.6	2.1	0.85[F]		MeCN	Ox alk	113, 114
549		H	F	H	2.37	2.37	u		MeCN	Ox alk	112
550		H	H	F	2.40	2.40	0.15[F]		MeCN	Ox alk	112

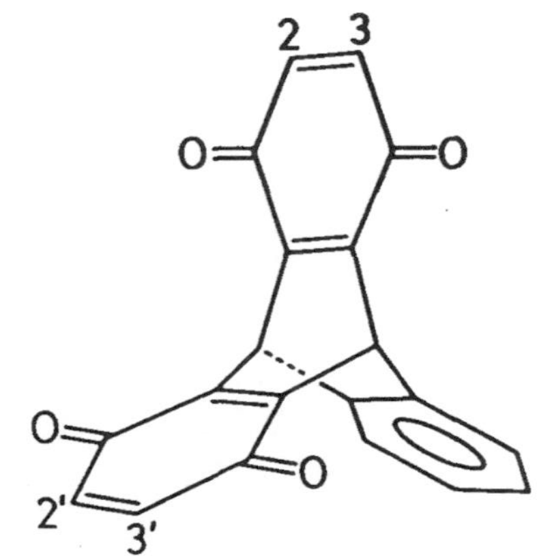

				a_2	a_3	$a_{2'}$	$a_{3'}$				
551	Triptycene bis(quinone)	$1^{\cdot-}$	Monoradical[b]	1.10[b]	1.10	1.10	1.10	2.00506	DMSO	Re el	419
552		$\cdot1^{\cdot2-}$	Biradical[c]	2.45	2.45	2.45	2.45	2.00506	DMSO	Re el	419
				2.40	2.40	2.40	2.40	2.00506	MeCN	Re el	419
553		$1^{\cdot3-}$	Monoradical[d]	1.26	1.26	1.26	1.26		MeCN	Re el	419
				1.30	1.30	1.30	1.30	2.00495	DMSO	Re el	419

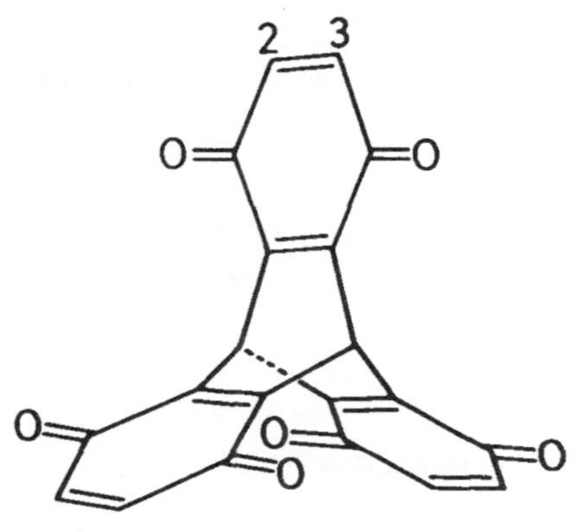

Table I.14 (continued)
SEMIQUINONE RADICALS DERIVED FROM TRIPTYCENE

No.	Radical		R′ R″ R‴	Splitting constants/Gauss: a_2	a_3	Others	g-Factor	Solvent	Method	Ref.
554	Triptycene tris(quinone)	$2^{\cdot}$	Monoradical[c]	a_H = 0.66 G (6H)			2.00507	DMSO	Re el	419
555		$\cdot 2^{3\cdot}$	Triradical[f]	a_H = 2.60 G (2H)				DMSO	Re el	419
556		$2^{5\cdot}$	Monoradical[g]	a_H = 0.68 G (6H)				DMSO	Re el	419

[a] For other triptycene semiquinone spectra, see Reference 153.
[b] A 1:4:6:4:1 quintet observed with a further splitting of 0.05 G from two hydrogens. The hydrogen atoms of the bis(quinone) rings are time averaged by electron migration with an energy barrier of $\Delta H^{\ddagger}$ = 6.2 kcal/mol ($\Delta S^{\ddagger}$ = −7 eu).
[c] Biradical giving rise to a 1:2:1 triplet spectrum only obtained after continued electrolysis.
[d] Extensive electrolytic reduction of 1 leads to a 1:4:6:4:1 spectrum assumed to be from the trianion radical 1^{3-}.
[e] Initially observed spectrum with a seven-line pattern. Reduction of 2 by KI in DMSO in the presence of Hg or by traces of Me_3COK in DMSO, MeCN, or DMF also yield $2^{=}$, and no line broadening is observed in DMF at −90°C.
[f] Reduction of 2 at higher potentials yields a 1:2:1 triplet assigned to the triradical trianion with no exchange of the electrons.
[g] Reduction of 2 at still higher potentials produces a second heptet from an unstable species, which may be $:\dot{2}^{4-}$ or $:2^{5-}$.

Table I.15.1
NAPHTHOSEMIQUINONES WITH TWO KEY OXYGENS

Anions

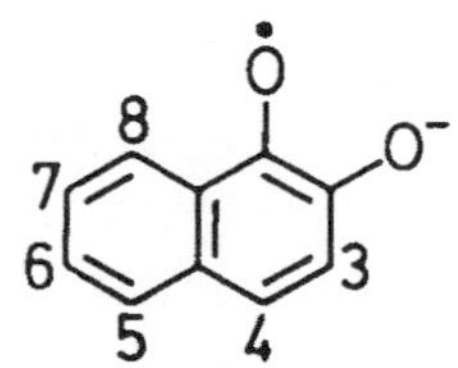

No.	Radical	Substituent(s)	a_3	a_4	a_5	a_6	a_7	a_8	g-Factor	Solvent	Method	Ref.
			Splitting constants/Gauss									
557	[C] 1,2-Naphthosemiquinone		0.56	4.61	0.27	1.45	0.17	1.36	2.00442	w EtOH	Re alk	8
			0.42	4.46	0.28	1.42	0.14	1.30		50% w/MeOH	Ox alk	136
			0.43	4.03	u	1.53	u	1.00		DMF	Re el	123
			0.34	4.00	u	1.47	u	1.14	2.0050	THF 0.05 *M* DCCw	Ox 0.01 *M* KO_2	7
			0.38	4.05	0.75	1.52	0.38	1.15		HMPA	Ox NaOMe[d]	140
558	-,4-sulfonato-	$-SO_3^-Na^+$	0.30	—	0.30	1.12	0.24	1.12	2.00449	w/EtOH	Ox alk	8, 3
			(12 Lines spectrum observed)						2.0049	w pH = 6.80	Photolysis	250

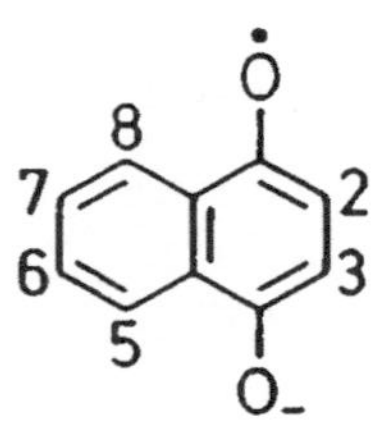

Table I.15.1 (continued)
NAPHTHOSEMIQUINONES WITH TWO KEY OXYGENS

Anions

No.	Radical	Substituent(s)	Splitting constants/Gauss a_3	a_4	a_5	a_6	a_7	a_8	g-Factor	Solvent	Method	Ref.
			a_2	a_3	a_5	a_6	a_7	a_8				
559	[C.I] 1,4-Naphthosemiquinone		3.11	3.11	0.55	0.63	0.63	0.55		w	Re alk	1, 79
			3.21	3.21	0.558	0.658	0.658	0.558		50% w/EtOH	Ox/Re alk	47
			3.26	3.26	0.54	0.68	0.68	0.54	2.00437	w alcohol	Re alk	3, 8
			3.23	3.23	0.513	0.655	0.655	0.513		w EtOH	Re alk	135
			3.24	3.24	0.58	0.65	0.65	0.58		50% w/MeOH	Ox alk	136
			3.25	3.25	0.51	0.64	0.64	0.51		EtOH/neut	Re phot	137, 138
			3.25	3.25	0.31	0.62	0.62	0.31		DMSO	Re KO_2	55
			3.31	3.31	0.300	0.633	0.633	0.300		DMSO	Re el	36
			3.31	3.31	0.300	0.633	0.633	0.300		MeCN	Re el	36
			3.35	3.35	0.32	0.64	0.64	0.32		DMF	Re el	95, 139
			3.27	3.27	0.26	0.57	0.57	0.26		DMF	Re el	77
			3.28	3.28	0.23	0.64	0.64	0.23		HMPA	Re K[b]	37, 89
			3.30	3.30	0.25	0.65	0.65	0.25		HMPA	Ox NaOMe[a]	140, 4
560	-,2-methyl-	–Me	3.01[l]	2.34	0.59	0.73	0.66	0.66		w	Ox alk	1, 79
			3.00[l]	2.35	0.62	0.76	0.68	0.68		50% w MeOH	Ox alk	136
			3.01[l]	2.38	0.64	0.64	0.64	0.64		50% w EtOH[c]	Re el	77, 3, 47
			2.98[l]	2.43	0.54	0.71	0.71	0.54	2.00429	w/EtOH	Re alk	8
			2.911[l]	2.467	0.48	0.78	0.70	0.56		EtOH	Re K[d]	76
			2.80[l]	2.51	0.40	0.69	0.69	0.40		10% w DMF	Re el	77
			2.69[l]	2.69	0.22	0.78	0.61	0.36		DMF	Re el	77
			2.69[l]	2.69	0.22	0.76	0.62	0.37		DMSO	Re el	77
561	-,2-ethyl-	–Et	2.28[l]	2.64	0.34	0.68	0.68	0.34		DMF	Re el	139
562	-,2-benzyl-	$–CH_2Ph$	2.45[l]	2.58	0.55	0.70	0.70	0.55	2.00434	w EtOH	Re alk	8
563	-,5-methyl-	–Me	2.49	3.80	u	u	u	u		w	Ox alk	35
			2.87	3.53	u	u	u	u		DMF	Ox alk	35
564	-,6-methyl-	–Me	3.17	3.35	0.55	0.69[l]	0.69	0.55		w/EtOH	Ox alk	8

565	-,2,3-dimethyl-	–Me	2.573[l]	2.573[l]	0.561	0.740	0.740	0.561		EtOH	Re K	76
			2.51[l]	2.51[l]	0.55	0.73	0.73	0.55		w/EtOH	Re alk	8
			2.51[l]	2.51[l]	0.32	0.73	0.73	0.32		MeCN	Re el	141
566	-,2,5-dimethyl-	–Me	2.27[l]	2.92	u	u	u	u		w	Ox alk	35
			2.27[l]	2.92	u	u	u	u		DMF	Ox alk	35
567	-,2-methyl-3-phytyl-	–Me,–R[c]	2.2[l]	1.1[l]	0.55	0.55	0.55	0.55		w alcohol	Re Zn/alk	3
			2.541[l]	1.269[l]	0.521	0.698	0.698	0.521		EtOH	Re el	76
			2.63[l]	1.39[l]	0.30	0.74	0.74	0.26		DMF	Re el	77
				1.23[l]								
			2.63[l]	1.41[l]	0.30	0.74	0.74	0.26		DMSO	Re el	77
				1.22[l]								
568	-,2-methoxymethyl-	$-CH_2OMe$	2.49[l]	2.90	(0.53	0.61	0.61	0.69)		85% w/DMF	Re alk	8
569	-,2-ethoxymethyl-	$-CH_2OEt$	2.92[l]	2.92	(0.52	0.60	0.60	0.71)	2.00432	w/EtOH	Re alk	8
570	-,2-acetyl-	–COMe	0.60[g]	5.87	0.15	0.64	0.50	0.32		DMF	Re el	139
571	-,2-amino-	$-NH_2^{\alpha}$	1.37[h]	(u	0.14	0.34	1.71	2.04)[i,j]		50% w/EtOH	Re el	47
			0.78[n,k]							(pH = 11)		
			1.75[h]	0.30	0.10	0.815	0.0	0.75		w/DMF	Re el	62
			1.75[n]									
			1.07[h]	(u	u	0.22	0.22	1.68)[j]		MeCN	Re el	143
			1.07[n]									
572	-,2-methylamino-	$-NH^{\alpha}CH_3^{\beta}$	1.89[h]	1.45	0.06	1.45	0.21	0.40		Pr[i]OH	Re alk	143
			1.89[n]									
			2.44[β]									
573	-,2-ethylamino-	$-NH^{\alpha}CH_2^{\beta}Me$	1.88[h]	1.46	0.06	1.46	0.21	0.41		Pr[i]OH	Re alk	143
			1.88[n]									
			2.37[β]									
574	-,2-butylamino-	$-NH^{\alpha}CH_2^{\beta}CH_2CH_2Me$	1.84[h]	1.52	0.06	1.52	0.21	0.39		Pr[i]OH	Re alk	143
			1.84[n]									
			2.38[β]									
575	-,2-isopropylamino-	$-NH^{\alpha}CH^{\beta}Me_2$	1.84[h]	1.51	0.06	1.51	(0.20	0.39)		Pr[i]OH	Re alk	143
			1.84[n]									
			1.51[β]									
576	-,2-*t*-butylamino-	$-NH^{\alpha}CMe_3$	1.87[h]	1.47	0.06	1.68	(0.20	0.40)		Pr[i]OH	Re alk	143
			1.87[n]									
577	-,2-chloro-	–Cl	—	4.05	0.33	0.70	0.75	0.33	2.00445	w/EtOH	Re alk	8
578	-,2,3-difluoro-	–F	5.31[F]	5.31[F]	0.37	0.74	0.74	0.37		15% MeCN/DME	Re el	66
579	[I]-,2,3-dichloro-	–Cl	—	—	0.59	0.77	0.77	0.59	2.00440	w/EtOH	Re alk	8, 104
580	-,2-trisdimethylaminophosphinium-	$-P^+(NMe_2)_3$	5.78[P]	6.72	0.57	0.57	0.57	0.57		CH_2Cl_2	TDAP[r]	168
			0.40[h]									
581	-,2-phenylthio-	–SPh	—	2.36	u	u	u	u		EtOH	Re NaOEt	74

Table I.15.1 (continued)
NAPHTHOSEMIQUINONES WITH TWO KEY OXYGENS

Anions

No.	Radical	Substituent(s)	a_3	a_4	a_5	a_6	a_7	a_8	g-Factor	Solvent	Method	Ref.
	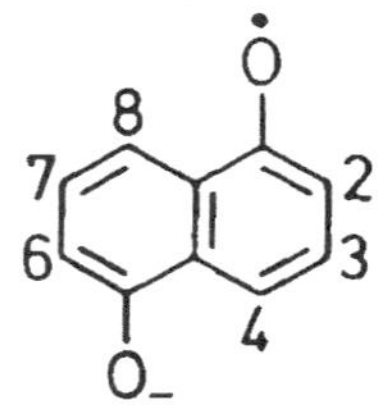		a_2	a_3	a_4							
582	1,5-Naphthosemiquinone		3.80	0.55	5.25	3.80	0.55	5.25		HMPA	Ox NaOMe[a]	140, 3
	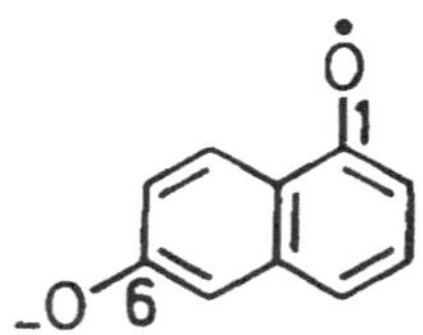											
583	1,6-Naphthosemiquinone		(Spectrum not analyzed)						2.00411	w/EtOH	Ox alk	8, 3

	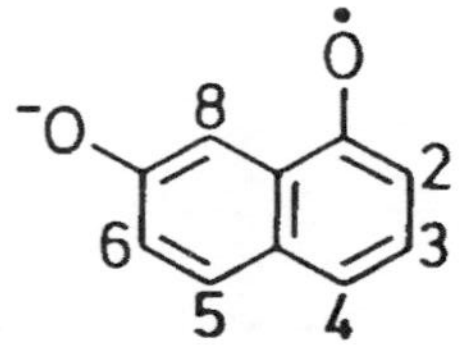										
					a_5	a_6					
584	1,7-Naphthosemiquinone	0.20	3.45	1.35	0.0	1.20	9.30		HMPA	Ox NaOMe[a]	140
	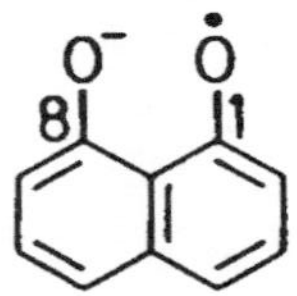										
585	1,8-Naphthosemiquinone	(Weak spectrum observed)							HMPA	Ox NaOMe[a]	140
	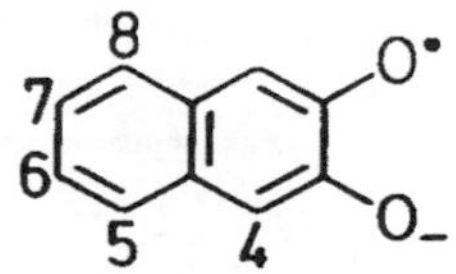										
586	2,3-Naphthosemiquinone	(~0.41 (2H), ~1.9(2H))						2.00447	w/EtOH	Ox alk	8
	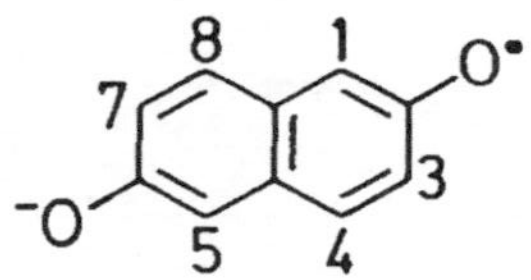										
		a_1				a_7					
587	2,6-Naphthosemiquinone	4.25	1.30	0.70	4.25	1.30	0.70		HMPA	Ox NaOMe[a]	140

Table I.15.1 (continued)
NAPHTHOSEMIQUINONES WITH TWO KEY OXYGENS

Anions

No.	Radical	Substituent(s)	Splitting constants/Gauss a_3	a_4	a_5	a_6	a_7	a_8	g-Factor	Solvent	Method	Ref.
	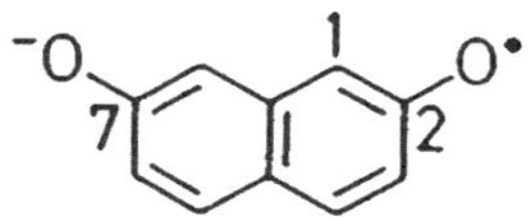											
588	2,7-Naphthosemiquinone		(21 Components of low resolution observed)						2.00428	w/alcohol	Ox alk	3

[a] KOBu^t also applied.
[b] Splitting constants remain constant with changing K^+ concentration. For a change in g-factor see Reference 37.
[c] Apparent pH = 8.
[d] ENDOR experiment at 223 K. Radical generated by K-metal reduction in DME and dried residue dissolved in absolute EtOH.
[e] $R = CH_2CH = C(CH_2CH_2CH_2CH-)_3Me$, with Me on the C and on each CH (as drawn: Me above C(and above CH–).
[f] Splitting from nonequivalent methylene protons.
[g] Splitting from methyl group protons.
[h] Splitting from nitrogen.
[i] Splittings in brackets unassigned.
[j] Splitting constants do not sum up to the spectral total width of 8.5 G reported in Reference 47.
[k] Splitting confirmed by deuteration.
[l] Reaction between quinone and tris(dimethylamino)phosphine (TDAP).

Table I.15.2
NAPHTHOSEMIQUINONES WITH THREE KEY OXYGENS

Anions

No.	Radical	Substituent(s)	Splitting constants/Gauss a_3	a_4	a_5	a_6	a_7	a_8	g-Factor	Solvent	Method	Ref.
	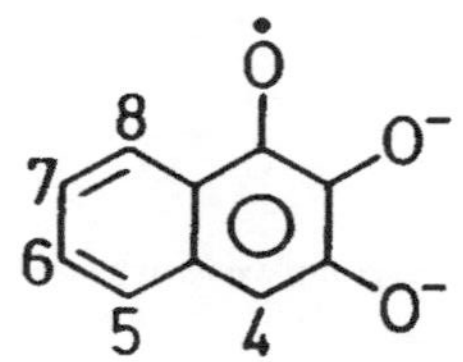											
589	3-Oxido-1,2-naphthosemiquinone		—	0.41	0.00	1.99	0.41	1.56	2.00425	w/EtOH	Ox alk	8
			—	0.39	0.00	2.00	0.39	1.84		50% w/MeOH	Ox alk	136
	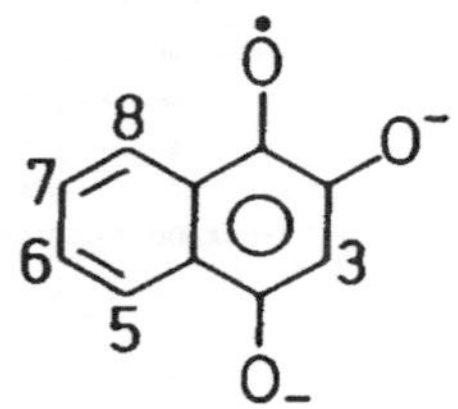											
590	4-Oxido-1,2-naphthosemiquinone		0.28	—	0.10	1.87	0.10	1.48		w	Ox KOH	144
			0.25	—	0.16	1.96	0.08	1.53		w	Ox alk	1, 105
			0.24	—	0.15	1.97	0.08	1.55	2.00425	w EtOH	Ox alk	8
			0.25	—	0.13	1.97	0.11	1.54		50% w MeOH	Ox alk	136
			0.10	—	0.38	2.14	0.0	1.48		50% Bu^{t}OH /DMSO	Re alk	105
			0.0	—	0.50	2.30	0.0	1.48		DMSO	Re alk	105
			0.35	—	0.60	2.45	0.60	1.55		HMPA	Ox NaOMe[a]	140
	-,2-methyl-	(2-Alkoxy- and 2-hydroxy-1,4-naphthosemiquinones, see nos. 610, 611, 612, this table)										

Table I.15.2 (continued) NAPHTHOSEMIQUINONES WITH THREE KEY OXYGENS

Anions

No.	Radical	Substituent(s)	Splitting constants/Gauss a_3	a_4	a_5	a_6	a_7	a_8	g-Factor	Solvent	Method	Ref.
591	-,3-methyl-	–Me	0.51[i]	—	0.12	1.97	0.12	1.54		w	Ox alk[h]	1, 144, 145
			0.38[i]	—	0.26	2.08	0.13	1.51		50% w/MeOH	Ox alk	136
592	-,3-(4-methyl-2-butene)-	–$CH_2CH{=}CMe_2$	0.24[i]	—	0.35	2.07	0.07	1.57	2.00429	w/EtOH	Re alk	8
593	-,3-phenyl-	–Ph	—	—	0.18	1.98	0.0	1.56		w	Re alk	105
			—	—	0.52	2.34	0.0	1.57		DMSO	Re alk	105
594	-,3-(2-oxidophenyl)-	–Ph–O (*o*)	—	—	0.38	2.05	0.0	1.30		DMSO[c]	Re alk	105
595	-,3-(2-methoxyphenyl)-	–Ph–OMe (*o*)	—	—	0.56	2.10	0.0	1.54		DMSO[c]	Re alk	105
596	-,3-(4-oxidophenyl)-	–Ph–O (*p*)	—	—	0.47	1.98	0.0	1.48		DMSO[c]	Re alk	105
597	-,3-(4-methoxyphenyl)-	–Ph–OMe (*p*)	—	—	0.55	2.10	0.0	1.52		DMSO[c]	Re alk	105
598	-,3-(1-oxido-4-methoxy-β-naphthyl)-	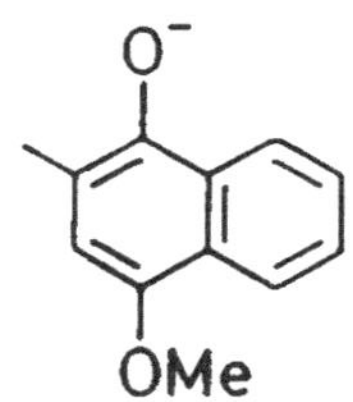	—	—	0.10	1.81	0.10	1.42		w	Ox alk	105
			—	—	0.39	2.18	0.39	1.51		DMSO	Ox alk	105
599	-,3-(1,4-dimethoxy-β-naphthyl)-	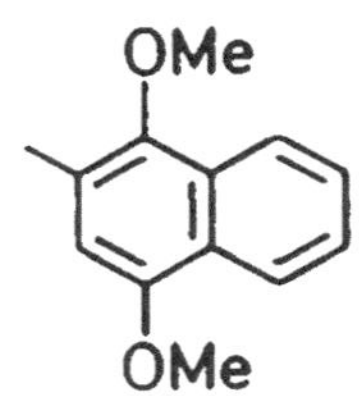	—	—	0.55	2.07	0.0	1.50		DMSO	Re alk	105
600	-,3-nitro-	–NO_2	—	—	0.54	2.20	0.18	1.64		DMSO	Re alk	105

601	-,3-chloro-	–Cl	0.11[Cl]	—	0.11	1.85	0.11	1.48		w	Re alk	105
			0.12[Cl]	—	0.12	1.90	0.12	1.56		w	Ox KOH	144
			0.0[Cl]	—	0.40	2.16	0.0	1.46		DMSO	Re alk	105
602	-,6-bromo-3-phenyl-	-Br,-Ph	—	—	0.25	—	0.0	1.45		w	Re alk	105
			—	—	0.54	—	0.0	1.52		DMSO	Re alk	105
603	-,7-bromo-3-phenyl-	–Br,–Ph	—	—	0.31	2.08	—	1.57		w	Re alk	105
			—	—	0.62	2.22	—	1.58		DMSO	Re alk	105
604	-,8-bromo-3-phenyl-	–Br,–Ph	—	—	0.26	2.20	0.0	—		w	Re alk	105
			—	—	0.68	2.74	0.28	—		DMSO	Re alk	105

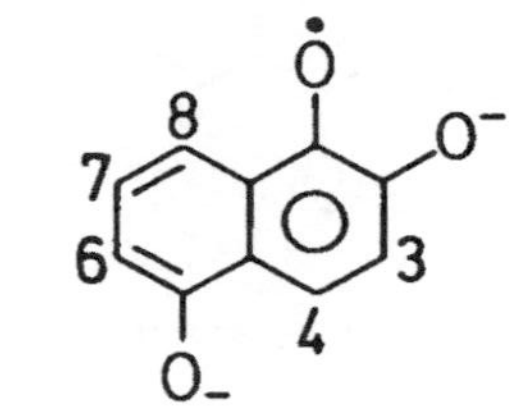

605	5-Oxido-1,2-naphthosemiquinone		0.98	4.94	—	2.50	0.16	2.35		50% w/MeOH	Ox H_2O_2/alk	136

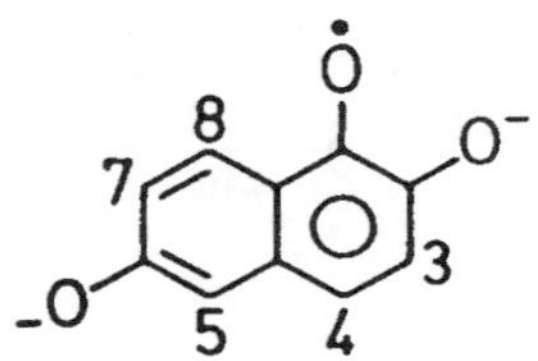

606	6-Oxido-1,2-naphthosemiquinone		0.58	5.02	1.40	—	0.36	1.40		50% w/MeOH	Ox alk	136
			0.60	5.20	1.48	—	0.35	1.48	2.00429	w/EtOH	Ox H_2O_2/alk	8
			0.35	5.25	1.10	—	0.10	1.90		HMPA	Ox NaOMe[a]	140

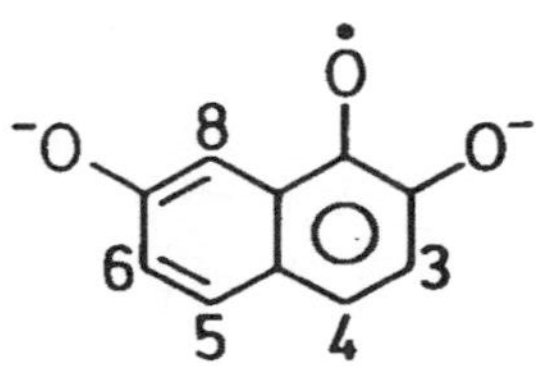

Table I.15.2 (continued)
NAPHTHOSEMIQUINONES WITH THREE KEY OXYGENS

Anions

No.	Radical	Substituent(s)	Splitting constants/Gauss a_3	a_4	a_5	a_6	a_7	a_8	g-Factor	Solvent	Method	Ref.
607	7-Oxido-1,2-naphthosemiquinone		0.48	4.97	0.48	2.43	—	1.02		50% w/MeOH	Ox alk	136
			0.48	4.60	0.0	2.68	—	0.85		HMPA	Ox NaOMe[a]	140
608	-,8-formyl-	–CHO	0.75	4.70	0.83	0.99	0.0	—	2.00462	w/EtOH	Ox alk	8

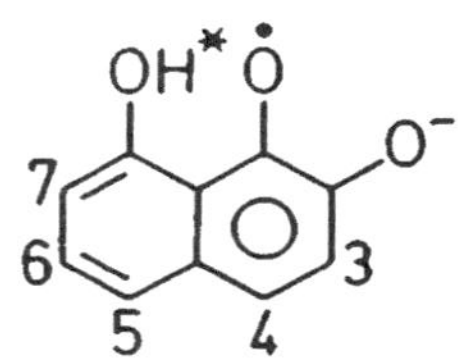

No.	Radical	Substituent(s)	a_3	a_4	a_5	a_6	a_7	a_8	g-Factor	Solvent	Method	Ref.
609	8-Hydroxy-1,2-naphthosemiquinone		0.75	4.60	1.00	1.50	1.00	0.52*		50% w/MeOH	Ox H_2O_2/alk	136
			0.59	4.37	0.97	1.51	1.00	0.52*	2.00428	w/EtOH	Ox H_2O_2/alk	8

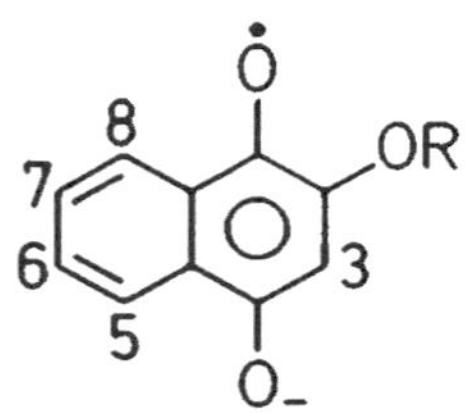

No.	Radical	Substituent(s)	a_2	a_3	a_5	a_6	a_7	a_8	g-Factor	Solvent	Method	Ref.
610	2-Hydroxy-1,4-naphthosemiquinone	(R = H*)[d]	0.24*	0.45	u	2.29	0.10	1.42	2.00452	MeCN	Re el	146
611	2-Methoxy-1,4-naphthosemiquinone	(R = CH_3)	0.75*	0.48	0.48	1.65	0.29	1.17		DMF	Re el	95, 139
			0.73*	0.43	0.73	1.62	0.0	1.17	2.00485	MeCN	Re el	146

612	2-Ethoxy-1,4-naphthosemiquinone	(R = CH_2Me)	0.88*	0.42	0.69	1.61	0.0	1.16	2.00483	MeCN	Re el	146

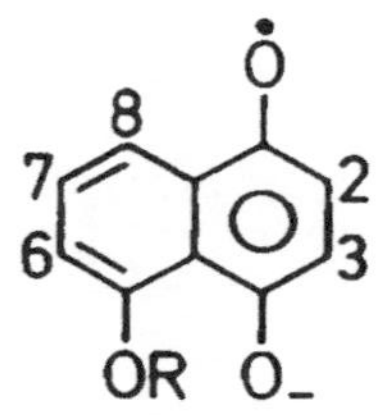

613	5-Hydroxy-1,4-naphthosemiquinone	(R = H*)	3.30	3.05	0.30*	1.27	0.68	1.27	2.00425[c]	w/EtOH	Ox alk	16, 79
			3.90	2.70	0.30*	1.12	0.80	0.80[l]		50% w/MeOH	Ox alk	136
			4.08	2.69	0.33*	1.13	0.82	0.82	2.00449	MeCN	Re el	146
			4.01	2.64	0.34*	1.05	0.78	0.78		DMF	Re el	139
			4.20	2.45	0.35*	0.85	0.65	1.00		HMPA	Ox NaOMe	140
614	-,2-methyl-	–Me	3.05[l]	2.33	0.28*	1.23	0.68	1.34	2.00416	w/EtOH	Ox alk	16, 79
615	-,2-benzyl-	$-CH_2Ph$	2.63[l]	2.48	0.30*	1.23	0.73	1.27	2.00420	w/EtOH	Re alk	8, 79
616	-,3-benzyl-	$-CH_2Ph$	2.80	2.24[l,g]	0.35*	1.28	0.75	1.28	2.00417	w/EtOH	Re alk	8, 79
	-,5-methyl-	(5-Methoxy-, see from no. 619, this table)										
617	-,6-methyl-	–Me	3.422	3.067	0.349*	1.288[l]	0.509	1.387		w/EtOH	Re alk	8
618	-,7-methyl-	–Me	3.11	3.11	0.28*	1.10	0.71[l]	1.39	2.00422	w/EtOH	Ox alk	16
619	5-Methoxy-1,4-naphthosemiquinone	(R = CH_3)	2.464	3.727	0.037*	0.913	0.727	1.102	2.00440	w/EtOH	Ox alk	8
			2.58	3.50	u	0.90	0.50	0.70		DMF	Re el	95, 139
620	-,2-benzyl-	$-CH_2Ph$	1.79[l]	3.03	u	0.81	0.81	1.08	2.00436	w/EtOH	Re alk	8
621	-,3-benzyl-	$-CH_2Ph$	1.80	2.86[l]	u	0.90	0.70	1.15	2.00434	w/EtOH	Re alk	8

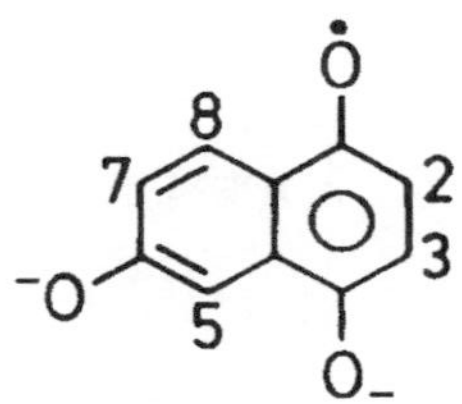

622	6-Oxido-1,4-naphthosemiquinone		4.42	2.78	0.60	—	1.64	0.0		50% w/MeOH	Ox alk	136
			4.27	2.73	0.71	—	1.73	0.0	2.00419	w/EtOH	Ox alk	8

Table I.15.2 (continued)
NAPHTHOSEMIQUINONES WITH THREE KEY OXYGENS

Anions

No.	Radical	Substituent(s)	Splitting constants/Gauss						g-Factor	Solvent	Method	Ref.
			a_3	a_4	a_5	a_6	a_7	a_8				
	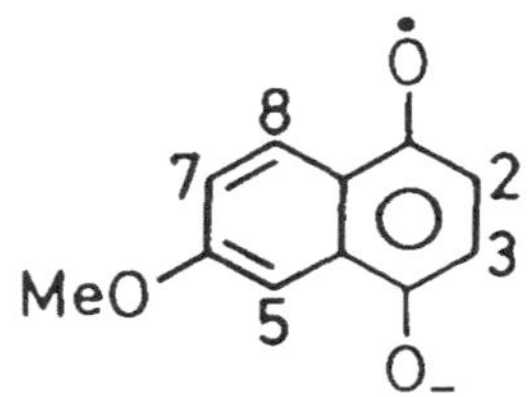											
623	6-Methoxy-1,4-naphthosemiquinone		3.30	2.20	0.48	u	0.96	0.48		HMPA	Ox NaOMe[a]	140

[a] KOBu^t also applied.
[b] Radical obtained as secondary product by hydroxylation of the 1,4-quinone at C-2.
[c] Similar splitting constants observed in 50% Bu^{t}OH/DMSO.
[d] Hydroxyl proton splitting identified by deuteration; radical structure may be 4-hydroxy-1,2-naphthosemiquinone.
[e] g-Factors in Reference 16 are 0.00004 too large.
[f] Hydroxyl proton splitting in Reference 136 was assigned the value 0.80 G, which we believe is incorrect.
[g] Reference 79 states the value 2.8 G for the methylene protons, which is a misprint.

Table I.15.3
NAPHTHOSEMIQUINONES WITH FOUR TO EIGHT KEY OXYGENS

Anions

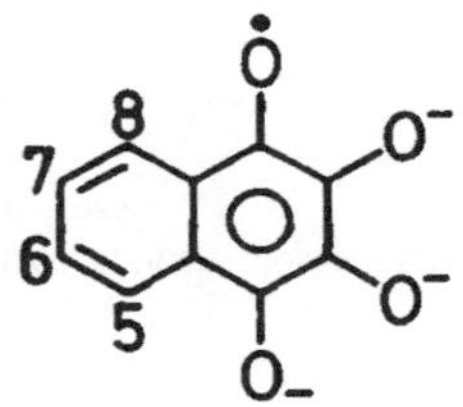

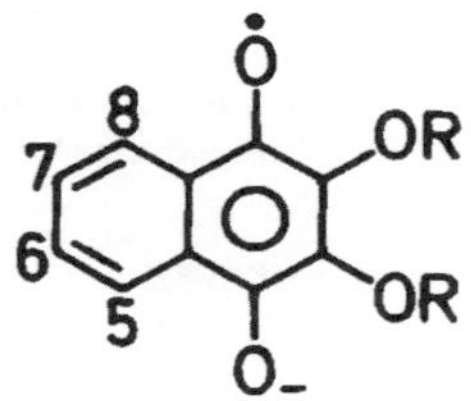

No.	Radical	Substituent(s)	Splitting constants/Gauss a_2	a_3	a_5	a_6	a_7	a_8	g-Factor	Solvent	Method	Ref.
624	2,3-Dioxido-1,4-naphthosemiquinone		—	—	0.73	1.03	1.03	0.73		50% w/MeOH	Ox H_2O_2/alk	136
			—	—	0.95	1.22	1.22	0.95		HMPA	Ox NaOMe	140
625	2,3-Dihydroxy-1,4-naphthosemiquinone	(R = H*)[a]	0.26*	0.26*	0.46	1.02	1.02	0.46	2.00466	MeCN	Re el	146
626	2,3-Dimethoxy-1,4-naphthosemiquinone	(R = Me)	u	u	0.36	0.79	0.79	0.36	2.00470	MeCN	Re el	146

Table I.15.3 (continued)
NAPHTHOSEMIQUINONES WITH FOUR TO EIGHT KEY OXYGENS

Anions

No.	Radical	Substituent(s)	Splitting constants/Gauss a_2	a_3	a_5	a_6	a_7	a_8	g-Factor	Solvent	Method	Ref.
627	2,3-Diphenoxy-1,4-*naphthosemiquinone*	(R = Ph)	u	u	0.36	0.67	0.67	0.36	2.00488	MeCN	Re el	146

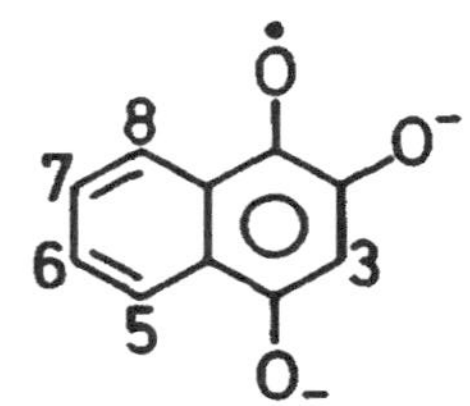

No.	Radical	Substituent(s)	a_2	a_3	a_5	a_6	a_7	a_8	g-Factor	Solvent	Method	Ref.
590	4-Oxido-1,2-naphthosemiquinone											
628	-,5-hydroxy-	–OH*	—	0.28	0.39*	2.60	0.10	2.14		50% w/MeOH	Ox alk	136
628D		–OD	—	u	u[b]	2.58	0.54	2.17		DMF/D_2O	Re el	139
629	-,6-ethyl-5-hydroxy-	–Et,–OH*	—	0.48	u*	2.10[i]	0.61	2.10		DMF	Re el	139
630	-,6-oxido-	$–O^-$	—	0.40	0.40	—	0.84	1.50		50% w/MeOH	Ox alk	136
631	-,7-oxido-	–O	—	0.60	0.10	2.00	—	1.64		50% w/MeOH	Ox alk	136
632	-,7-methoxy-	–OMe	—	0.227	0.055	1.815	u	1.665		w/EtOH	Ox alk	8
633	-,8-hydroxy-	–OH*	—	0.70	u	2.78	0.75	u*		DMF	Re el	139

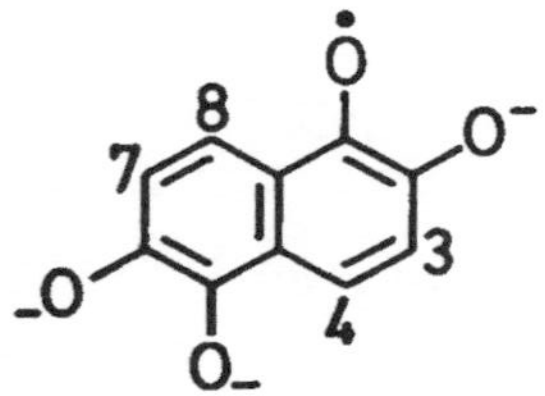

634	5,6-Dioxido-1,2-naphthosemiquinone	—	0.33	—	—	0.33	1.50		50% w/MeOH	Ox alk	136

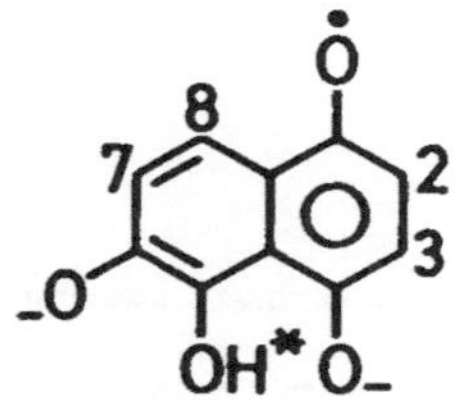

635	5-Hydroxy-6-oxido-1,4-naphthosemiquinone	4.13	2.55	0.51*	—	0.70	2.21		w/EtOH	Ox alk	8

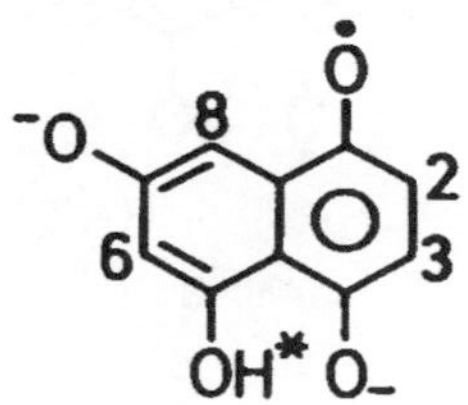

636	5-Hydroxy-7-oxido-1,4-naphthosemiquinone	3.20	4.35	0.33*	0.33	—	2.52	2.00401	w/EtOH	Ox alk	8
		3.12	4.41	u^D	0.33	—	2.52		D_2O/EtOD	Ox NaOD	8

Table I.15.3 (continued)
NAPHTHOSEMIQUINONES WITH FOUR TO EIGHT KEY OXYGENS

Anions

No.	Radical	Substituent(s)	Splitting constants/Gauss a_2	a_3	a_5	a_6	a_7	a_8	g-Factor	Solvent	Method	Ref.
	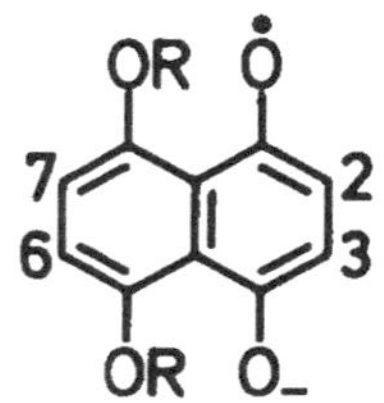											
637	Naphthazarin semiquinone[u]	(R = H*)	2.356	2.356	0.587*	2.356	2.356	0.587*		w	Re el	147, 148
			2.37	2.37	0.6*	2.37	2.37	0.6*		w/NH_3	Re[h]	149
			2.35	2.35	0.50*	2.35	2.35	0.50*		DMSO	Re KO_2	55, 147
			2.38	2.38	0.57*	2.38	2.38	0.57*		C_5H_5N	Re	150
			2.395	2.395	0.534*	2.395	2.395	0.534*		3.8% w/DMF	Re el	147
			2.390	2.390	0.526*	2.390	2.390	0.526*		DMF	Re el	147
			2.39	2.39	0.52*	2.39	2.39	0.52*	2.00410	DMF	Re el	139, 290
			2.406	2.406	0.520*	2.406	2.406	0.520*		THF	Re el	147
			2.30	2.30	0.50*	2.30	2.30	0.50*	2.00443	MeCN	Re el	146
637D		(R = D)	2.335	2.335	0.079[D]	2.335	2.335	0.079[D]		D_2O	Re el	147, 148
			2.362	2.362	0.070[D]	2.362	2.362	0.070[D]		5% D_2O/DMSO	Re el	147
			2.368	2.368	0.069[D]	2.368	2.368	0.069[D]		3.8% D_2O/DMF	Re el	147

			2.376	2.376	0.068[D]	2.376	2.376	0.068[D]		THF	Re el	147
			2.27	2.27	u[D]	2.27	2.27	u[D]		D_2O/ND_3	Re[h]	149
		(R = H*)										
638	-,2-ethyl-	–Et	(1.82[f]	2.35	0.55*	2.16	2.16)[e]	0.55*		DMF	Re el	139
639	-,2,3-dimethyl-	– Me	1.90[f]	1.90[f]	0.61*	2.60	2.60	0.61*	2.00426	DMF	Re el	151, 290
640	-,2,3,6,7-tetramethyl-	–Me	2.10[f]	2.10[f]	0.64*	2.10[f]	2.10[f]	0.64*	2.00403	DMF	Re el	290, 151
641	-,2-(1-hydroxy-4-methyl-3-pentene)-	OH | $–CHCH_2CH{=}CMe_2$	(0.105	1.389[d]	0.598*	2.417	4.508)[d]	0.598*		w/EtOH	Re alk	8
	-,2-hydroxy-	(Hydroxy and alkoxy substituents, see after no. 646)										
642	-,2-chloro-	–Cl	—	2.97	0.52*	1.92	1.59	0.52*		DMF	Re el	139
			—	3.00	0.53*	1.88	1.61	0.53*	2.00418	DMF	Re el	290
643	-,2,3-dichloro-	–Cl	0.19[Cl]	0.19[Cl]	0.53*	1.56	1.56	0.53*		DMF	Re el	139
644	-,2,6-dichloro-	–Cl	—	2.18	0.53*	—	2.18	0.53*	2.00437	DMF	Re el	290
645	-,2,3,6-trichloro-	–Cl	—	—	0.40*	—	1.70	0.40*	2.00435	DMF	Re el	290
646	-,tetrachloro-	–Cl	(Single broad line observed)						2.00483	DMF	Re el	290, 139
647	-,2-hydroxy-	–OH*	0.51*	1.27	0.51*	2.72	1.50	0.51*		DMF	Re el	139
648	-,3-chloro-2-hydroxy-	–Cl, –OH*	0.48*	—	0.48*	2.45	1.22	0.48*		DMF	Re el	139
			u[D]	—	u[D]	2.60[e]	1.30[e]	u[D]		D_2O/DMF	Re el	139
649	-,2-methoxy-	–OMe	u	1.519	0.491*	2.596	2.026	0.525*		DMF	Re el	8
			(0.74	2.77	u	2.30	0.43	0.57*)[e]		DMF	Re el	139, 95
650	-,3-chloro-2-methoxy-	–Cl, –OMe	u	—	0.52*	1.90	1.58	0.52*		DMF	Re el	95, 139
			u	—	u[D]	2.05[f]	1.66[f]	u[D]		D_2O/DMF	Re el	139
651	-,2-acetoxy-	O ‖ –OCMe	—	1.58	0.49*	2.59	2.03	0.51*		DMF	Re el	139
652	-,2,3-dihydroxy-	–OH*	0.27*	0.27*	0.42*	2.02	2.02	0.42*		DMF	Re el	139
653	-,6,7-dichloro-2,3-dioxido-	$–Cl, –O^-$	—	—	0.88*	u	u	0.88*		DMF	Re el	139
654	-,3-methoxy-2-oxido-	$–OMe, –O^-$	—	u	0.37*	3.40	1.20	0.74*		DMF	Re el	139
655	-,2,3-dimethoxy-	–OMe	u	u	0.52*	1.89	1.89	0.52*		DMF	Re el	95, 139
			u	u	0.550*	1.814	1.814	0.550*		DMF	Re el	8
656	-,6,7-dichloro-2,3-dimethoxy-	–Cl, –OMe	(Single broad line observed)							DMF	Re el	95, 139
657	-,2,6-dimethoxy-	–OMe	0.65[g]	1.00	0.65*	0.65[g]	1.00	0.65*		DMF	Re el	95, 139
658	-,2,7-dihydroxy-	–OH*	0.30*	1.82	0.70*	1.82	0.30*	0.39*		DMF	Re el	139

Table I.15.3 (continued)
NAPHTHOSEMIQUINONES WITH FOUR TO EIGHT KEY OXYGENS

Anions

No.	Radical	Substituent(s)	Splitting constants/Gauss a_2	a_3	a_5	a_6	a_7	a_8	g-Factor	Solvent	Method	Ref.
659	-,2,7-dimethoxy-	–OMe	0.61	1.79	u^h	1.79	0.61	u^h		DMF	Re el	95, 139
660	-,2,7-dimethoxy-3-ethyl-	–OMe, –Et	u^g	1.24^i	0.50*	2.40	0.76^g	0.50*		DMF	Re el	139, 95
661	-,2,3-dimethoxy-7-ethyl-6-hydroxy-	–OMe, –Et, –OH*	u	u	0.67*	0.50*	0.67^i	0.67*		DMF	Re el	95, 139
662	-,7-ethyl-2,3,6-trimethoxy-	–Et, –OMe	u	u	0.62*	u	0.84^i	0.62*		DMF	Re el	95, 139
663	-,tetrahydroxy-	–OH*	0.16*	0.16*	0.56*	0.16*	0.16*	0.56*		DMF	Re el	139
664	-,tetramethoxy-	–OMe	u	u	0.47*	u	u	0.47*		DMF	Re el	95, 139
665	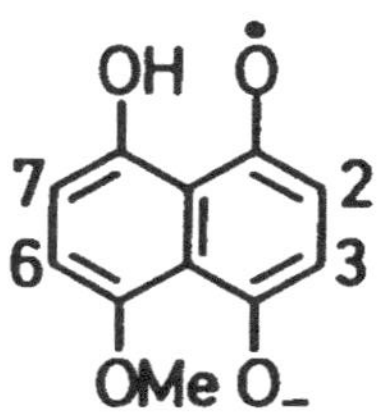 8-Hydroxy-5-methoxy-1,4-naphthosemiquinone											
666	-,2-chloro-	–Cl	—	2.11	u	1.35	1.14	0.41*	2.00462	DMF	Re el	290
667	-,6-chloro-	–Cl	1.933	4.069	u	—	1.271	0.488*	2.00460	DMF	Re el	152, 290
668	-,2-acetyl-6-methoxy-3-oxido-	–COMe, –OMe, –O	—	—	u	u	2.52	u^i		DMF	Re el	139, 95

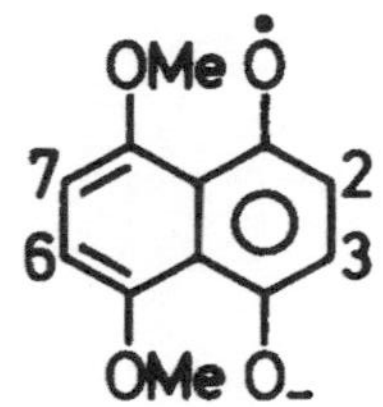

669	5,8-Dimethoxy-1,4-naphthosemiquinone		2.87	2.87	u	1.10	1.10	u		DMF	Re el	95, 139
			2.85	2.85	u	1.09	1.09	u	2.00478	DMF	Re el	290
670	-,2-chloro-	–Cl	—	3.15	u	1.12	0.92	u	2.00480	DMF	Re el	290
671	-,6-chloro-	–Cl	2.49	2.90	u	—	1.20	u	2.00443	DMF	Re el	290

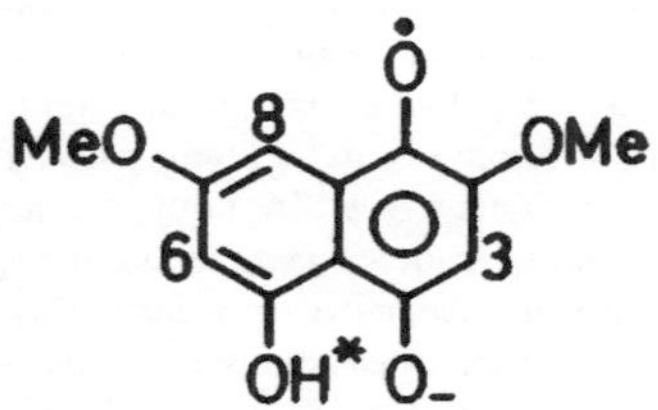

672	2,7-Dimethoxy-5-hydroxy-1,4-naphthosemiquinone											
673	-,3-oxido-	–O	u	—	0.58*	0.58	0.15[g]	0.58		DMF	Re el	95
674	-,3-methoxy-	–OMe	u	u	0.25*	1.38	u	0.90		DMF	Re el	95, 139

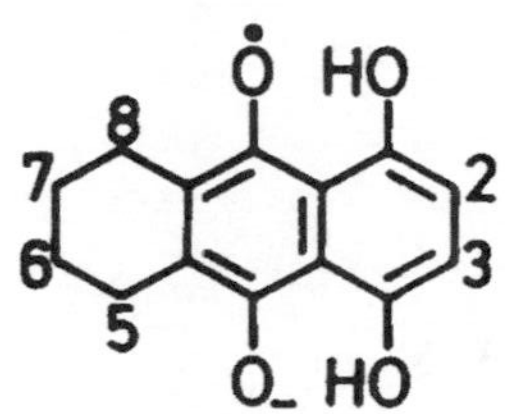

Table I.15.3 (continued)
NAPHTHOSEMIQUINONES WITH FOUR TO EIGHT KEY OXYGENS

Anions

No.	Radical	Substituent(s)	Splitting constants/Gauss a_2	a_3	a_5	a_6	a_7	a_8	g-Factor	Solvent	Method	Ref.
675	5,6,7,8-Tetrahydroquinizarin anion radical		2.37	2.37	2.85[i]	u[j]	u	2.85[i]		C_5H_5N/ $N(CH_2CH_2OH)_3$ (4:1)		150

[a] Hydroxyl proton splitting identified by deuteration.
[b] Naphthazarin dissolved in concentrated ammonia (0.88 g $m\ell^{-1}$).
[c] Assignment tentative, agreement with simulated spectrum not achieved.
[d] Parameters in brackets not assigned apart from the one with a*.
[e] Obtained from Figure 4, Reference 139.
[f] Obtained from Figure 5, Reference 139.
[g] Splitting from three methoxyl protons.
[h] Peri-hydroxyls replaced by deuterium.
[i] Hydroxyl proton at C-8 claimed to be dissociated.
[j] A triplet splitting of 0.55 G is observed from two hydroxyl protons.

Table I.15.4
BICYCLONAPHTHOSEMIQUINONES

No.	Radical	R′	R″	R‴	R⁗	Splitting constants/Gauss $a_{R'}$	$a_{R''}$	$a_{R'''}$	$a_{R''''}$	$a_{5,8}$	$a_{6,7}$	Solvent	Method	T [K]	Ref.
	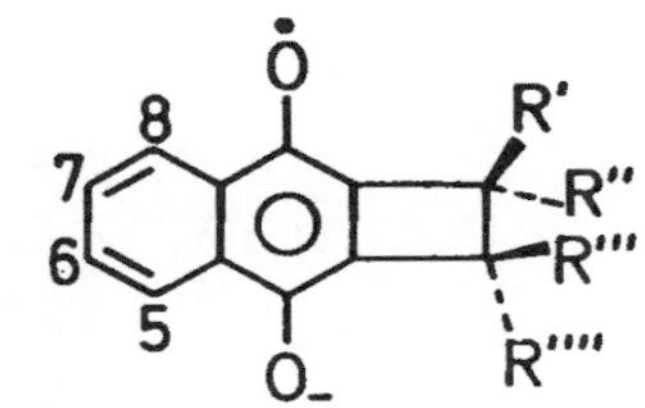														
676	*p*-Naphthocyclobutane-semiquinone	H	H	H	H	2.92	2.92	2.92	2.92	0.44	0.75	MeCN	Re el	248	141, 418
677		H	Me*	H	Me*	2.25	0.13*	2.25	0.13*	0.26	0.52	DMSO	Re el	RT	417, 418
678		H	H	H	Ph	2.99	2.78	2.65	—	0.36	0.72	DMSO	Re el	RT	417, 418
679		H	Ph	H	Ph	2.73	—	2.73	—	0.39	0.70	DMSO	Re el	RT	417, 418
679D		D	Ph	D	Ph	0.41^{b}	—	0.41^{b}	—	0.35	0.70	DMSO-d_6	Re el	RT	417
	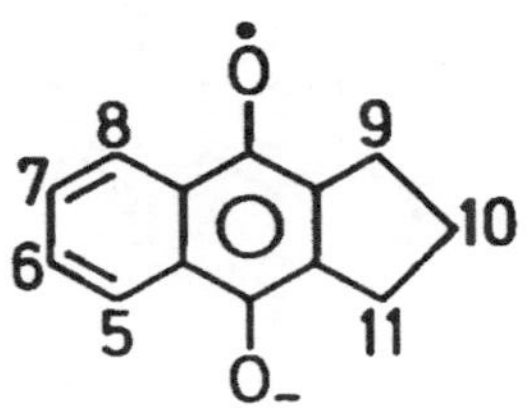					a_9	a_{10}	a_{11}	a_{12}						
680	*p*-Naphthocyclopentane-semiquinone					2.73^{a}	0.18^{a}	2.73^{a}	—	0.36	0.74	MeCN	Re el	248	141

Table I.15.4 (continued)
BICYCLONAPHTHOSEMIQUINONES

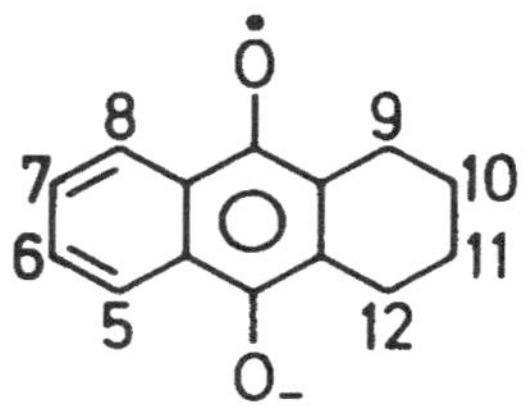

No.	Radical	R′	R″	R‴	R⁗	Splitting constants/Gauss $a_{R'}$	$a_{R''}$	$a_{R'''}$	$a_{R''''}$	$a_{5,8}$	$a_{6,7}$	Solvent	Method	T [K]	Ref.
681	*p*-Naphthocyclohexane-semiquinone					3.28[a]	u	u	3.28[a]	0.36	0.78	MeCN	Re cl	248	141

[a] Splitting from two equivalent methylene protons.

ANTHRASEMIQUINONE ANIONS

Table I.16.1
ANTHRASEMIQUINONES WITH TWO KEY OXYGENS

Anions

No.	Radical	Subst.	Splitting constants/Gauss								g-Factor	Solvent	Method	Ref.
			a_2	a_3	a_5	a_6	a_7	a_8	a_9	a_{10}				
	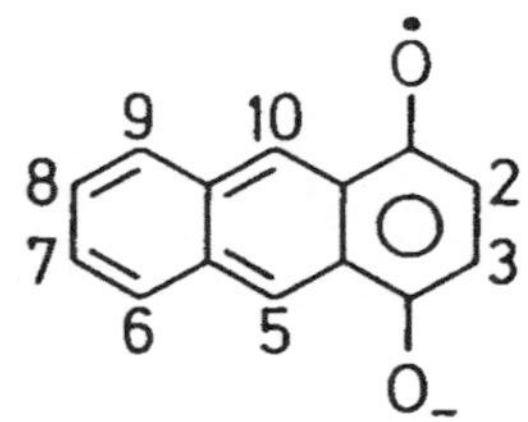													
682	1,4-Anthrasemiquinone		2.6	2.6	1.1	0.4	0.4	0.4	0.4	1.1[a]		w/MeOH	Re Zn/alk	153
	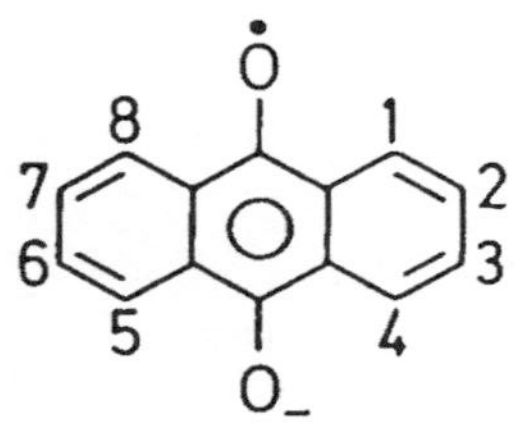													
			a_1	a_2	a_3	a_4	a_5	a_6	a_7	a_8				
683D	9,10-Anthrasemiquinone-d_1[b]		~0.1[D]	0.97	0.97	0.55	0.55	0.97	0.97	0.55		w/EtOH	Re DTN/alk	8
683	[C.1]9,10-Anthrasemiquinone		0.651	0.962	0.962	0.651	0.651	0.962	0.962	0.651	2.00448	w/alk	Re glucose	428, 437
			0.651	1.042	1.042	0.651	0.651	1.042	1.042	0.651		w/KOH	Re glucose	428, 437

Table I.16.1 (continued)
ANTHRASEMIQUINONES WITH TWO KEY OXYGENS

Anions

No.	Radical	Subst.	Splitting constants/Gauss								g-Factor	Solvent	Method	Ref.
			a_2	a_3	a_5	a_6	a_7	a_8	a_9	a_{10}				
			0.563	0.963	0.963	0.563	0.563	0.963	0.963	0.563		w/EtOH	Re DTN/alk	8, 78
			0.54	0.97	0.97	0.54	0.54	0.97	0.97	0.54		EtOH/abs	Re KOH[c]	104
			0.56	1.00	1.00	0.56	0.56	1.00	1.00	0.56		50% w/DMF	Re el	159[c]
			0.53	1.01	1.01	0.53	0.53	1.01	1.01	0.53		40% w/DMF	Re el	159[c]
			0.51	0.99	0.99	0.51	0.51	0.99	0.99	0.51		30% w/DMF	Re el	159[c]
			0.45	0.94	0.94	0.45	0.45	0.94	0.94	0.45		20% w/DMF	Re el	159[c]
			0.41	1.00	1.00	0.41	0.41	1.00	1.00	0.41		10% DMF	Re el	159[c]
			0.30	1.02	1.02	0.30	0.30	1.02	1.02	0.30	2.0048	DMF	Re el	160, 158
			0.33	1.01	1.01	0.33	0.33	1.01	1.01	0.33	2.0043	DMSO	Re KOH	294
			0.303	0.986	0.986	0.303	0.303	0.986	0.986	0.303		DMSO	Re el	36
			0.30	0.97	0.97	0.30	0.30	0.97	0.97	0.30		DMSO	Re KO_2	55
			0.271	0.974	0.974	0.271	0.271	0.974	0.974	0.271		DME	Re el[e]	39
			0.269	0.980	0.980	0.269	0.269	0.980	0.980	0.269	2.00442	p,p'-Azoxy-anisole	Re el[d]	161, 424[c]
			0.244	0.998	0.998	0.244	0.244	0.998	0.998	0.244		THF/173K	Re Na	162
			0.24	1.00	1.00	0.24	0.24	1.00	1.00	0.24		HMPA	Ox NaOMe	140
			0.191	0.923	0.923	0.191	0.191	0.923	0.923	0.191	2.00462	p,p'-Azoxy-anisole	Re el[f]	161, 424[c]
684	-,1-methyl-	–Me	0.498[i]	1.285	0.860	0.703	0.750	0.680	1.213	0.233		Pr^iOH	Re $KOPr^i$	154, 155
685	-,2-methyl-	–Me	1.05	0.95[i]	0.95	0.52	0.52	0.73	0.90	0.61	2.00404	w/EtOH	Re DTN/alk	8, 118
			1.052	0.950[i]	0.950	0.541	0.541	0.759	0.858	0.652		w/EtOH	Re DTN/alk	8[c]
			1.07	0.923[i]	0.928	0.449	0.445	0.660	0.923	0.567		Pr^iOH	Re $KOPr^i$	154
686	-,1,4-dimethyl-	–Me	0.837[i]	1.193	1.193	0.837[i]	0.433	0.931	0.931	0.433		EtOH/abs	Re el	156
			0.512[i]	1.179	1.179	0.512[i]	0.210	0.902	0.902	0.210		DME	Re el	156
687	-,2,3-dimethyl-	–Me	1.029	1.029[i]	1.029[i]	1.029	0.576	0.696	0.696	0.576		w/EtOH	Re DTN/alk	8
688	-,2,6-dimethyl-	–Me	(0.71	0.87[i]	0.47	0.47	0.71	0.87[i]	0.47	0.47)[g]		Pr^iOH	Re K or Na	155
689	-,2,7-dimethyl-	–Me	(0.69	0.75[i]	0.56	0.31	0.31	0.56	0.75[i]	0.69)[g]		Pr^iOH	Re K or Na	155
690	-,1,4,5,8-tetramethyl-	–Me	0.329[i]	1.153	1.153	0.329[i]	0.329[i]	1.153	1.153	0.329[i]		DME	Re el	156
691	-,1-carbamoyl-	$-CONH_2$	(Spectrum shown in Reference 157 without EPR data)									DMF	Re el	157

692	-,2-carboxy-	–COOH		(u	—	1.64	0.81	0.81	1.78	1.78	1.78)[h]		DMF	Re el	158
693	-,2-chloroformyl-	–COCl		(0.12	—	1.79	0.89	0.89	0.25	0.45	0.45)[h]		DMF	Re el	158
694	-,1-amino-	$-NH_2$		(1.45^N / 0.80*	0.51, 0.25, 0.14, 0.14)[h]								DMF	Re el	158
695	-,1,4-diamino-	$-NH_2$		(1.32^N / 0.44*	0.22	0.22	1.32^N / 0.44*	u	u	u	u)[h]		DMF	Re el	158
696	-,1-nitro-	$-NO_2$		(1.40^N	0.32	0.64	0.32	0.64	0.64	0.64	0.64)[h]		DMF	Re el	158, 425
	-,1-hydroxy-	(Hydroxy- and alkoxyanthrasemiquinones, see Table I.16.2 and no. 701 below)													
697	-,1-chloro-	–Cl	A[i]	—	1.111	1.099	0.647	0.697	0.597	1.109	0.307		w/EtOH	Re DTN/alk	8
			B[i]	—	1.099	1.111	0.647	0.697	1.109	0.597	0.307				8
698	-,2-chloro-	–Cl		(0.78	—	0.05	0.78	0.78	0.28	0.28	0.78)[h]		DMF	Re el	158
699	-,1,5-dichloro-	–Cl	A[i]	—	0.778	1.247	0.344	—	0.778	1.247	0.344		w/EtOH	Re DTN/alk	8
			B[i]	—	1.247	0.778	0.344	—	1.247	0.778	0.344				8
			A[i]	—	0.80	1.26	0.35	—	0.80	1.26	0.35		EtOH/abs	Re KOH	104
			A[i]	—	0.74	1.30	0.70	—	0.74	1.30	0.70		DMF	Re el	158
700	-,1,8-dichloro-	–Cl	A[i]	—	1.328	0.727	0.831	0.831	0.727	1.328	—		w/EtOH	Re DTN/alk	8
			B[i]	—	0.727	1.328	0.831	0.831	1.328	0.727	—				8
			A[i]	—	1.26	0.88	0.64	0.64	0.88	1.26	—		DMF	Re el	158
701	-,1,4-dichloro-5,8-dihydroxy-	–Cl, –OH*		—	1.22	1.22	—	0.52*	1.75	1.75	0.52*		w/DMF	Re Zn/KOH	164
		–Cl, –OD		—	1.22	1.22	—	u^D	1.66	1.66	u^D		D_2O/DMF	Re Zn/KOH	164
702	-,1,4,5,8-tetrachloro-	–Cl		—	1.073	1.073	—	—	1.073	1.073	—		w/EtOH	Re DTN/alk	8, 118
703	-,2,6-dichloro-	–Cl		(~1.1	—	~1.1	0.54	~1.1	—	~1.1	0.54)[j]		60% w/Pr^iOH	Re Phot/alk	165
704	-,1-sulfonato-	$-SO_3^-Na^+$		—	(12 Lines spectrum, a_H^{min} = 0.45 G)							2.0039	w/alk	Re Phot	166
705	-,2-sulfonato-[k]	$-SO_3^-Na^+$		0.27	—	1.01	0.53	0.69	0.78	1.21	0.53	2.00402	w/EtOH	Re DTN/alk	8, 167
706	-,1,5-disulfonato-[k]	$-SO_3^-Na^+$		—	0.37	0.89	0.37	—	0.37	0.89	0.37	2.0038	w/alk	Re Phot	166
707	-,1,8-disulfonato-[k]	$-SO_3^-K^+$		—	0.57	1.12	0.72	0.72	1.12	0.57	—	2.00415	w/EtOH	Re DTN/alk	8
708	-,2,6-disulfonato-[k]	$-SO_3^-Na^+$		0.39	—	1.19	0.43	0.39	—	1.19	0.43	2.00408	w/EtOH	Re DTN/alk	8, 166

[a] Assignment tentative.

[b] The correct assignment of 9,10-anthrasemiquinone, i.e., $a_1 < a_2$, is established by this experiment with deuterium at C-1. Some papers in the past contain the incorrect assignment $a_2 < a_1$.

[c] ENDOR investigation.

[d] Result from the isotropic phase of the liquid crystal at 418 K.

[e] ENDOR study in nematic and smetic phases with information of isotropic and anisotropic hyperfine contributions.

[f] Result from the nematic phase of the liquid crystal at 393 K.

[g] The data, obtained from low-resolution spectra, may be erroneous; cf. Reference 154.

[h] Set of data of questionable correctness, assignment seems improbable. Simulations based on set of data do not fit published spectra for all set in Reference 158.

[i] Two different assignments, A and B, suggested by means of additivity rules; cf. Reference 78.

[j] Spectrum of low resolution, assignment tentative.

[k] Assignment tentative.

Table I.16.2
HYDROXY- AND ALKOXYANTHRASEMIQUINONES

Anions

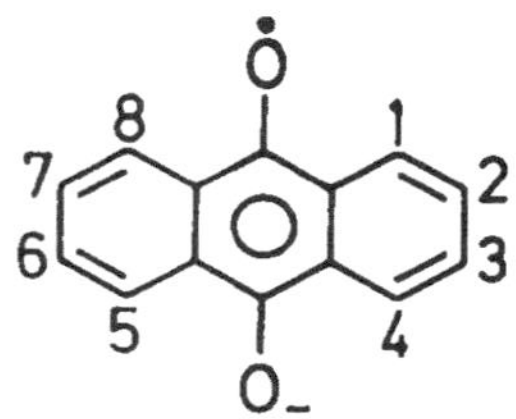

9,10-Anthrasemiquinone

No.	Radical	Substituent(s)	Splitting constants/Gauss[a] a_1	a_2	a_3	a_4	a_5	a_6	a_7	a_8	g-Factor	Solvent	Method	Ref.
709	-,1-hydroxy-	–OH*	0.32*	0.157	1.17	1.09	0.50	1.04	0.86	0.65		w/EtOH	Re alk	78, 439
			0.319*	1.586	1.243	1.083	0.534	1.041	0.853	0.690		w/EtOH	Re alk	78
			0.36*	1.59	1.59	0.94	0.36	0.63	0.36	0.36		DMF	Re el	158, 438
709D		–OD	0.035[D]	1.55	1.18	1.11	0.50	1.06	0.87	0.65		D_2O/EtOD	Re NaOD	8
710	-,2-oxido-	–O^-	1.96	—	0.54	0.27	0.15	1.61	0.62	1.03	2.00392	w/EtOH[b]	Re alk	78
711	-,2-methoxy-	–OCH_3^*	1.081	<0.015*	0.785	0.310	0.338	1.237	0.737	0.788		w/EtOH	Re alk	8
712	-,1,2-dihydroxy-	–OH*	0.59*	—	1.07	0.59	0.13	1.67	0.13	1.67	2.00385	w/EtOH	Re alk	8, 430
713[c]		–OH*(2-O^-?)	0.580*	—	1.075	0.593	0.098	2.234	0.108	1.054		w/EtOH	Re alk	8
712D		–OD	0.10[D]	—	1.01	0.59	0.10	1.64	0.10	1.64		D_2O/EtOD	Re NaOD	8
712		–OH*	0.32*	—	0.96	0.32	0.32	0.96	0.96	0.32	2.0042	DMF	Re NaH	433, 434
714[d]	-,1,3-dihydroxy-	–OH*(3-O^-?)	0.27*	0.59	—	3.25	1.01	0.59	1.56	0.17	2.00382	w/EtOH	Re alk	8
714D		–OD	u[D]	0.60	—	3.27	1.05	0.60	1.65	0.18		D_2O/EtOD	Re NaOD	8
715	-,1,4-dihydroxy-	–OH*	0.520*	2.084	2.084	0.520*	0.554	0.905	0.905	0.554	2.00379	w/EtOH	Re alk	78, 427
			0.526*	2.104	2.104	0.526*	0.526	0.918	0.918	0.526		EtOH	Re el	147
			0.483*	2.045	2.045	0.483*	0.425	0.934	0.934	0.425		4.1% w/ DMF	Re el	147, 164
			0.476*	2.034	2.034	0.476*	0.404	0.941	0.941	0.404		DMF	Re el	147, 158, 439

715D		–OD	0.07^{D}	2.08	2.08	0.07^{D}	0.55	0.90	0.90	0.55	2.00379	D_2O/EtOD	Re NaOD	8
			0.065^{D}	1.999	1.999	0.065^{D}	0.435	0.932	0.932	0.435		4% w/DMF	Re el	147, 164
716	-,1,4-dihydroxy-2-hydroxymethyl-	–OH*,–CH_2OH	0.532*	1.902^{1}	1.983	0.532*	0.534	0.882	0.943	0.589		w/EtOH	Re alk	8
717	-,1,4-dihydroxy-5,8-dichloro-	–OH*,–Cl	0.52*	1.75	1.75	0.52*	—	1.22	1.22	—		w/DMF		164
717D		–OD,–Cl	u^{D}	1.66	1.66	u^{D}	—	1.22	1.22	—		w/DMF		164
718	-,1-hydroxy-4-methoxy-	–OH*,–OCH_3	0.454*	(2.177	$1.954)^{c}$	0.026*	0.396	(0.891	$0.958)^{c}$	0.142		w/EtOH	Re alk	8
719	-,1,5-dihydroxy-	–OH*	0.30*	1.71	1.35	0.94	0.30*	1.71	1.35	0.94		w/MeOH	Re alk	427, 78
			0.304*	1.699	1.160	0.929	0.304*	1.699	1.160	0.929		DMSO	Re el	147^{-}, 439
			0.3*	1.70	1.16	0.94	0.3*	1.70	1.16	0.94		w/DMF		164
719D		–OD	0.39^{D}	1.661	1.101	0.936	0.039^{D}	1.661	1.101	0.936		4% D_2O/DMF	Re el	147^{-}, 158, 164
720	-,1,6-dihydroxy-	–OH	(Spectrum not obtained)									w/EtOH	Re alk	8
721	-,1-hydroxy-6-methoxy-	–OH*,–OCH_3	0.282*	1.829	0.862	1.440	1.039	0.028*	0.666	0.434		w/EtOH	Re alk	8
722	-,1-hydroxy-7-oxido-	–OH*,–O	0.416*	1.272	1.740	0.526	0.301	0.766	—	2.092		w/EtOH	Re alk	8
733	-,1-hydroxy-7-methoxy-	–OH*,–OCH_3	0.352*	1.352	1.374	0.999	0.280	0.859	0.015*	1.188		w/EtOH	Re alk	8
724	-,1-hydroxy-3-methyl-7-methoxy-	–OH*,–Me,–OCH_3	0.346*	1.247	1.370^{1}	1.210	0.329	0.799	0.024*	1.123		w/EtOH	Re alk	8
725	-,1,8-dihydroxy-	–OH*	0.196*	1.591	1.202	0.989	0.989	1.202	1.591	0.196*		w/EtOH	Re alk	8, 78, 427
			0.202*	1.509	1.064	0.515	0.515	1.064	1.509	0.202*		DMSO	Re el	147, 439
725D		–OD	0.028^{D}	1.447	1.131	0.575	0.575	1.131	1.447	0.028^{D}		3.8% D_2O/DMF	Re el	147
			u^{D}	1.423	1.147	0.598	0.598	1.147	1.423	u^{D}		5% D_2O/DMF	Re el	147, 158
726	-,1,8-dihydroxy-3-methyl-	–OH*,–Me	0.19*	1.70	0.95^{1}	1.42	0.99	1.25	1.42	0.19*		w/EtOH	Re alk	8
727^{b}	-,2-hydroxy-3-oxido-	–OH,–O^-	1.350	—	—	1.350	0.602	1.133	1.133	0.602	2.00392	w/EtOH	Re alk	8, 78
728^{b}	-,2,3-dioxido-	–O^-	1.661	—	—	1.661	0.618	1.178	1.178	0.618		w/EtOH	Re alk	8
729	-,2,3-dimethoxy-	–OMe	0.790	—	—	0.790	0.592	1.026	1.026	0.592		w/EtOH	Re alk	8
730	-,2,6-dioxido-	–O (or –OH)	1.586	—	<0.025	0.924	1.586	—	<0.025	0.924	2.00393	w/EtOH	Re alk	8, 78, 432
731	-,1,2-dihydroxy-3-oxido-	–OH*,–O	0.367*	—	—	2.736	0.518	1.218	1.036	0.685		w/EtOH	Re alk	8, 431, 434
732	-,2,3-dioxido-1-hydroxy-	–O^-,–OH*	0.437*	—	—	2.940	0.613	1.147	1.185	0.583		w/EtOH	Re alk	8

Table I.16.2 (continued)
HYDROXY- AND ALKOXYANTHRASEMIQUINONES

Anions

No.	Radical	Substituent(s)	Splitting constants/Gauss[a] a_1	a_2	a_3	a_4	a_5	a_6	a_7	a_8	g-Factor	Solvent	Method	Ref.
733	-,1,4-dihydroxy-2-oxido-[illegible]	–OH*,–O⁻	0.81*	—	1.62	0.46*	0.09	1.04	0.59	0.33	2.00380	w/EtOH	Re alk	8
733D		–OD,–O⁻	0.13[D]	—	1.60	~0.06[D]	~0.06	1.01	0.57	0.30	2.00376	D_2O/EtOD	Re NaOD	8
734	-,1,5-dihydroxy-2-oxido-[illegible]	–OH*,–O⁻	0.280*	—	1.617	0.635	0.269*	2.176	0.665	1.885		w/EtOH	Re alk	8
735	-,1,2,6-trihydroxy-	–OH*	0.48*	—	0.33	1.55	2.05	—	0.00	0.89		w/EtOH	Re alk	8
736	-,2,7-dioxido-1-hydroxy-[illegible]	–O⁻, –OH*	0.592*	—	1.250	0.380	0.209	0.804	—	2.264		w/EtOH	Re alk	8
737	-,1,3,8-trihydroxy-		(Spectrum not obtained)									w/EtOH	Re alk	8
738	-,1,3,8-trihydroxy-6-methyl-	–OH*,–Me	0.14*	0.72	—	3.12	1.30	0.58[l]	2.59	0.14*		w/EtOH	Re alk	8
739	-,1,4,5-trihydroxy-	–OH*	0.50*	2.06	1.97	0.40*	0.18*	1.65	1.30	0.85		w/EtOH	Re alk	8, 78
		–OD	u[D]	2.00	1.95	u[D]	u[D]	1.65	1.30	0.85		D_2O/EtOD	Re NaOD	78
740	-,1,5-dimethoxy-4-hydroxy-	–OCH_3^*,–OH*	0.031*	1.932	1.539	0.439*	u	1.427	0.877	0.531		w/EtOH	Re alk	8
741	-,1,4-dihydroxy-6-oxido-[illegible]	–OH*,–O⁻	0.58*	2.72	1.69	0.43*	1.99	—	0.58	0.17		w/EtOH	Re alk[b]	8, 462
741D		–OD,–O⁻	~0.09[D]	2.73	1.68	~0.08[D]	2.02	—	0.60	~0.18		D_2O/EtOD	Re NaOD	8, 462
742	-,1,4,6-trihydroxy-	–OH*	0.55*	2.46	1.83	0.47*	1.18	0.09*	0.63	0.39		w/EtOH	Re alk[b]	8, 462
742D		–OD	~0.08[D]	2.40	1.80	~0.08[D]	1.20	—	0.68	0.40		D_2O/EtOD	Re NaOD	8, 462
743	-,1,4-dihydroxy-6-methoxy-	–OH*,–OMe	0.55*	2.35	1.81	0.47*	1.05	u	0.71	0.34		w/EtOH	Re alk	8, 462
744	-,1,2,3,4-tetrahydroxy-	–OH*(–O⁻?)	2.1*	—	—	2.1*	0.5	0.5	0.5	0.5		w/EtOH	Ox	431, 434
745	-,1,2,5,6-tetrahydroxy-	–OH*	0.46*	—	0.30	1.96	0.46*	—	0.30	1.96	2.00383	w/EtOH	Re alk	8, 78
745D		–OD	0.075[D]	—	0.30	1.92	0.075[D]	—	0.30	1.92		D_2O/EtOD	Re NaOD	8, 78
746	-,1,2,5,8-tetrahydroxy-	–OH*	0.48*	—	1.71	0.38	0.38*	2.46	1.72	0.27*	2.00374	w/EtOH	Re alk	8, 78
746D		–OD	u[D]	—	1.27	0.40	u[D]	2.25	1.85	u[D]	2.00379	D_2O/EtOD	Re NaOD	8, 78

747	-,1,3,5,8-tetrahy-droxy-2,7-dimethyl-	–OH,–Me	(One exchange narrowed line of width = 4.11 G)								2.0037	Powder spectrum		441, 442
748	-,1,3,6,8-tetrahydroxy-	–OH*	~0.12*	1.15	—	3.31	3.31	—	1.15	~0.12*	2.00368	w/EtOH	Re alk	8,78
748D		–OD	u^D	1.15	—	3.30	3.30	—	1.15	u^D		D_2O/EtOD	Re NaOD	8, 78
749	-,1,4,5,8-tetrahydroxy-	–OH*	0.36*	1.85	1.85	0.36*	0.36*	1.85	1.85	0.36*	2.00373	w/EtOH	Re alk	78
			0.38*	2.02	2.02	0.38*	0.38*	2.02	2.02	0.38*		w/MeOH	Re alk	435
			u	1.96	1.96	u	u	1.96	1.96	u		DMF		436
750	-,1,2,3,5,6,7-hexahydroxy-	–OH*	0.38*	—	—	2.70	0.38*	—	—	2.70		w/EtOH	Re alk	8, 431
750D		–OD	u^D	—	—	2.10	u^D	—	—	2.10		D_2O/EtOD	Re NaOD	8

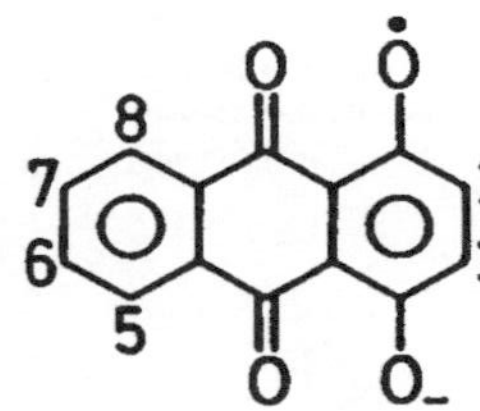

751	1,4,9,10-Anthracenetetrone radical anion		(0.52 (2H), a_C = 3.50)	DMSO	Ox KOBu^t	164
752	-,5,8-dichloro-	–Cl	(0.69 (2H), a_C = 3.25)	DMSO	Ox KOBu^t	164

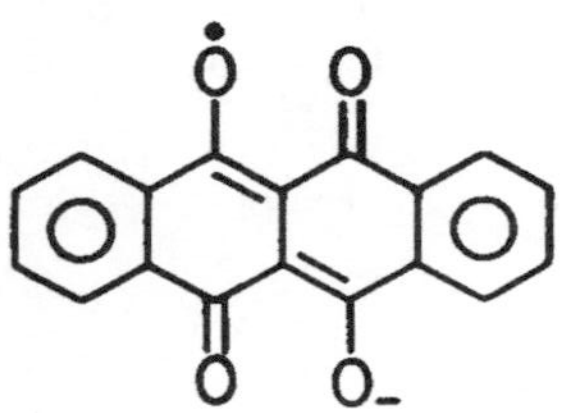

753	5,6,11,12-Naphthacenetetrone radical anion		(0.073 (8H), a_C = 3.5)	DMSO	Ox KOBu^t	164

Table I.16.2 (continued)
HYDROXY- AND ALKOXYANTHRASEMIQUINONES

Anions

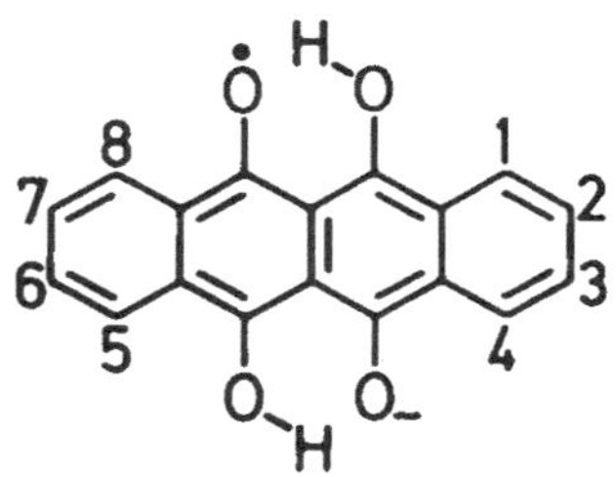

No.	Radical	Substituent(s) a^{OH} or a^{OD}	Splitting constants/Gauss[a] a_1	a_2	a_3	a_4	a_5	a_6	a_7	a_8	g-Factor	Solvent	Method	Ref.
754	Dihydro-9,10,11,12-naphthacenetetrone radical anion	0.74	0.31	0.94	0.94	0.31	0.31	0.94	0.94	0.31		w/DMF	Ox KOBu^t	164, 443
		0.11[D]	0.31	0.94	0.94	0.31	0.31	0.94	0.94	0.31		D_2O/DMF	Ox KOBu^t	164, 147
		0.752	0.319	0.958	0.958	0.319	0.319	0.958	0.958	0.319		DMF	Re el	147

[a] The assignments of splitting constants in this table are based mainly on additivity relations and must be considered as tentative. In particular, assignments for β-hydroxylated compounds are ambiguous, since the degree of protonation at β-positions (2,3,6,7) is often unknown. In contrast, α-hydroxylated compounds are fully protonated throughout the relevant pH-range, making complete assignments based on additivity relations feasible. For a preliminary discussion, see Reference 78.

[b] For pH values, at which the radicals are observed, see References 78 and 462.

[c] The structure of no. 713 is probably 1-hydroxy-2-oxido-9,10-anthrasemiquinone.

[d] The structure of no. 714 is probably 1-hydroxy-3-oxido-9,10-anthrasemiquinone.

[e] Assignments within the pairs (a_2,a_3), (a_5,a_8), and (a_6,a_7) unknown.

PHENANTHRASEMIQUINONE ANIONS

Table I.17
PHENANTHRENESEMIQUINONES

Anions

No.	Radical	Substituent(s)	Splitting constants/Gauss								g-Factor	Solvent	Method	Ref.
			a_2	a_3	a_5	a_6	a_7	a_8	a_9	a_{10}				
	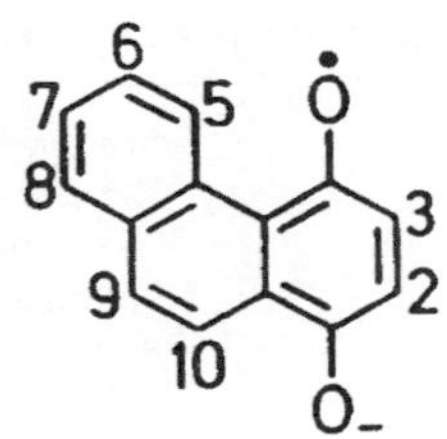													
755	1,4-Phenanthrenesemiquinone		(Spectrum not analyzed)									w/EtOH	Re alk	8
	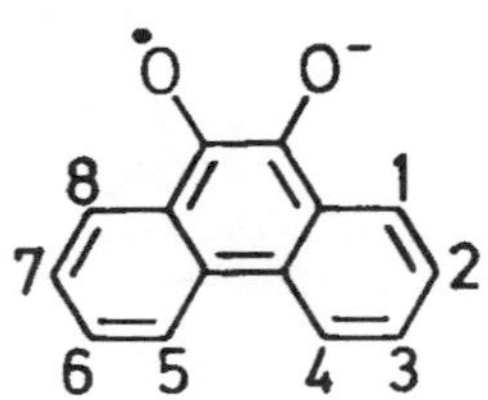													
			a_1	a_2	a_3	a_4	a_5	a_6	a_7	a_8				
756	[C]9,10-Phenanthrenesemi-quinone		1.69	0.41	1.73	0.41	0.41	1.73	0.41	1.69	2.0049	w/DEA	Re[a]	121
			1.66	u	1.66	u	u	1.66	u	1.66		w/EtOH	Re KOH	118[b]

Table I.17 (continued)
PHENANTHRENESEMIQUINONES

Anions

No.	Radical	Substituent(s)	Splitting constants/Gauss								g-Factor	Solvent	Method	Ref.
			a_2	a_3	a_5	a_6	a_7	a_8	a_9	a_{10}				
			1.62	0.36	1.75	0.42	0.42	1.75	0.36	1.62		EtOH/abs	Re KOH[c]	104
			1.35	0.20	1.67	0.42	0.42	1.67	0.20	1.35	2.00485[d]	DMF	Re el	117, 255
			1.37	0.22	1.65	0.42	0.42	1.65	0.22	1.37		DMSO	Re KO_2	55
			1.40	u	1.60	u	u	1.60	u	1.40	2.0049	THF/DCCw	Re KO_2	7
757	-,3-isobutyl-	$-CH_2CH(Me)_2$	1.52	u	u	u	u	1.52	u	1.52		w/EtOH	Re KOH	118
758	-,3-cyclohexyl-	$-C_6H_{11}$	1.65	u	u	u	u	1.65	u	1.65		w/EtOH	Re KOH	118
759	-,3-acetyl-	–COMe	1.33	u	—	u	u	1.33	u	1.33		w/EtOH	Re KOH	118
760	-,2-nitro-	$-NO_2$	1.60	u[e]	1.60	u	u	1.60	u	1.60		w/EtOH	Re KOH	118
			0.97	0.00[e]	1.94	0.44	0.44	1.60	0.00	0.97		DME	Re K	120, 119
761	-,2,7-dinitro-	$-NO_2$	0.97	0.00[e]	1.94	0.44	0.44	1.60	0.00[e]	0.97		DME	Re K	120, 119
762	-,4-nitro-	$-NO_2$	1.65	u	1.65	u[e]	u	1.65	u	1.65		w/EtOH	Re KOH	118
763	-,2-chloro-	–Cl	1.58	—	1.58	u	u	1.58	u	1.58		w/EtOH	Re KOH	118
764	-,3-chloro-	–Cl	1.60	u	—	u	u	1.60	u	1.60		w/EtOH	Re KOH	118
765	-,2-sulfonato-	$-{}^{CH}_{H}{}_2$	(1.66	—	1.38	0.38	0.28	1.25	0.28	1.25)[f]		w/DEA	Re[a]	121

[a] Spontaneous transfer of electron from diethylamine (DEA) in deaerated water to the quinone.
[b] Reference 118 has spectra of low resolution (±0.1 G).
[d] From an ENDOR experiment. Temperature 213 to 293 K.
[d] g-Factor from Reference 255.
[e] Splitting from nitrogen not observed.
[f] In Reference 121 spectrum published only. Constants obtained by simulation and tentatively assigned by editor.

Table I.18.1
TROPONE ANION RADICALS AND TROPOLONYL RADICALS

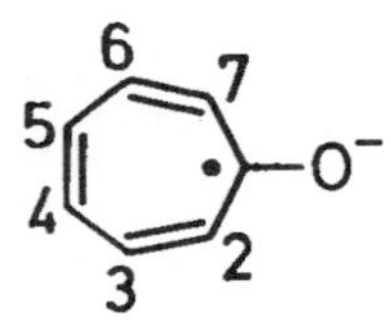

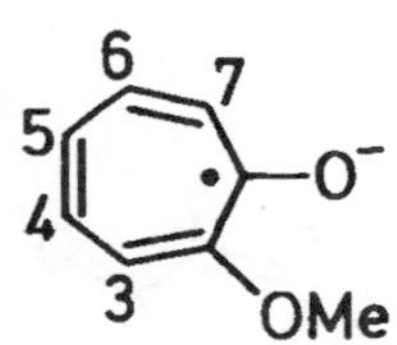

No.	Radical	Substituent(s)	Splitting constants/Gauss a_3	a_4	a_5	a_6	a_7	g-Factor	Solvent	Method	Ref.
766	Tropone anion radical		0.10	5.050	5.050	0.10	8.578[a]		DMF	Re el	421
767	-,2,7-dimethyl-	–Me	0.48	4.77	4.77	0.48	8.76[f,b]		DMF	Re el	423
768	2-Methoxytropone anion radical		u	4.9	4.9	u	8.3		DMF	Re el	422
769	-,5-methyl-	–Me	1.1	5.2	4.1[1]	u	8.3		DMF	Re el	422
770	-,5-isopropyl-	$-CHMe_2$	1.12	5.40	0.87[1]	u	8.70		DMF	Re el	422
771[c]	-,5-nitro-	$-NO_2$	1.10	6.42	3.36[N]	0.12	1.46		DMF	Re el	422, 127

Table I.18.1 (continued)
TROPONE ANION RADICALS AND TROPOLONYL RADICALS

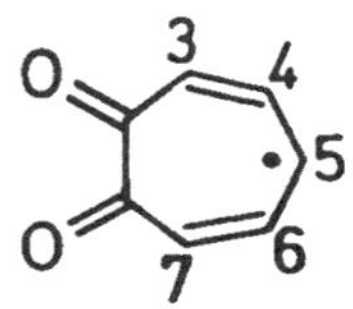

No.	Radical	Substituent(s)	Splitting constants/Gauss a_3	a_4	a_5	a_6	a_7	g-Factor	Solvent	Method	Ref.
772	Tropolonyl radical		7.1	2.5	10.3	2.5	7.1		w	Ox CeIV	126
773	-,4-isopropyl-	$-CHMe_2$	6.25	1.0^{I}	10.5	2.5	7.25		w	Ox CeIV	89
774	-,3-carboxy-4-carboxymethyl-	–COOH,–CH_2COOH	—	1.13^{I}	10.5	2.75	7.25		w	Ox CeIV	89
775	-,4-carboxy-3,6-dicarboxymethyl-	–COOH, $-CH_2COOH$ –COOH	—	1.17^{I}	10.0	—	7.7		w	Ox CeIV	89
776	-,3,4,6-tricarboxy-	–COOH	—	—	10.0	—	7.5		w	Ox CeIV	89
777	-,3-nitro-	$-NO_2$	7.19^{N}	1.76	4.51	0.95	3.98	2.0039	DMF	Re el	127^{+}
778	-,5-nitro-	$-NO_2$	2.84	1.42	8.04^{N}	1.42	2.84	2.0046	DMF	Re el	127^{-}
779	-,3,5-dinitro-	$-NO_2$	0.78^{N}	1.94	6.78^{N}	1.08	3.92	2.0043	DMF	Re el	127^{+}
780	-,3-oxido-	$-O^-$	—	1.09	2.05	2.05	1.09		HMPA	Ox NaOMe	420
781	-,5-oxido-	$-O^-$	1.20	1.99	—	1.99	1.20		HMPA	Ox NaOMe	420, 89
781D		$(3,7\text{–}d_2)$	0.18^{D}	1.96	—	1.96	0.18^{D}		HMPA	Ox NaOMe	420
782	-,4-isopropyl-5-oxido-	$-CHMe_2$, $-O^-$	1.75	0.65^{I}	—	2.25	1.08		HMPA	Ox NaOMe	89

[a] $a_2 = a_7 = 8.578$ G.
[b] $a_2^{Me} = a_7^{Me} = 8.76$ G.
[c] A splitting of 0.75 G observed from three protons of the OMe-Group for radical no. 771.

Table I.18.2
BENZTROPOLONYL RADICALS

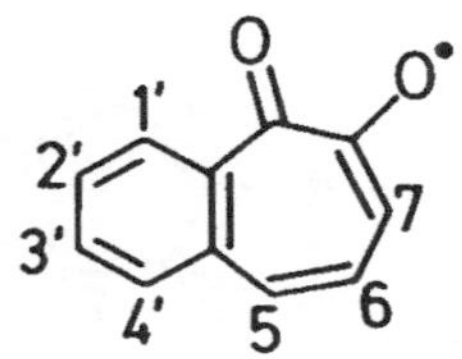

No.	Radical	Substituent(s)	a_5	a_6	a_7	$a_{1'}$	$a_{2'}$	$a_{3'}$	$a_{4'}$	Solvent	Method	Ref.
			Splitting constants/Gauss									
783	3,4-Benztropolonyl radical		10.5	2.5	8.5	0.0	2.5	1.0	2.5	w	Ox CeIV	89, 83
784	-,1′,2′-dihydroxy-6-carboxy-	–OH,–COOH	10.0	—	7.5	—	—	0.65	0.65	w	Ox CeIV	89
785	-,1′,2′-dihydroxy-6-ethoxycarbonyl-	–OH,–COOEt	10.25	—	7.0	—	—	0.81	0.81	w	Ox CeIV	89
786	-,1′,2′-dihydroxy-6-carboxy-4′-methyl-	–OH,–COOH,–Me	10.25	—	7.5	—	—	0.86	0.86[1]	w	Ox CeIV	89
787	-,1′,2′,3′-trihydroxy-	–OH	9.0	2.5	7.0	—	—	—	8.5	w	Ox CeIV	89, 83
788	-,1′,2′,3′-trihydroxy-6-carboxy-	–OH,–COOH	9.25	—	7.5	—	—	—	8.5	w	Ox CeIV	89

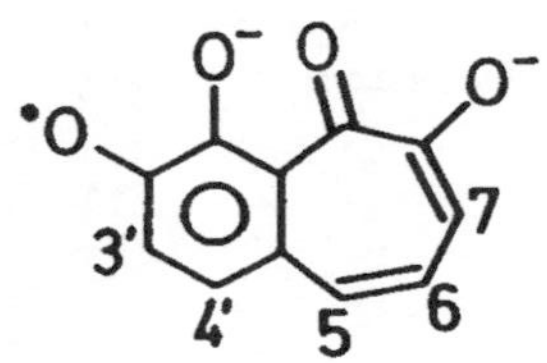

789 1′,2′-Dioxidobenztropolonyl radical

Table I.18.2 (continued)
BENZTROPOLONYL RADICALS

No.	Radical	Substituent(s)	Splitting constants/Gauss a_5	a_6	a_7	$a_{1'}$	$a_{2'}$	$a_{3'}$	$a_{4'}$	Solvent	Method	Ref.
789	1′,2′-Dioxidobenztropolonyl radical											
790[a]	-,6-carboxylato-	–COO⁻	0.5	—	0.13	—	—	0.45	3.58	w	Ox alk	89, 83
791	-,6-ethoxycarbonyl-	–COOEt	0.5	—	0.13	—	—	0.45	3.58	w	Ox alk	89
792	-,6-carboxylato-3′-methyl-	–COO⁻,–Me	0.13	—	0.13	—	—	0.13[1]	4.6	w	Ox alk	89
793	-,6-carboxylato-4′-methyl-	–COO⁻,–Me	0.85	—	0.3	—	—	0.0	0.49[1,b]	w	Ox alk	89
794	-,3′-oxido-	–O⁻	0.75	2.8	0.55	—	—	—	1.1	w	Ox alk	89, 83
			1.7	4.3	1.7	—	—	—	1.5	w/neut	Ox CeIV	83
795	-,3′-oxido-6-carboxylato-	–O⁻,–COO⁻	0.5	—	0.5	—	—	—	1.45	w	Ox alk	89
796[c]	-,4′-oxido-6-carboxylato-	–O⁻,–COO⁻	2.35	—	1.35	—	—	0.5	—	w	Ox alk	89
797	4,5-Benztropolonyl radical		10.4	0.0	1.5	4.8	0.6	5.6	1.1	w	Ox alk	89

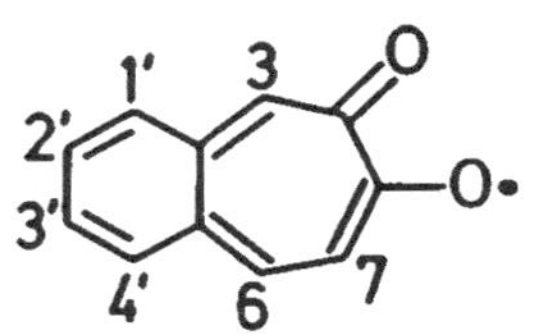

[a] The radicals of the nos. 790 to 793 are best described as *o*-benzosemiquinones.

[b] This small splitting from the methyl group protons does not make sense; 4.9 G seems appropriate.

[c] Radical obtained as a hydroxylation product of compound no. 790; structure tentatively assigned.

Table I.19
DIPHENOSEMIQUINONES

Anions

No.	Radical	Substituents(s)	Splitting constants/Gauss $a_2 = a_{2'}$	$a_3 = a_{3'}$	$a_5 = a_{5'}$	$a_6 = a_{6'}$	Solvent	Method	Ref.
798	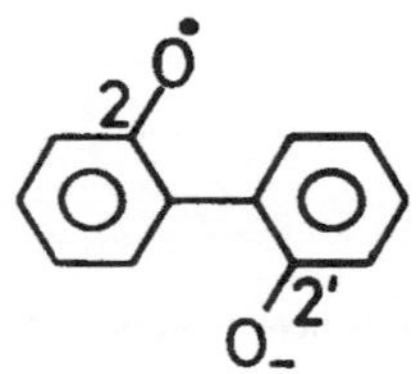 2,2'-Diphenosemiquinone		(Spectra not obtained; see Reference 15)						
799	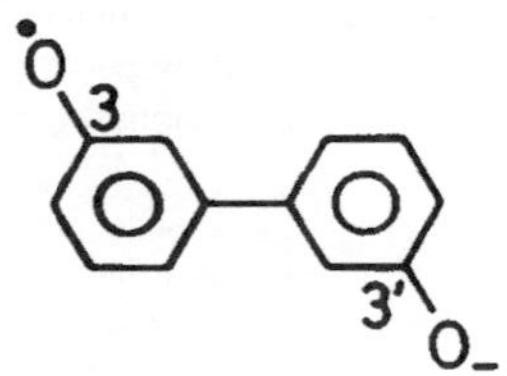 3,3'-Diphenosemiquinone		(EPR studies not reported)						
800	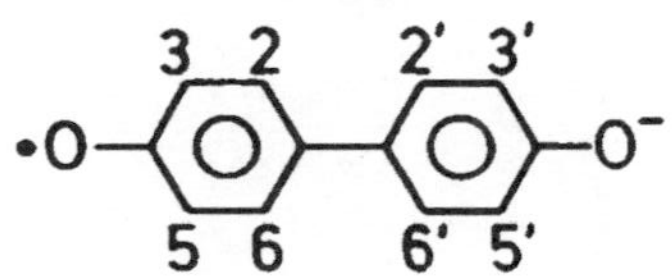 4,4'-Diphenosemiquinone		0.56	2.36	2.36	0.56	MeCN	Re el	142

Table I.19 (continued)
DIPHENOSEMIQUINONES

Anions

No.	Radical	Substituents(s)	Splitting constants/Gauss $a_2 = a_{2'}$	$a_3 = a_{3'}$	$a_5 = a_{5'}$	$a_6 = a_{6'}$	Solvent	Method	Ref.
801	-,3,3′-dimethyl-	–Me	0.48	1.96[1]	2.36	0.78	MeCN	Re el	142
802	-,3,3′-di-*t*-butyl-	–CMe_3	0.40	u	2.30	0.62	MeCN	Re el	142
803	-,3,3′,5,5′-tetramethyl-	–Me	0.78	1.56[1]	1.56[1]	0.78	50% w/C_5H_5N	Re KOH	87
			0.54	1.88[1]	1.88[1]	0.54	MeCN	Re el	142
			0.52	1.92[1]	1.92[1]	0.52	DMF	Re el	231
804	-,3,3′,5,5′-tetraisopropyl-	–$CHMe_2$	0.56	1.11[1]	1.11[1]	0.56	MeCN	Re el	142
805	-,3,3′,5,5′-tetra-*t*-butyl-	–CMe_3	0.46	u	u	0.46	MeCN	Re el	142
806	-,3,3′-dimethyl-5,5′-di-*t*-butyl-	–Me,–CMe_3	0.76	1.69[1]	u	0.76	w/MeOH	Ox alk[a]	444
			0.56	2.00[1]	u	0.42	MeCN	Re el	142
807	-3,3′-diisopropyl-5,5′-di-*t*-butyl-	–$CHMe_2$,–CMe_3	0.56	1.00[1]	u	0.56	MeCN	Re el	142

[a] Radical generated by oxidation of parent phenol with potassium hexacyanoferrate (III); g = 2.0039 measured.

Table I.20
SEMIQUINONES DERIVED FROM DIQUINONES

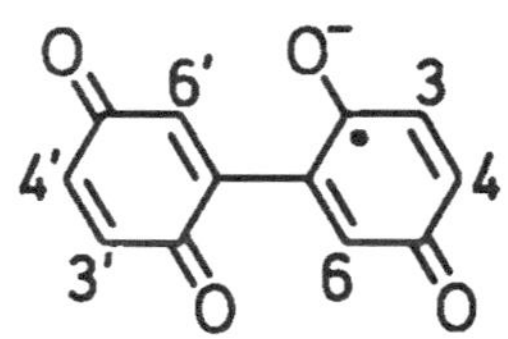

No.	Radical	Substituent(s)	Splitting constants/Gauss $a_3 = a_{3'}$	$a_4 = a_{4'}$	$a_6 = a_{6'}$	g-Factor	Solvent	Method	Ref.
808	2,2′,5,5′-Diquinone radical anion		1.13	0.82	2.34		DMSO	Re el	125

809	-,3,3′-di-*t*-butyl-	$-CMe_3$	u	0.63	1.90	5% w/DMSO	Re KOH	105
810	-,4,4′-di-*t*-butyl-	$-CMe_3$	0.96	u	2.42	5% w/DMSO	Re KOH	105
811	-,4,4′-dioxido-[a]	$-O^-$	0.22	—	1.48	w	Ox alk	130
812	-,6,6′-dimethyl-4,4′-dioxido-	$-Me, -O^-$	0.32	—	0.64[i]	w	Ox KOH	43
813	-,4,4′-dimethoxy-	$-OMe$	0.35	0.35[b]	1.55	w	Re el	450
			0.44	0.30[b]	1.84	56% w/DMF	Re el	450
			0.58	0.15[b]	3.10	5% w/DMF	Re el	450
			0.60	0.13[b]	3.41	DMF	Re el	450
			0.54	0.18[b]	3.75	5% w/DMSO	Re KOH	105

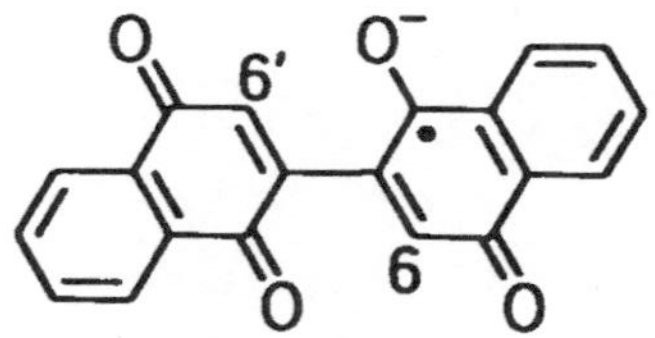

814	2,2′,5,5′-Dinaphthoquinone radical anion		—	—	3.45[c]	DMSO	Re KOH	105

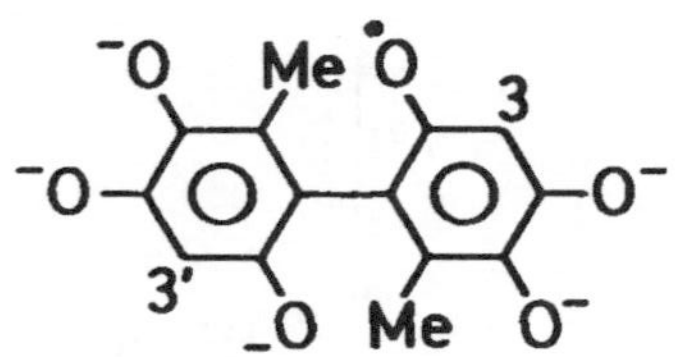

815	6,6′-Dimethyl-2,2′,4,4′,5,5′-hexaoxidodiphenyl radical anion[d]		0.32	—	0.32[i]	w	Ox KOH	43

Table I.20 (continued)
SEMIQUINONES DERIVED FROM DIQUINONES

No.	Radical	Substituent(s)	Splitting constants/Gauss $a_3 = a_{3'}$	$a_4 = a_{4'}$	$a_6 = a_{6'}$	g-Factor	Solvent	Method	Ref.
	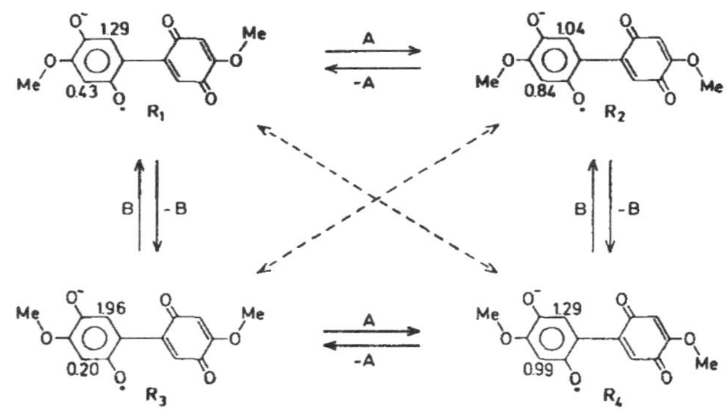			a_3	a_H^{OMe}	a_6			
816			R_1	0.43	0.78	1.29	w	Ox alk	130
817			R_2	0.84	0.78	1.04	w	Ox alk	130
818			R_3	0.20	0.78	1.96	w	Ox alk	130
819			R_4	0.99	0.78	1.29	w	Ox alk	130

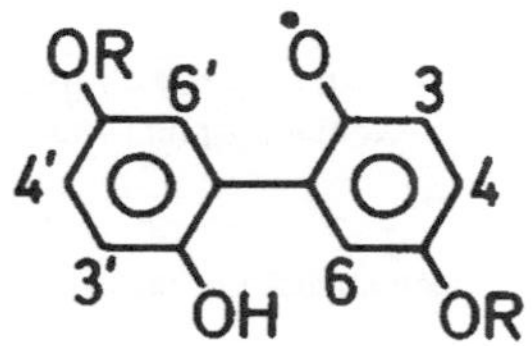

			a_2^{OH}[f]	$a_3 = a_{3'}$	$a_5^{OR} = a_{5'}^{OR}$				
820	3,3'-di-*t*-butyl	(R = Me)	0.50*	u	0.83[b]	2.0043	w	Re Ag_2O	451
821		(R = Et)	0.50*	u	0.92[g]	2.0042	w	Re Ag_2O	451
822	4,4'-di-*t*-butyl	(R = Me)	0.50*	1.38	0.90[b]	2.0042	w	Re Ag_2O	451
823		(R = Et)	0.49*	1.38	1.18[g]	2.0042	w	Re Ag_2O	451
823D			u[D]	1.50	1.01[g]		D_2O	Re Ag_2O	451

[a] Radical generated by oxidation of the corresponding hexahydroxy compound. Structure tentatively assigned: it is unknown whether some oxido groups are protonated.

[b] Septet splitting from six equivalent methoxyl-group protons.

[c] A quintet splitting of 0.30 G was observed from two equivalent pairs of unidentified hydrogens in the aromatic rings.

[d] Structure tentatively assigned; cf. the results for nos. 811 and 812.

[e] R_1 to R_4 are four radical conformers with lack of coplanarity (see scheme) observed simultaneously when 4,4'-dimethoxydiquinone is oxidized in alkaline water. The result shows a semiquinone in which electron transfer between rings is slow. It is unknown whether R_1 to R_4 should be described as tri-anion radicals, cf. Reference 105.

[f] Splitting observed from one hydroxyl proton only.

[g] Quintet splitting from two pairs of equivalent methylene protons.

Table I.21
DIARYLFURAN SEMIQUINONE DERIVATIVES

Anions

No.	Radical	Substituent(s)	Splitting constants/Gauss a_1	a_2	a_3	a_4	a_6	a_7	a_8	a_9	Solvent	Method	Ref.
	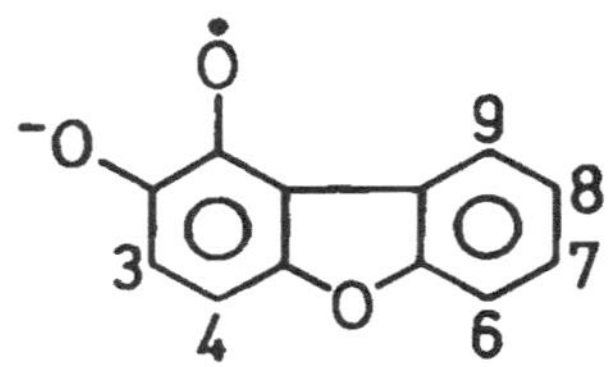												
824	Dibenzofuran-1,2-semiquinone		—	—	2.02	3.51	(0.20	0.10	0.20	0.10)[a]	DMSO/w	Re alk	105
825	-,4-oxido-	–O⁻	—	—	0.87	—	(0.87	0.08	0.22	u)[a]	DMSO/w	Re alk	105
	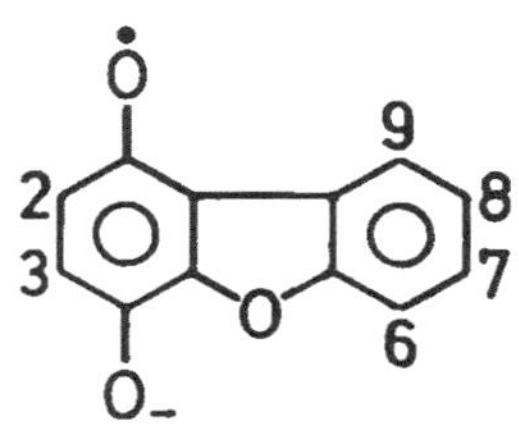												
826	Dibenzofuran-1,4-semiquinone		—	3.41	2.25	—	u	u	u	u	DMF/w	Re alk	5
			—	3.21	2.15	—	0.13	0.13	0.13	0.13	DMSO/w	Re alk	105
827	-,2,6-di-*t*-butyl-8-methoxy-	–Bu^t,–OMe	—	u	1.75	—	u	u	—	u	DMF/w	Re alk	452
828	-,3,7-di-*t*-butyl-8-methoxy-	–Bu^t,–OMe	—	2.64	u	—	u	u	—	u	DMF/w	Re alk	5

829	-,2-methoxy-	–OMe	—	—	0.62	—	(0.26	0.09	0.26	0.09)[a]	DMSO/ w	Re alk	105
830	-,3,7-dimethoxy-8-oxido-	–OMe,–O⁻	—	0.98	0.91[b]	—	0.16	0.16[b]	—	0.04	w	Ox alk	130
831	-,3,7-diethoxy-8-oxido-	–OEt,–O	—	1.07	1.04[c]	—	0.13	0.15[c]	—	0.13	w/EtOH	Ox alk	8

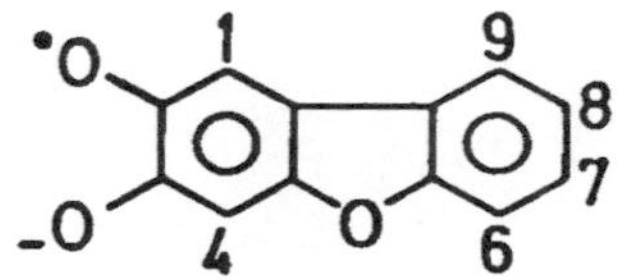

832	Dibenzofuran-2,3-semiquinone		(No data published)										
833	-,4-isopropyl-9-methyl-7-oxido-	–CHMe$_2$, –Me,–O⁻	0.56	—	—	u	0.13	—	0.13	1.22[f]	w/alk	—[d]	453
833D$_1$		(1-d)[e]	u[D]	—	—	u	u	—	u	1.22[f]	w/alk	—[d]	453
833D$_2$		(6,8-d$_2$)[e]	0.56	—	—	u	u[D]	—	u[D]	1.22[f]	w/alk	—[d]	453

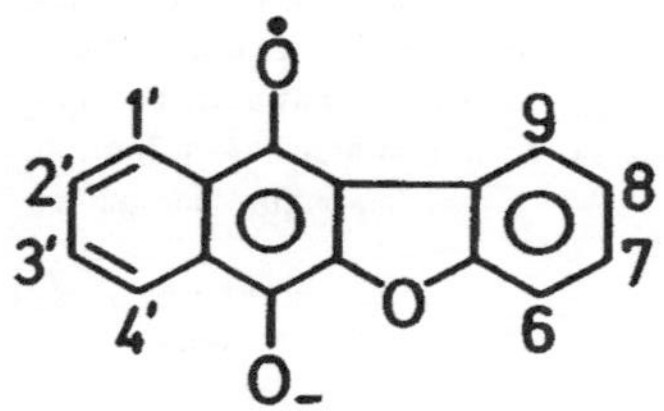

			$a_{1'}$	$a_{2'}$	$a_{3'}$	$a_{4'}$							
834	2,3-Benzo(dibenzofuran)-1,4-semiquinone		(0.35	1.40	u	1.02)[a]	0.19	0.19	0.19	0.19	DMSO/ w	Re alk	105
835	2,3;6,7-Dibenzo(dibenzofuran)-1,4-semiquinone		(No data published)										
836	-,8-methoxy	–OMe	(Complex spectrum, not analyzed)										
											DMSO/ w	Re alk	105

Table I.21 (continued)
DIARYLFURAN SEMIQUINONE DERIVATIVES

Anions

No.	Radical	Substituent(s)	Splitting constants/Gauss								Solvent	Method	Ref.
			a_1	a_2	a_3	a_4	a_6	a_7	a_8	a_9			
837	2,3;8,9-Di-benzo(dibenzofuran)-1,4-semiquinone		(No data published)										
838	-,7-oxido-	$-O^-$	(0.57	1.15	0.09	0.90	0.23	—	0.34	0.38)[a]	w/EtOH	Re alk	8

[a] Assignment unknown.
[b] Splitting from three equivalent OMe-group protons.
[c] Splitting from two equivalent OCH_2Me-group protons.
[d] Compound and radical obtained by reacting an equimolar solution of 3-isopropylcatechol and orcinol (0.01 *M*) with 1 *M* NaOH in the presence of air.
[e] Deuterated compound.

Table I.22.1
3′,4′-DIOXIDOFLAVONOID ANION RADICALS

B-Ring Semiquinones

No.	Radical	Trivial name parent compound	Splitting constants/Gauss $a_{2'}$	$a_{5'}$	$a_{6'}$	a_2	a_3		g-Factor	Solvent	Method	Ref.
	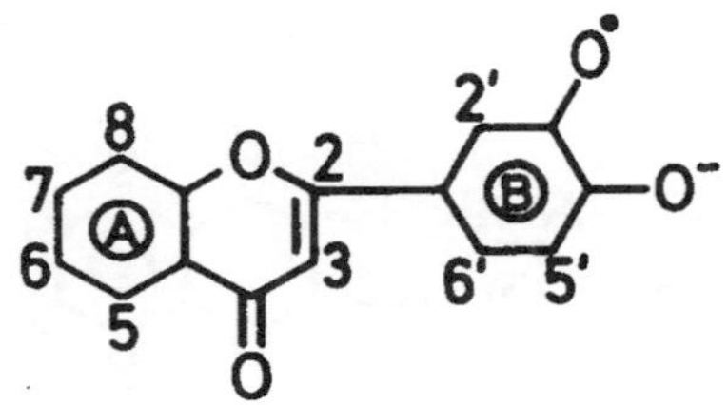											
839	3′,4′-Dioxidoflavone anion radical											
840	-,5,7-dihydroxy-[a]	Luteolin	1.15	1.40	2.63	—	1.25	0.11[b]	2.00463	20% w/DMSO	Ox alk	455, 454
			0.95	1.50	2.80	—	1.25		2.00454	20% w/EtOH	Ox alk	455
841	-,5,7-dihydroxy-6-glucoside-	Isoorientin	1.25	1.40	2.65	—	1.50		2.00458	20% w/DMSO	Ox alk	455
			1.20	1.40	2.75	—	1.15		2.00458	20% w/EtOH	Ox alk	455
	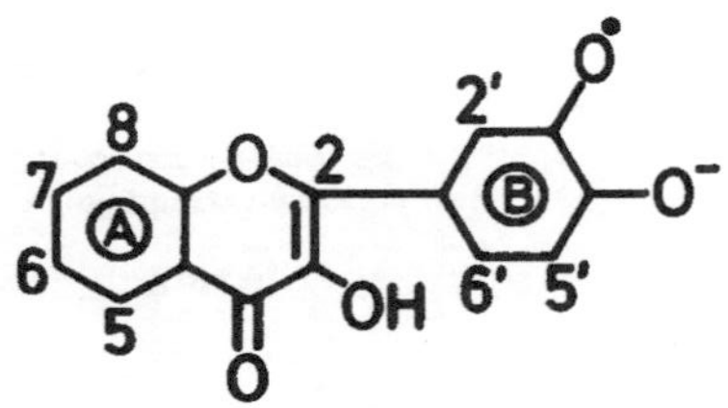											
842	3′,4′-Dioxidoflavonol anion radical											
843	-,7-hydroxy-	Fisetin	0.75	1.40	2.70	—	—	0.30[b]	2.00451	20% w/DMSO	Ox alk	455, 454
			0.80	1.40	2.55	—	—	0.25[b]	2.00450	10% w/EtOH	Ox alk	455
844	-,5,7-dihydroxy-	Quercetin	0.65	1.48	2.88	—	—			DMSO	Ox alk	454
			0.70	1.40	2.70	—	—		2.00451	20% w/DMSO	Ox alk	454, 455
			0.78	1.40	2.58	—	—			50% w/DMSO	Ox alk	454
			0.80	1.40	2.60	—	—		2.00448	20% w/EtOH	Ox alk	455

Table I.22.1 (continued)
3′,4′-DIOXIDOFLAVONOID ANION RADICALS

B-Ring Semiquinones

No.	Radical	Trivial name parent compound	Splitting constants/Gauss					g-Factor	Solvent	Method	Ref.
			$a_{2'}$	$a_{5'}$	$a_{6'}$	a_2	a_3				
845	-,5,7-dihydroxy-3-methoxy-	3-Methyl-quercetin	1.40	1.40	2.80	—	—		20% w/DMSO	Ox alk	454
846	-,5,7-dihydroxy-3-rutinoside-	Rutin	1.08	1.50	2.80	—	—	2.00457	20% w/DMSO	Ox alk	455
			0.95	1.25	2.90	—	—	2.00454	20% w/EtOH	Ox alk	455
847	-,5,7-dihydroxy-3-rhamnoside-	Quercitrin	1.25	1.40	2.70	—	—	2.00461	20% w/DMSO	Oa alk	455
			1.15	1.15	2.83	—	—	2.00455	20% w/EtOH	Ox alk	455
			0.96	0.96	2.94	—	—		w	Ox alk	130
848	-,3,7-dihydroxy-5-methoxy-	5-Methyl-quercetin	0.75	1.40	2.75	—	—		20% w/DMSO	Ox alk	454
849	-,3,5-dihydroxy-7-methoxy-	Rhamnetin	0.78	1.45	2.68	—	—	2.00454	20% w/DMSO	Ox alk	455, 454
			0.75	1.43	2.55	—	—	2.00454	20% w/EtOH	Ox alk	455
850	3′,4′-Dioxidoflavanone anion radical										
851	-,5,7-dihydroxy-	Eriodictyol	1.23	1.38	3.33	2.18	u		DMSO	Ox alk	454
			1.08	1.20	3.40	2.25	u		20% w/DMSO	Ox alk	454
			0.85	1.03	3.43	2.35	u		50% w/DMSO	Ox alk	454

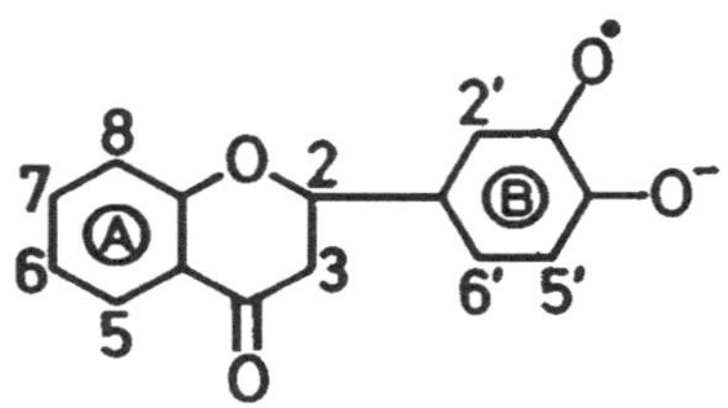

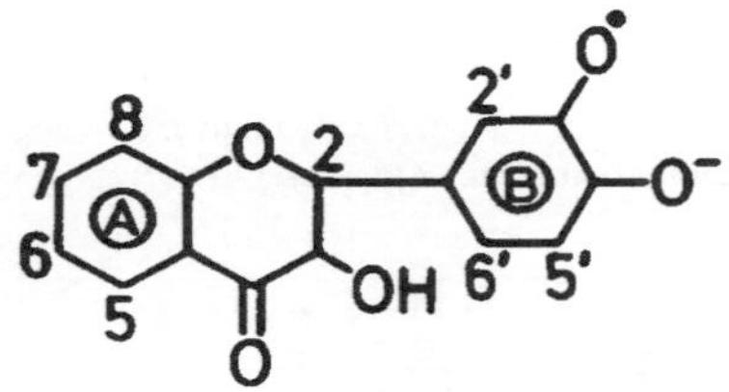

852	3′,4′-Dioxido-2,3-dihydro-flavonol anion radical										
853	-,7-hydroxy-	Dihydrofisetin	1.05	1.13	3.40	1.15	u	2.00455	20% w/DMSO	Ox alk	454, 455
			0.95	0.95	3.40	1.25	0.10	2.00451	20% w/EtOH	Ox alk	455
			0.94	0.94	3.41	0.94	0.10		w	Ox alk	130
854	-,5,7-dihydroxy-	Dihydroquercetin	1.15	1.33	3.40	1.38	u		DMSO	Ox alk	454
		(Taxifolin)	1.08	1.15	3.40	1.20	u	2.00450	20% w/DMSO	Ox alk	454, 455
			0.98	1.08	3.43	1.08	u		50% w/DMSO	Ox alk	454
			0.85	0.85	3.45	0.85	0.13		w	Ox alk	130

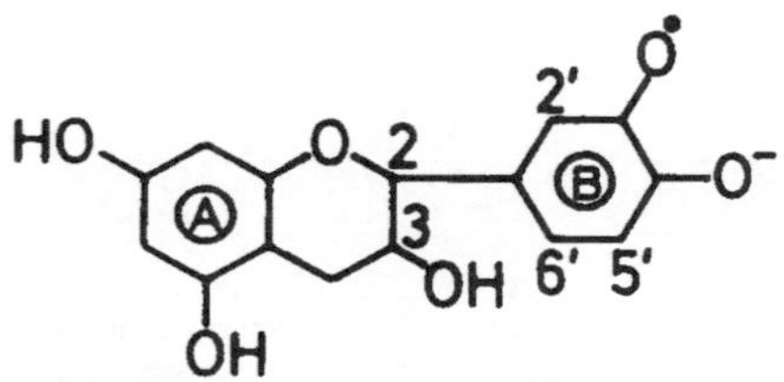

855	(+) Catechin anion radical	Catechin	1.23	1.48	3.30	2.15	u		DMSO	Ox alk	454
			1.08	1.23	3.45	1.90	u		20% w/DMSO	Ox alk	454, 456

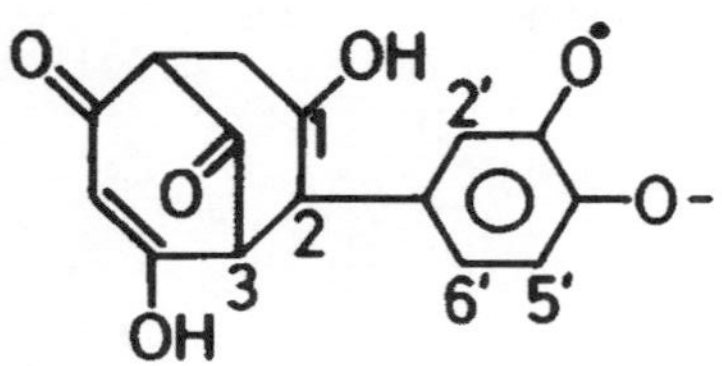

Table I.22.1 (continued)
3′,4′-DIOXIDOFLAVONOID ANION RADICALS

B-Ring Semiquinones

No.	Radical	Trivial name parent compound	Splitting constants/Gauss $a_{2'}$	$a_{5'}$	$a_{6'}$	a_2	a_3	g-Factor	Solvent	Method	Ref.
856	Catechinic acid anion radical	Catechinic acid	0.65	0.90	3.75	2.20	u	2.00445	33% w/DMSO	Ox alk	456
			0.65	0.90	3.75	2.20	u	2.00445	50% w/EtOH	Ox alk	456

[a] OH-substituents in A-ring are given as hydroxy since their state of protonation is unknown.
[b] Assignment of this doublet splitting unknown.

Table I.22.2
2′,3′,4′- and 3′,4′,5′-TRIOXIDOFLAVONOID ANION RADICALS

B-Ring Semiquinones

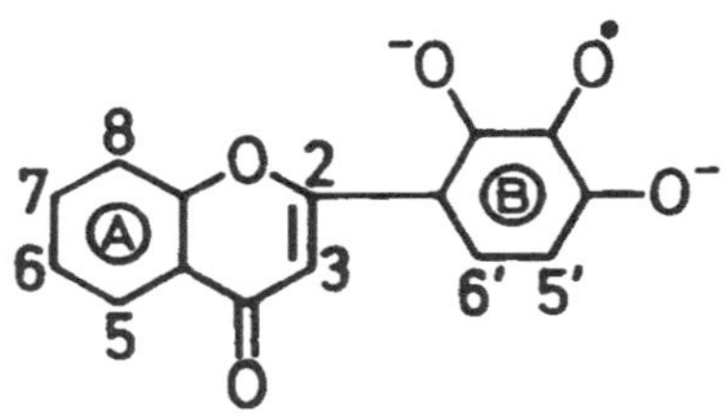

No.	Radical	Splitting constants/Gauss $a_{2'}$	$a_{5'}$	$a_{6'}$	a_2	a_3	g-Factor	Solvent	Method	Ref.
857	2′,3′,4′-Trioxidoflavone anion radical									
858	-,5,7-dihydroxy-	—	0.85	4.75	—	0.25	2.00463	20% w/DMSO	Ox alk	455
		—	0.88	5.05	—	0.24	2.00458	20% w/EtOH	Ox alk	455

859 -,5,7-dihydroxy-6-glucoside-	—	0.85	4.65	—	0.25	2.00462	20% w/DMSO	Ox alk	455
	—	0.85	4.95	—	0.25	2.00458	20% w/EtOH	Ox alk	455

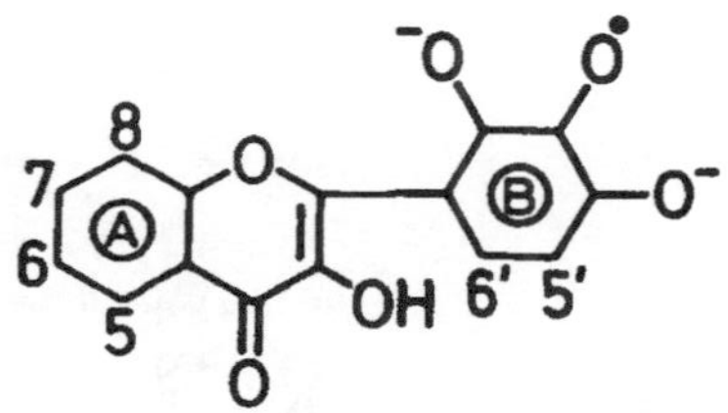

860 2′,3′,4′-Trioxidoflavonol anion radical									
861 -,7-hydroxy-	—	0.75	4.95	—	—	2.00462	20% w/DMSO	Ox alk	455
	—	0.75	4.95	—	—		w/EtOH	Ox alk	455
862 -,5,7-dihydroxy-	—	0.70	4.50	—	—	2.00463	50% w/DMSO	Ox alk	455
	—	0.70	4.55	—	—	2.00459	33% w/EtOH	Ox alk	455
863 -,5,7-dihydroxy-3-rutinoside-	—	0.88	5.00	—	—	2.00457	50% w/DMSO	Ox alk	455
	—	0.90	5.15	—	—	2.00456	50% w/EtOH	Ox alk	455
864 -,5,7-dihydroxy-3-rhamnoside-	—	0.88	4.98	—	—	2.00459	20% w/DMSO	Ox alk	455
	—	0.90	5.10	—	—	2.00458	20% w/EtOH	Ox alk	455

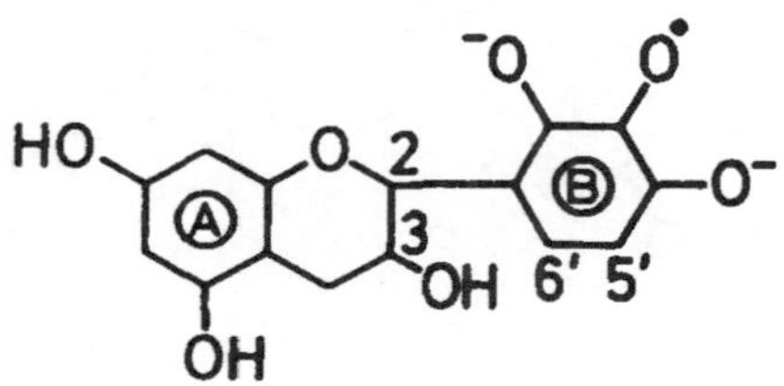

865 2′-Oxidocatechin anion radical	—	0.55	3.70	0.27	0.23	2.00439	25% w/DMSO	Ox alk	456
866 2′-Oxidocatechinic acid anion radical[b]	—	0.60	5.85		0.15 (3H)[c]	2.00437	33% w/DMSO	Ox alk	456

Table I.22.2 (continued)
2′,3′,4′- and 3′,4′,5′-TRIOXIDOFLAVONOID ANION RADICALS

B-Ring Semiquinones

No.	Radical	$a_{2'}$	$a_{5'}$	$a_{6'}$	a_2	a_3	g-Factor	Solvent	Method	Ref.
		Splitting constants/Gauss								
	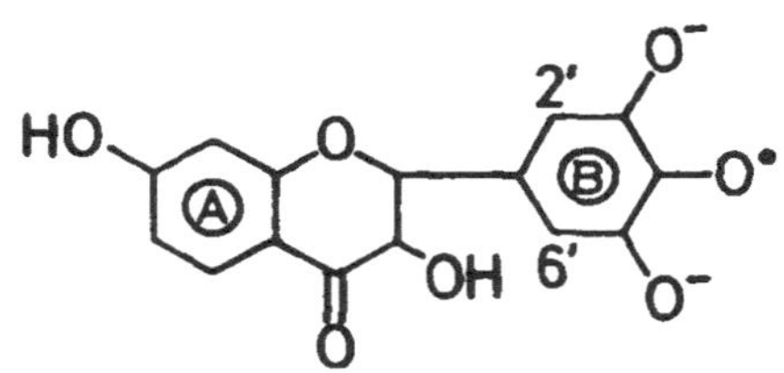									
867	(+) Dihydrorobinetin anion radical	0.825	—	0.825	1.55	—		20% w/DMSO	Ox alk	454
	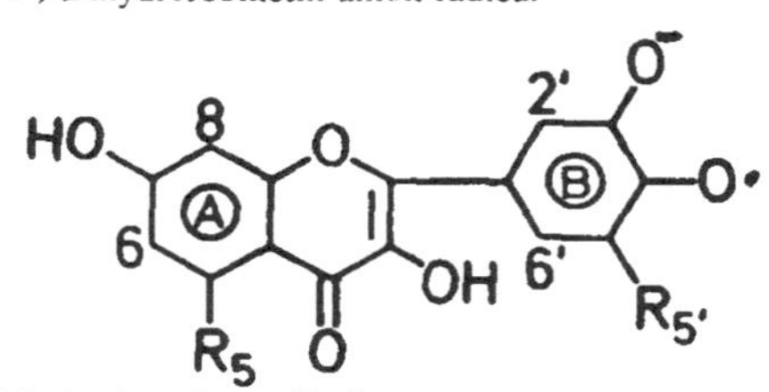 Myricetin anion radical									
868	($R_{5'}$ = O^-, R_5 = OH)[d]	0.98	—	0.98	—	—	2.00455	20% w/DMSO	Ox alk	454, 455
869	($R_{5'}$ = OH*, R_5 = OH)[d]	0.85	0.30*	1.70	—	—	2.00452	20% w/DMSO	Ox alk	455
	Robinetin anion radical									

				a_5	a_6	a_8			
870 ($R_{5'}$ = O^-, R_5 = H)	0.92	—	0.92	0.20	1.12	1.08	20% w/DMSO	Ox alk	454

[a] The compound nos. 858 to 864 obtained as secondary products by 2′-hydroxylation of the parent compounds of Table I.22.1.
[b] For radical structure, see Table I.22.1, no. 856.
[c] 1:3:3:1 Quartet might tentatively be assigned to the protons at C-1, C-2, and C-3 (Table I.22.1, No. 856).
[d] pK_3 (B-ring) ~12.3.[453]

Table I.22.3
3′,4′,6′-TRIOXIDOFLAVONOID ANION RADICALS[a]

B-Ring Semiquinones

No.	Radical	Splitting constants/Gauss $a_{2'}$	$a_{5'}$	a_5	a_6	a_8	g-Factor	Solvent	Method	Ref.
	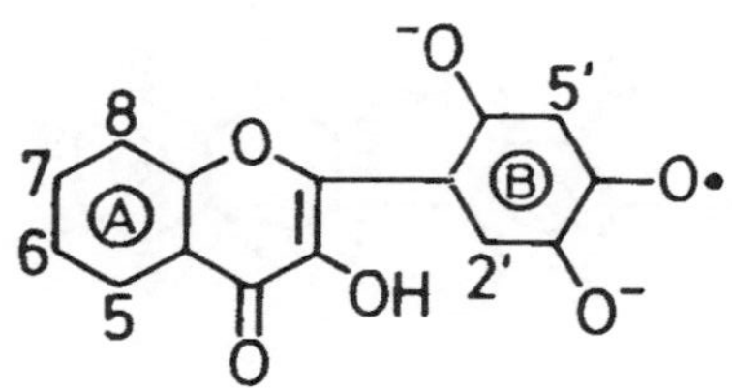									
871	3′,4′,6′-Trioxidoflavonol anion radical									
872	-,7-hydroxy-	1.77	0.27	(0.11	0.24	1.40)[b]		50% w/EtOH	Ox alk	455
873	-,5,7-dihydroxy-	2.15	0.35	—	u	u	2.00445	50% w/DMSO	Ox alk	455

Table I.22.3 (continued)
3′,4′,6′-TRIOXIDOFLAVONOID ANION RADICALS[a]

B-Ring Semiquinones

No.	Radical	Splitting constants/Gauss $a_{2'}$	$a_{5'}$	a_5	a_6	a_8	g-Factor	Solvent	Method	Ref.
	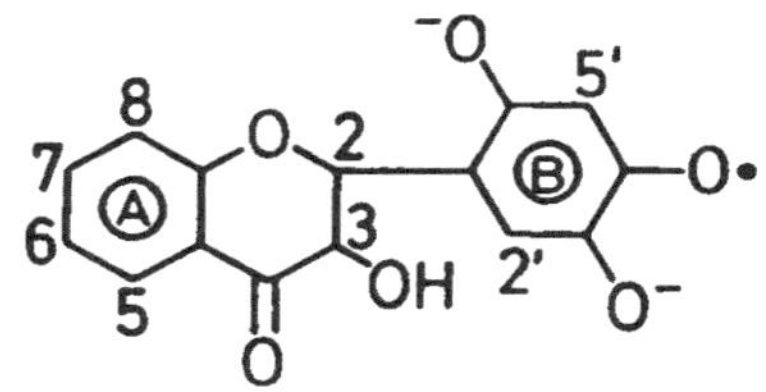									
874	3′,4′,6′-Trioxido-2,3-dihydroflavonol anion radical									
				a_2	a_3					
875	-,7-hydroxy-	1.55	0.55	3.20	0.20		2.00449	20% w/DMSO	Ox alk	455
		1.55	0.55	2.65	0.20		2.00447	20% w/EtOH	Ox alk	455
876	-,5,7-dihydroxy-[d]	1.55	0.55	3.15	0.20		2.00453	20% w/DMSO	Ox alk	455
		1.55	0.55	2.35	0.15		2.00449	20% w/EtOH	Ox alk	455

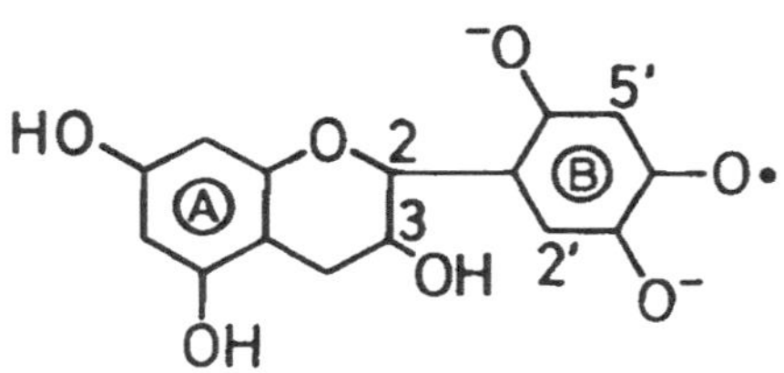

877	6′-Oxidocatechin anion radical	1.30	0.60	3.75	0.30	2.00439	25% w/DMSO	Ox alk	456
878	6′-Oxidocatechinic acid anion radical[d]	1.13	0.45	2.25	0.15(3H)[c]	2.00438	50% w/DMSO	Ox alk	456
		1.15	0.50	2.20	0.15(3H)[c]	2.00437	50% w/EtOH	Ox alk	456

[a] All compounds in Table I.22.3 obtained as secondary products by 6′-hydroxylation of parent compounds of Table I.22.1.
[b] Assignment unknown for values in the bracket.
[c] 1:3:3:1 quartet which might tentatively be assigned to the protons at C-1, C-2, and C-3 (Table I.22.1, no. 856).
[d] For radical structure see Table I.22.1, no. 856.

Table I.22.4
RADICALS OF SOME FLAVONES

A-Ring Radicals

No.	Radical	Trivial name parent compound	Splitting constants/Gauss a_2	a_3	a_6	a_8	Solvent	Method	Ref.

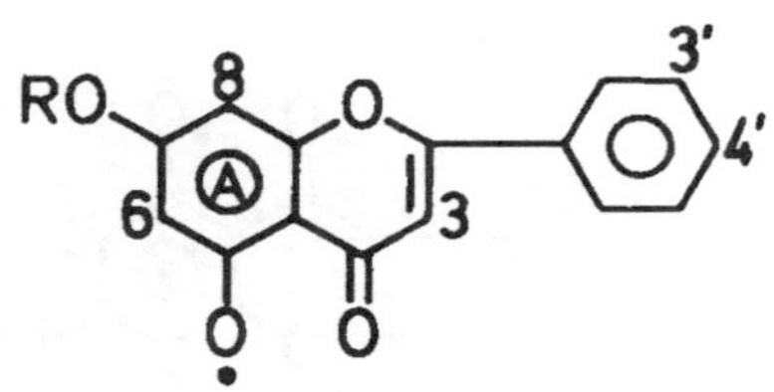

No.	Radical	Trivial name parent compound	a_2	a_3	a_6	a_8	Solvent	Method	Ref.
879	7-Hydroxy-5-oxidoflavone radical	Chrysin (R = H)[a]	—	1.0	8.2	11.25	w/acidic	Ox CeIV[b]	457
880	3′-Hydroxy-4′-methoxy-5-oxido-7-rutinosylflavanone radical	Hesperidin[c] (R = rutinose)	—	1.5	9.0	11.5	w/acidic	Ox CeIV	457

Table I.22.4 (continued)
RADICALS OF SOME FLAVONES

A-Ring Radicals

No.	Radical	Trivial name parent compound	Splitting constants/Gauss a_2	a_3	a_6	a_8	Solvent	Method	Ref.
	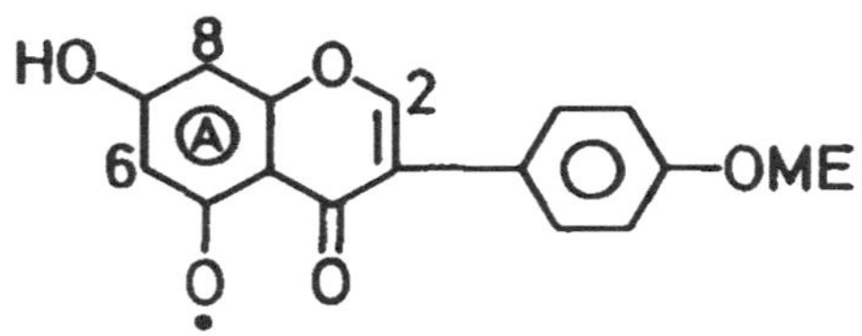								
881	7-Hydroxy-4′-methoxy-5-oxido-isoflavone radical	Biochanin A	0.0	—	7.8	10.75	w/acidic	Ox CeIV	457

[a] Calculations indicate the 7-oxygen to be protonated.
[b] Oxidation also performed with the H_2O_2/TiIII system.
[c] Hesperidin is a flavanone; the starting material in Reference 457 is named a flavone (?).

Table I.23
OXIDOCOUMARINE RADICALS

Anions and Neutral Species

No.	Radical	Trivial name parent compound	Splitting constants/Gauss a_3	a_4	a_5	a_6	a_7	a_8	Solvent	Method	Ref.
			Anions								
	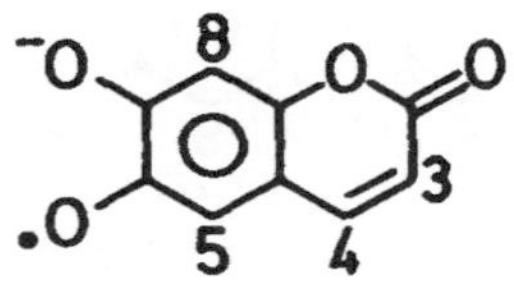										
882	6,7-Dioxidocoumarine semiquinone	Aesculetin	3.38	1.06	2.34	—	—	0.47	50% w/EtOH	Ox alk	17, 91
883	-,4-methyl-	4-Methylaesculetin	2.78	0.58[1]	2.20	—	—	0.45	50% w/EtOH	Ox alk	17, 91
	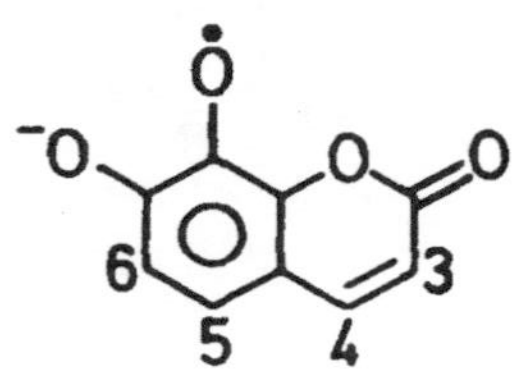										
884	7,8-Dioxidocoumarine semiquinone	Daphnetin									
885	-,4-methyl-	4-Methyldaphnetin	1.52	0.33[1]	3.08	0.51	—	—	50% w/EtOH	Ox alk	17

Table I.23 (continued)
OXIDOCOUMARINE RADICALS

Anions and Neutral Species

No.	Radical	Trivial name parent compound	Splitting constants/Gauss a_3	a_4	a_5	a_6	a_7	a_8	Solvent	Method	Ref.
			Anions								
			Neutral Species								
	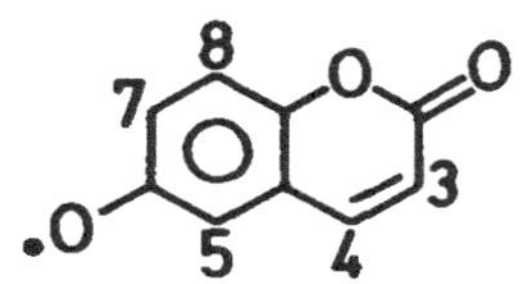										
886	6-Oxidocoumarine radical										
887	-,4-methyl-		0.4	0	5.2	—	7.3	1.1	w/acidic	Ox CeIV[a]	457
	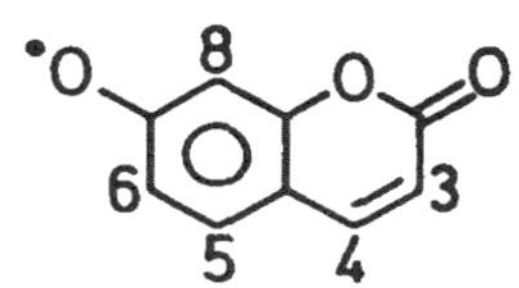										
888	7-Oxidocoumarine radical	Umbelliferone	5.6	1.9	2.1	5.6	—	6.8	w/acidic	Ox CeIV	457
889	-,4-methyl-		5.6	1.9[1]	1.9	5.6	—	6.3	w/acidic	Ox CeIV	457

	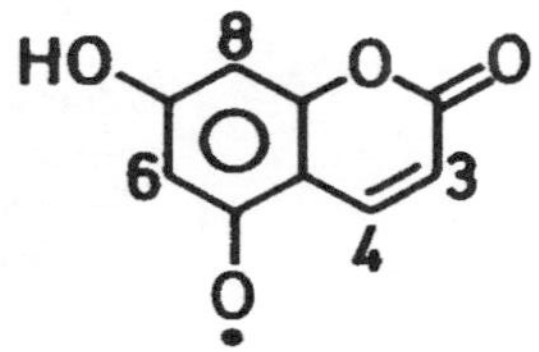										
890	**7-Hydroxy-5-oxidocoumarine radical**[b]										
891	-,4-methyl-		5.45	1.30[1]	—	4.25	—	10.5	w/acidic	Ox CeIV	457
	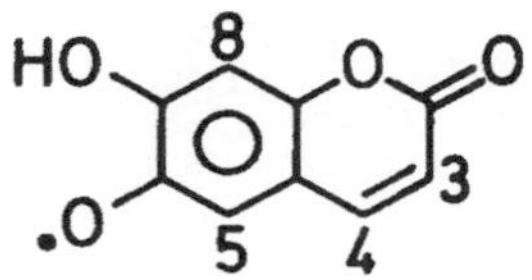										
892	7-Hydroxy-6-oxidocoumarine radical	Aesculetin	2.62	1.0	2.62	—	—	0.25	w/acidic	Ox CeIV	457
893	-,4-methyl-	4-Methylaesculetin	2.4	0.8[1]	2.4	—	—	0.0	w/acidic	Ox CeIV	457
	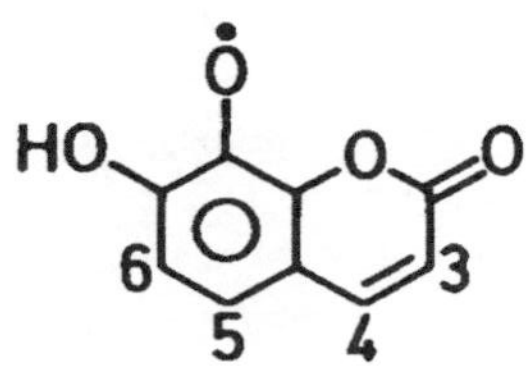										
894	**7-Hydroxy-8-oxidocoumarine radical**	**Daphnetin**									
895	**-,4-methyl-**	**4-Methyldaphnetin**	0.0	0.0	6.6	1.3	—	—	w/acidic	Ox CeIV	457

[a] Oxidation also performed using the TiIII/H_2O_2 system.
[b] A calculation indicates radicals protonated at 5- or 7-0 in equilibrium give rise to the observed spectrum.

II. NEUTRAL SPECIES

Table II.1
PHENOXYL RADICALS

Neutral Species

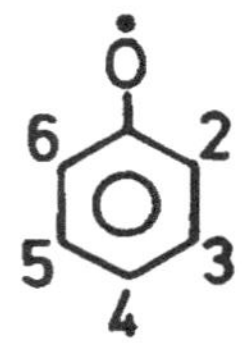

No.	Radical	Substituent(s)	Splitting constants/Gauss a_2	a_3	a_4	a_5	a_6	g-Factor	Solvent	Method	Ref.
896	Phenoxyl radical		6.61	1.85	10.22	1.85	6.61	2.00461	w/pH = 12.2	Radiolysis[a]	18
			6.65	1.85	10.20	1.85	6.65	2.00461	w/H_2SO_4	Ox CeIV	85, 173
			6.60	1.88	10.10	1.88	6.60		C_6H_6/PhCl	Photolysis[b]	176
897	-,2-methyl-	–Me	6.0[1]	1.9	11.5	1.9	6.0		w/H_2SO_4	Ox CeIV	232
			7.5[1]	2.0	9.8	1.5	6.0	2.00429	w/H_2SO_4	Ox CeIV	85, 19
898	-,2-isopropyl-	–$CHMe_2$	4.1[1]	2.0	10.2	2.0	6.1		w/H_2SO_4	Ox CeIV	232
899	-,4-methyl-	–Me	6.1	1.4	12.7[1]	1.4	6.1	2.00432	w/H_2SO_4	Ox CeIV	85
			6.0	1.45	11.95[1]	1.45	6.0		w/H_2SO_4	Ox CeIV	173
			5.93	1.47	11.86[1]	1.47	5.93		C_6H_6/PhCl	Photolysis[b]	176
900	-,4-ethyl-	–Et	6.0	1.5	10.15[1]	1.5	6.0		w/H_2SO_4	Ox CeIV	173
901	-,4-propyl-	–CH_2CH_2Me	6.1	1.45	8.7[1]	1.45	6.1		w/H_2SO_4	Ox CeIV	173
902	-,4-benzyl-	–CH_2Ph	6.1	1.45	8.7[1]	1.45	6.1		w/H_2SO_4	Ox CeIV	173
903	-,4-isopropyl-	–$CHMe_2$	6.0	1.2	4.5[1] 0.4[2]	1.2	6.0		w/H_2SO_4	Ox CeIV	19
904	-,4-*s*-butyl-	–CH(Me)CH_2Me	6.0	≤1.4	4.3[1]	≤1.4	6.0		w/H_2SO_4	Ox CeIV	173
905	-,4-*t*-butyl-	–CMe_3	6.0	≤1.4	u	≤1.4	6.0		w/H_2SO_4	Ox CeIV	173

			6.25	1.65	0.42[2]	1.65	6.25		C_6H_6/PhCl	Photolysis[b]	176
906	-,2,4-dimethyl-	–Me	6.5[1]	1.7	11.2[1]	1.7	6.5		C_6H_6	Ox Ag_2O/PbO_2	177
907	-,2-*t*-butyl-4-methyl-	–CMe_3, –Me	u	1.7	11.4[1]	1.7	6.6		C_6H_6	Ox Ag_2O/PbO_2	177
908	-,2,6-dimethyl-	–Me	6.5[1]	1.65	9.5	1.65	6.5[1]		w/H_2SO_4	Ox CeIV	173
			7.22[1]	1.99	10.13	1.99	7.22[1]		C_6H_6/PhCl	Photolysis[b]	176
909	-,2,6-diethyl-	–Et	5.7[1]	1.65	9.35	1.65	5.7[1]		w/H_2SO_4	Ox CeIV	173
910	-,2,6-diisopropyl-	–$CHMe_2$	3.7[1]	1.8	9.5	1.8	3.7[1]		w/H_2SO_4	Ox CeIV	173
			3.7[1]	1.85	9.4	1.85	3.7[1]		C_6H_6	Ox Ag_2O/PbO_2	177
911	-,2,6-di-*t*-butyl-	–CMe_3	u	2.0	9.65	2.0	u		C_6H_6	Ox Ag_2O/PbO_2	177, 188
			u	1.95	9.62	1.95	u		C_6H_6/PhCl	Photolysis[b]	176
912	-,6-*t*-butyl-2-methyl-	–CMe_3, –Me	7.0[1]	1.9	9.5	1.9	u		C_6H_6	Ox Ag_2O/PbO_2	177
913	-,3,5-dimethyl-	–Me	6.24	1.69[1]	10.0	1.69[1]	6.24		C_6H_6/PhCl	Photolysis[b]	176
			6.30	1.60[1]	10.70	1.60[1]	6.30	2.00438	w/H_2SO_4	Ox CeIV	241
914	-,3,5-di-*t*-butyl-	–CMe_3	6.60	u	10.0	u	6.60		C_6H_6/PhCl	Photolysis[b]	176
915	-,2,4,6-trimethyl-	–Me	6.0[1]	1.4	11.95[1]	1.4	6.0[1]		w/H_2SO_4	Ox CeIV	173
			6.06[1]	1.65	10.54[1]	1.65	6.06[1]		C_6H_6	Photolysis[b]	237
			6.25[1]	1.7	10.9[1]	1.7	6.25[1]		C_6H_6	Ox Ag_2O/PbO_2	177
916	-,2,6-dimethyl-4-ethyl-	–Me, –Et	6.0[1]	1.45	9.1[1]	1.45	6.0[1]		w/H_2SO_4	Ox CeIV	173
			6.25[1]	1.7	8.1[1]	1.7	6.25[1]		C_6H_6	Ox Ag_2O/PbO_2	177
917	-,2,6-dimethyl-4-propyl-	–Me, –$CH_2\dot{C}H_2Me$	6.0[1]	1.35	8.3[1]	1.35	6.0[1]		w/H_2SO_4	Ox CeIV	173
918	-,4-benzyl-2,6-dimethyl-	–CH_2Ph, –Me	6.0[1]	1.35	8.5[1]	1.35	6.0[1]		w/H_2SO_4	Ox CeIV	173
919	-,4-*t*-butyl-2,6-dimethyl-	–CMe_3, –Me	6.1[1]	1.8	u	1.8	6.1[1]		—	—	173
920	-,2,6-di-*t*-butyl-4-methyl-	–CMe_3, –Me	0.07[2]	1.69	11.05[1]	1.69	0.07[2]		CCl_4	Ox PbO	254
			0.07[2]	1.70	11.07[1]	1.70	0.07[2]		Hexane	Ox PbO	254
			0.07[2]	1.66	11.10[1]	1.66	0.07[2]		PhMe	Ox PbO	254
			u	1.68	11.20[1]	1.68	u		C_6H_6	Photolysis[b]	237
			u	1.73	11.08[1]	1.73	u		C_6H_6	Ox Pt	251
			0.07[2]	1.65	11.27[1]	1.65	0.07[2]		THF	Ox PbO	254
921	-,2,6-di-*t*-butyl-4-ethyl-	–CMe_3, –Et	0.06[2]	1.60	8.21[1]	1.60	0.06[2]		PhMe (–60°C)	Ox PbO	267
922	-,2,6-di-*t*-butyl-4-isopropyl-	–CMe_3, –$CHMe_2$	0.07[2]	1.66	4.05[1] 0.38[2]	1.66	0.07[2]		PhMe (–60°C)	Ox PbO	267
923	-,2,6-di-*t*-butyl-4-phenyl-	–CMe_3, –Ph	u	1.74	1.74[o] 0.74[m] 1.74[p]	1.74	u	2.00415	PhMe	Ox PbO	268

Table II.1 (continued)
PHENOXYL RADICALS

Neutral Species

No.	Radical	Substituent(s)	Splitting constants/Gauss a_2	a_3	a_4	a_5	a_6	g-Factor	Solvent	Method	Ref.
924	-,2,4,6-tri-*t*-butyl-	$-CMe_3$	0.075[2]	1.74	0.39[2]	1.74	0.075[2]		MeOH	Ox PbO	254
			0.076[2]	1.72	0.385[2]	1.72	0.076[2]		PhMe	Ox PbO	254
			0.073[2]	1.72	0.376[2]	1.72	0.073[2]		CS_2	Ox PbO	254
			0.069[2]	1.70	0.378[2]	1.70	0.069[2]		C_6H_6	Ox PbO	254, 238
			0.074[2]	1.69	0.378[2]	1.69	0.074[2]		THF	Ox PbO	254
925	-,3,4,5-trimethyl-	–Me	5.91	1.69[1]	11.79[1]	1.69[1]	5.91		C_6H_6/PhCl	Photolysis[b]	176
926	-,2,3,5,6-tetramethyl-	–Me	6.93[1]	1.8[1]	9.89	1.18[1]	6.93[1]	2.0048	PhMe	Ox PbO_2	191
927	-,2-formyl-	–CHO	—	1.7	10.0	2.0	7.1		w/H_2O_4	Ox CeIV	19
928	-,3-formyl-	–CHO	7.1	—	9.8	2.0	6.75		w/H_2SO_4	Ox CeIV	19
929	-,4-formyl-	–CHO	6.8	2.2	u	2.2	6.8		w/H_2SO_4	Ox CeIV	173, 232
930	-,2-cyano-	–CN	0.7[d]	1.4	10.1	2.1	7.1		w/alk	Ox FeIII	26
931	-,3-cyano-	–CN	7.25	0.2[d]	9.9	2.05	6.8		w/H_2SO_4	Ox CeIV	26
932	-,4-cyano-	–CN	6.75	2.25	1.45[d]	2.25	6.75		w/H_2SO_4	Ox CeIV	26
933	-,2-acetyl-	–COMe	u	1.50	10.10	2.15	7.20	2.00497	w/H_2SO_4	Ox CeIV	241, 19
934	-,3-acetyl-	–COMe	7.10	u	9.90	1.90	6.70	2.00480	w/H_2SO_4	Ox CeIV	241, 19
935	-,4-acetyl-	–COMe	6.75	2.20	0.10[3]	2.20	6.75	2.00520	w/H_2SO_4	Ox CeIV	241, 173
936	-,4-benzoyl-	–COPh	7.0	2.2	u	2.2	7.0		w/H_2SO_4	Ox CeIV	173, 232
937	-,2-carboxylato-	$-COO^-$	—	1.84	9.98	1.78	6.39	2.00476	w/pH ≥ 3	Radiolysis[a]	18
938	-,2-carboxy-	–COOH	—	1.3	10.8	2.2	7.5		w/H_2SO_4	Ox CeIV	173
			—	1.23	10.20	2.09	7.11	2.00459	w/pH ≤ 3	Radiolysis[a]	18
939	-,3-carboxylato-	$-COO^-$	6.82	—	9.99	1.83	6.56	2.00472	w/pH = 12	Radiolysis[a]	18
940	-,3-carboxy-	–COOH	7.5	—	10.15	2.0	6.65		w/H_2SO_4	Ox CeIV	173
941	-,4-carboxylato-	$-COO^-$	6.53	1.93	—	1.93	6.53	2.00477	w/pH = 4.5—12	Radiolysis[a]	18

942	-,4,-carboxy-	–COOH	6.8	2.2	—	2.2	6.8		w/H_2SO_4	Ox CeIV	173, 232
943	-,4-carboxy-2,6-dimethyl-	–COOH, –Me	6.8[f]	2.2	—	2.2	6.8[f]		w/H_2SO_4	Ox CeIV	173, 232
944	-,3,5-dicarboxylato-	–COO^-	6.75	—	9.80	—	6.75	2.00481	w/pH = 12	Radiolysis[a]	18
945	-,3,5-dicarboxy-	–COOH	6.87	—	9.72	—	6.87	2.00494	w/pH = 3	Radiolysis[a]	18
946	-,3-ethoxycarbonyl-	–COOEt	7.80	—	11.33	1.91	6.95		C_6H_6/PhCl	Photolysis[b]	176
947	-,4-ethoxycarbonyl-	–COOEt	6.8	2.3	—	2.3	6.8		w/H_2SO_4	Ox CeIV	173
948	-,3,5-diethoxycarbonyl-	–COOEt	6.98	—	9.38	—	6.98		C_6H_6/PhCl	Photolysis[b]	176
949	-,2-trifluoromethyl-	–CF_3	4.00[c]	1.35	10.50	2.15	7.50	2.00474	w/H_2SO_4	Ox CeIV	241
950	-,3-trifluoromethyl-	–CF_3	7.10	1.50[c]	10.05	2.00	6.70	2.00490	w/H_2SO_4	Ox CeIV	241
951	-,4-trifluoromethyl-	–CF_3	7.20	2.20	11.20[c]	2.20	7.20	2.00491	w/H_2SO_4	Ox CeIV	241
952	-,2-amino-	–NH_2	4.76[d] 5.30[e]	<0.1	4.31	2.94	1.01	2.00372	w/pH = 7—9	Radiolysis[a]	18
			4.8[d] 5.25[e]	0.1	4.25	2.95	1.0	2.00372	w/H_2SO_4	Ox CeIV	85, 19
			4.5[d] 6.9[f] 5.5[f]	0.1	3.0	2.5	1.4		MeOH	Ox phot[g] (−61°C)	171
			4.43[d] 7.6[f] 5.8[f]	0.46	2.25	2.18	0.53		CH_2Cl_2/THF	Ox phot[g] (−75°C)	171
953	-,2-amino-3-carboxylato-	–NH_2, –COO	4.45[d] 5.14[e]	—	5.14	2.55	1.68	2.00369	w/pH = 7—9	Radiolysis[a]	18
954	-,2-amino-5-carboxylato-	–NH_2, –COO^-	4.51[d] 5.09[e]	<0.1	4.02	—	1.49	2.00389	w	Radiolysis[a]	18
955	-,2-amino-4-sulfonato-	–NH_2, –SO_3	5.08[d] 5.80[e]	0.49	—	2.30	1.27	2.00369	w/pH ≤11	Radiolysis[a]	18
956	-,3-amino-	–NH_2	(Data not reported)						w/H_2SO_4	Ox CeIV	19, 85
957	-,4-amino-	–NH_2	2.76	1.77	5.21[d] 5.53[e]	1.77	2.76	2.00377	w/pH = 3.5—11	Radiolysis[a]	18
958	-,4-amino-2-carboxylato-	–NH_2, –COO	—	1.62	5.07[d] 5.22[e]	1.44	2.94	2.00388	w/pH ~7	Radiolysis[a]	18
959	-,2,6-di-*t*-butyl-4-isocyano-	–CMe_3, –NC	—	0.88	5.06[d]	0.88	—		C_6H_6	Ox PbO_2	295
960	-,2-nitro-	–NO_2	1.25[d]	1.25	10.35	2.25	7.50	2.00502	w/H_2SO_4	Ox CeIV	241, 19
961	-,3-nitro-	–NO_2	7.40	0.35[d]	9.80	2.10	6.75	2.00503	w/H_2SO_4	Ox CeIV	241, 19
962	-,4-nitro-	–NO_2	7.20	2.40	2.40[d]	2.40	7.20	2.00545	w/H_2SO_4	Ox CeIV	241, 19, 173
963	-,2,6-di-*t*-butyl-4-nitro-	–CMe_3, –NO_2	—	2.14	2.14[d]	2.14	—		C_6H_6	Ox PbO_2	295

Table II.1 (continued)
PHENOXYL RADICALS

Neutral Species

No.	Radical	Substituent(s)	Splitting constants/Gauss					g-Factor	Solvent	Method	Ref.
			a_2	a_3	a_4	a_5	a_6				
	-,2-hydroxy-	(Hydroxy- and alkoxyphenoxyl radicals, see Tables II.2, II.3, II.4.1)									
964	-,2-fluoro-	–F	17.25[F]	1.40	10.00	2.15	5.80	2.00475	w/H_2SO_4	Ox CeIV	241, 173
965	-,2-chloro-	–Cl	1.05[i]	2.00	9.80	1.60	6.00	2.00535	w/H_2SO_4	Ox CeIV	241, 19
966	-,3-fluoro-	–F	5.00	5.80[F]	10.75	2.25	7.90	2.00456	w/H_2SO_4	Ox CeIV	241, 19
			7.9	5.8[F]	8.5	2.4	7.9	2.0052	Matrix[h]		240
967	-,3-chloro-	–Cl	6.20	0.25[i]	10.50	2.10	7.50	2.00450	w/H_2SO_4	Ox CeIV	241, 19
968	-,4-fluoro-	–F	6.50	1.45	27.50[F]	1.45	6.50	2.00503	w/H_2SO_4	Ox CeIV	241, 173
969	-,4-chloro-	–Cl	6.40	1.75	1.90[i]	1.75	6.40	2.00597	w/H_2SO_4	Ox CeIV	241, 19
970	-,2,3,5,6-tetrafluoro-	–F	16.7[F]	5.85[F]	10.0	5.85[F]	16.7[F]	2.0047	w	Ox CeIV	239
971	-,2,3,5,6-tetrafluoro-4-trifluoromethyl-	–F, $-CF_3$	18.0[F]	6.5[F]	9.5[c]	6.5[F]	18.0[F]	2.0048	w	Ox CeIV	239
972	-,pentafluoro-	–F	15.5[F]	5.2[F]	28.3[F]	5.2[F]	15.5[F]	2.0052	w	Ox CeIV	239
973	-,4-sulfo-	$-SO_3H$	7.25	2.1	—	2.1	7.25		w/H_2SO_4	Ox CeIV	252
974	-,4-sulfo-2-carboxy-	$-SO_3H$, –COOH	—	1.6	—	2.2	7.1		w/H_2SO_4	Ox CeIV	252
975	-,2,4-disulfo-	$-SO_3H$	—	1.75	—	2.25	6.3		w/H_2SO_4	Ox CeIV	252
976	-,2,4,6-trisulfo-	$-SO_3H$	—	2.0	—	2.0	—		w/H_2SO_4	Ox CeIV	252

[a] *In situ* radiolysis, aqueous solutions saturated with N_2O to convert e_{aq} into ·OH; see Reference 18.
[b] *In situ* photolysis of an α-diketone and the phenol.
[c] Splitting from three fluorine nuclei.
[d] Splitting from nitrogen nucleus of an NH_2, NO_2, or CN group.
[e] Splitting from amino protons.
[f] Proton splittings of nonequivalent NH_2 protons.
[g] Radicals produced by direct photooxidation in the microwave cavity; solution oxygen free.
[h] Radical obtained by X-irradiation of phenol in an adaman-tane matrix at room temperature.
[i] Splitting from chlorine nucleus.

Table II.2
2-HYDROXY- AND 2-ALKOXYPHENOXYL RADICALS

Neutral Species

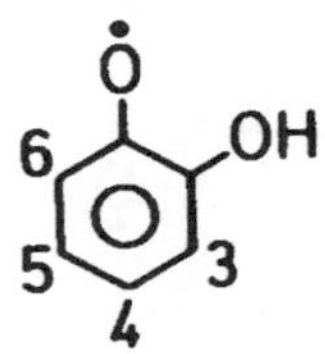

No.	Radical	Substituent(s)	Splitting constants/Gauss a_2	a_3	a_4	a_5	a_6	g-Factor	Solvent	Method	T [K]	Ref.
977	2-Hydroxyphenoxyl radical		1.51[a]	1.54	8.02	0.16	3.85		CCl_4	UV Phot[b]	295	169, 170, 244
			1.67[a]	1.69	8.18	0.16	3.96		CCl_4	UV Phot[b]	258	169
			1.70[a]	1.82	7.99	0.16	4.11		CCl_4	UV Phot[b]	171	169
			0.23[c]	1.82	7.90	0.16	4.04		CCl_4	UV Phot[b]	345	169
			0.24[c]	1.90	8.00	0.16	4.12		CCl_4	UV Phot[b]	291	169
978	-,4-methyl-	–Me	1.35[a]	1.84	9.68[f]	0.38	3.89		CH_2Cl_2/dioxane	UV Phot[b]	199	171
979	-,3,6-di-*t*-butyl-	$-CMe_3$	0.26[c]	u	7.86	u	u		PhMe	UV Phot	283	379
980	-,6-chloro-3,5-di-*t*-butyl-	–Cl, $-CMe_3$	1.5[a]	u	8.0	u	—		DEE	Phot	173	395
981	-,3-chloro-4,6-di-*t*-butyl-	–Cl, $-CMe_3$	1.5[a]	—	u	u	u		DEE	Phot	173	395
982	-,4-(2-carboxyvinyl)-	–CH=CHCOOH	u	1.75	4.8[1] 2.45[2]	0.15	3.5		w/H_2SO_4	Ox CeIV	RT	26
983	-,3-carboxylato-	–COO	0.57[a]	—	6.59	1.55	2.89	2.00438	w/pH = 3.5	Radiolysis	RT	18
984	-,3,5-dicarboxylato-	–COO	0.37[a]	—	5.94	—	3.26	2.00456	w/pH = 6.6	Radiolysis	RT	18
985	-,3,5-dicarboxy-	–COOH	0.25[a]	—	5.47	—	3.77	2.00463	w/pH = 3.0	Radiolysis	RT	18

Table II.2 (continued)
2-HYDROXY- AND 2-ALKOXYPHENOXYL RADICALS

Neutral Species

No.	Radical	Substituent(s)	Splitting constants/Gauss a_2	a_3	a_4	a_5	a_6	g-Factor	Solvent	Method	T [K]	Ref.
	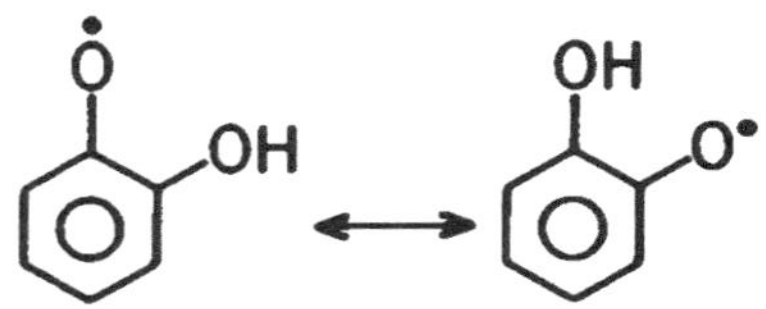											
986	2-Hydroxyphenoxyl radical		u	0.95[d]	4.1[d]	4.1	0.95	2.00428[c]	w/H_2SO_4	Ox CeIV	RT	84, 19, 85
987	-,3,6-di-*t*-butyl-	–CMe_3	1.62[a]	u	3.9[d]	3.9[d]	u		Heptane	Phot	273	174, 379
			0.26[c]	u	3.9[d]	3.9[d]	u		Heptane	Phot	273	174
988	-,6-chloro-3,5-di-*t*-butyl-	–Cl, –CMe_3	1.5[a]	u	3.3	u	—		DEE	Phot	373	395
	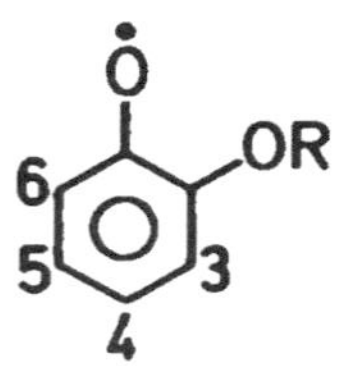											
989	2-Methoxyphenoxyl radical	(R = Me)	1.92[f]	1.92	8.70	u	4.30		w/H_2SO_4	Ox H_2O_2	RT	172, 244
			1.8[f]	1.9	8.5	0.0	4.3	2.00435	w/H_2SO_4	Ox CeIV	RT	85, 19
			1.8[f]	1.8	8.3	u	4.2		w/H_2SO_4	Ox CeIV	RT	173
			1.85[f]	1.85	8.5	0.25	4.30	2.00449	w/H_2SO_4	Ox CeIV	RT	256
990	-,3,6-di-*t*-butyl-	–CMe_3	1.37	u	8.65	1.0	u		C_6H_6	Ox Phot	263	331
			1.13	u	8.75	1.13	u		C_6H_6	Ox Phot	473	331

991	-,5-(2-carboxyvinyl)-	–CH:CHCOOH	1.9[f]	1.9	8.0	0.0[1] 0.0[2]	3.2		w/H_2SO_4	Ox CeIV	RT	26
992	-,4-formyl-	–CHO	2.2[f]	2.2	u	u	4.3		w/H_2SO_4	Ox CeIV	RT	173
993	-,5-formyl-	–CHO	1.8[f]	1.8	7.8	u	5.4		w/H_2SO_4	Ox CeIV	RT	173
994	2-Ethoxyphenoxyl radical	(R = Et)										
995	-,3,6-di-*t*-butyl-	$–CMe_3$	0.12[g]	u	14	2.2	0.5[2]		EtOH	Phot	283	379
996	2-Isopropoxyphenoxyl radical	($R = CHMe_2$)										
997	-,3,6-di-*t*-butyl-	$–CMe_3$	0.12[h]	u	14	2.2	0.5[2]		Pr^iOH	Phot	283	379
998	2-Phenoxyphenoxyl radical	(R = Ph)	—	1.90	7.65	1.90	4.40		w/alk	Ox $K_3Fe(CN)_6$		57

[a] Splitting from proton of OH group.
[b] Radicals produced by direct photodissociation in a flow system. Same constants observed in CH_2Cl_2 solution.
[c] Deuterium splitting; hydroxyl proton exchanged by deuteron.
[d] Average values; rapid proton exchange assumed.
[e] g-Factor from Reference 85.
[f] Splitting from protons of OCH_3 group.
[g] Splitting from CH_2-protons of OEt-group. Assignment suggested by the editor.
[h] Splitting from CH-proton of OPr^i-group. Assignment suggested by the editor.

Table II.3
3-HYDROXY- AND 3-ALKOXYPHENOXYL RADICALS

Neutral Species

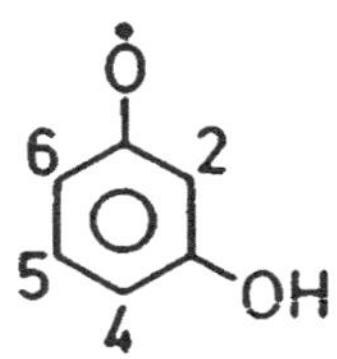

No.	Radical	Substituent(s)	Splitting constants/Gauss a_2	a_3	a_4	a_5	a_6	g-Factor	Solvent	Method	Ref.
999	3-Hydroxyphenoxyl radical		3.8	—	9.8	2.25	9.8		w/H_2SO_4	Ox CeIV	27
			3.75	—	10.0	2.3	10.0		0.5 *M* H_2SO_4	Ox CeIV	83, 19
			3.9	—	11.3[a]	2.3	9.7[a]	2.00435	w/H_2SO_4	Ox CeIV	85, 84
1000	-,2-methyl-	–Me	5.5[l]	—	9.0	1.9	9.0		w/H_2SO_4	Ox CeIV	83
			5.3[l]	—	8.9	1.85	8.9		w/H_2SO_4	Ox CeIV	27
1001	-,4-methyl-	–Me	3.20	—	12.30[l]	2.05	9.20	2.00406	w/H_2SO_4	Ox CeIV	241
1002	-,4-ethyl-	–Et	3.2	—	10.0[l]	2.05	10.0		w/H_2SO_4	Ox CeIV	27
1003	-,4-hexyl-	–$(CH_2)_5Me$	3.2	—	8.6[l]	2.0	8.6		w/H_2SO_4	Ox CeIV	27
1004	-,4-benzyl-	–CH_2Ph	3.2	—	8.15[l]	2.1	9.15		w/H_2SO_4	Ox CeIV	27
1005	-,5-methyl-	–Me	4.0	—	9.55	2.1[l]	9.55		w/H_2SO_4	Ox CeIV	27
			4.2	—	9.75	2.05[l]	9.75	2.00424	w/H_2SO_4	Ox CeIV	256
1006	-,2-acetyl-	–COMe	—	0.85	11.5	2.6	10.7	2.00421	w/H_2SO_4	Ox CeIV	256
1007	-,2-carboxy-	–COOH	—	—	10.65	2.5	10.65		w/H_2SO_4	Ox CeIV	27, 83
1008	-,4-carboxy-	–COOH	4.6	—	—	1.9	9.95		w/H_2SO_4	Ox CeIV	27
1009	-,5-carboxy-	–COOH	3.95	—	9.85	—	9.85		w/H_2SO_4	Ox CeIV	27
1010	-,2-nitro-	–NO_2	0.5[b]	—	10.7	2.6	10.7		w/H_2SO_4	Ox CeIV	27, 83

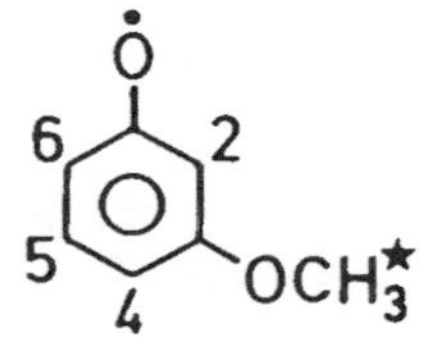

1011 3-Methoxyphenoxyl radical	3.5	0.6*	11.4	2.3	9.0	2.00429	w/H_2SO_4	Ox CeIV		85, 19

[a] Average values obtained only; values given calculated from trends in Reference 19.
[b] Splitting from nitrogen nucleus.

Table II.4.1
4-HYDROXY- AND 4-ALKOXYPHENOXYL RADICALS

Neutral Species

No.	Radical	Substituent(s)	Splitting constants/Gauss a_2	a_3	a_4	a_5	a_6	g-Factor	Solvent	Method	T [K]	Ref.
	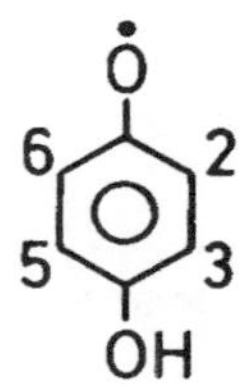											
1012	[1]4-Hydroxyphenoxyl radical		4.82	0.20	1.77[a]	0.20	4.82		$(CH_2OH)_2$	CIDEP	RT	38
			5.09	0.29	1.86[a]	0.29	5.09		EtOH	Phot	292	137, 199
			5.11	0.31	1.84[a]	0.31	5.11	2.00427[b]	EtOH	Phot	299	137, 85
			5.21	0.34	1.81[a]	0.34	5.21		EtOH	Phot	309	137

Table II.4.1 (continued)
4-HYDROXY- AND 4-ALKOXYPHENOXYL RADICALS

Neutral Species

No.	Radical	Substituent(s)	Splitting constants/Gauss a_2	a_3	a_4	a_5	a_6	g-Factor	Solvent	Method	T [K]	Ref.
			5.05	0.28	1.73[a]	0.28	5.05		EtOH	Phot	RT	175
			5.05	0.27	1.78[a]	0.27	5.05		MeOH	Phot	RT	175
			5.13	0.31	1.79[a]	0.31	5.13		Pr^iOH	Phot	RT	175, 198
			5.2	0.35	1.8[a]	0.35	5.2		$C_6H_6/PhCl$	Phot[c]	303	176
			5.24	0.38	1.76[a]	0.38	5.24		$C_5H^i_{11}OH$	Phot	RT	175
			5.54	0.68	1.63[a]	0.68	5.54		Bu_3PO_4	Phot	RT	175
			5.53	0.68	1.61[a]	0.68	5.53		DME	Phot	RT	175
			5.56	0.70	1.54[a]	0.70	5.56		THF	Phot	RT	175
			5.56	0.74	1.54[a]	0.74	5.56		Me_2CO	Phot	RT	175
			5.71	0.80	1.47[a]	0.80	5.71		Dioxane	Phot	RT	175
			5.64	0.85	1.40[a]	0.85	5.64	2.0046	CH_2Cl_2	Phot	307	137, 242, 346
			5.22[d]	0.20	2.10[a]	0.20	4.84[d]		EtOH	Phot	220	179
			5.62[d]	0.61	1.85[a]	0.61	5.32[d]		DME	Phot	200	182
			5.69[d]	0.57	1.90[a]	0.57	5.34[d]		THF	Phot	203	180
1013	-,2-methyl-	–Me	5.15[f]	0.68	1.77[a]	u	5.15		Pr^iOH	Phot	RT	346
			5.00[f]	0.71	1.72[a]	Small	5.00		EtOH	Phot	297	179, 137
			3.85[f]	0.28	u	1.24	3.85	2.00432	w/H_2SO_4	Ox CeIV	296	241
1014	-,2-*t*-butyl-	$–CMe_3$	—	0.5	2.0[a]	0.2	5.0		AcOH/ PhOH	Phot[c]	261	184
			—	0.52	1.8[a]	0.2	5.24		10% Pr^iOH/ PhMe	Phot[c]	253	184
1015	-,3-*t*-butyl-	$–CMe_3$	4.53	—	1.8[a]	0.2	5.24		10% Pr^iOH	Phot[c]	253	184
			4.3	—	2.0[a]	0.2	4.6		AcOH/ PhOH	Phot[c]	261	184
1016	-,2,3-dimethyl-	–Me	4.95[f]	1.10[f]	1.60[a]	0.48	5.68		DME	Phot	293	180

1017	-,2,5-dimethyl-	–Me	5.18[l]	0.71	1.70[a]	0.20[l]	4.32		EtOH	Phot	291	179, 137
			5.90[l]	1.10	1.37[a]	0.70[l]	4.93		DME	Phot	293	180
1018	-,2,5-di-*t*-butyl-	$-CMe_3$	—	0.5	1.8[a]	—	4.6		AcOH/ PhOH	Phot[c]	261	184
1019	-,2,6-dimethyl-	–Me	4.77[l]	0.42	1.65[a]	0.42	4.77[l]		EtOH	Phot	296	179
			4.70[l]	0.42	1.54[a]	0.42	4.70[l]		EtOH	Phot	RT	183
			4.70[l]	0.42	1.58[a]	0.42	4.70[l]		MeOH	Phot	RT	183
			4.78[l]	0.49	1.59[a]	0.49	4.78[l]		Pr^iOH	Phot	RT	183
			4.84[l]	0.53	1.50[a]	0.53	4.84[l]		$C_5H^t_{11}OH$	Phot	RT	183, 31
			5.05[l]	0.74	1.44[a]	0.74	5.05[l]		Bu_3PO_4	Phot	RT	183
			5.19[l]	0.81	1.37[a]	0.81	5.19[l]		DME	Phot	RT	183, 31
			5.12[l]	0.81	1.33[a]	0.81	5.12[l]		THF	Phot	RT	183
			5.17[l,d]	0.61	1.70[a]	0.61	4.78[l,d]		THF	Phot	203	180
			5.34[l]	0.84	1.30[a]	0.84	5.34[l]		Dioxane	Phot	RT	183
1020	-,2,6-diisopropyl-	$-CHMe_2$	3.1[l]	1.2	u	1.2	3.1[l]		C_6H_6	Ox P_bO_2	RT	243
1021	-,2,6-di-*t*-butyl-	$-CMe_3$	—	0.54	1.6[a]	0.54	—		AcOH/ PhOH	Phot[c]	261	184
			—	0.68	1.36[a]	0.68	—		$C_5H^t_{11}OH$	Phot	RT	183
			—	0.85	1.34[a]	0.85	—		$Bu^t_2O_2$	Phot	RT	57
			—	0.82	1.65[a]	0.82	—		$Bu^t_2O_2$	Phot	258	188
			—	1.75	u	1.75	—		C_6H_6	Ox PbO_2	RT	243
1022	-,tetramethyl-	–Me	4.80[l]	0.62[l]	0.62[a]	0.62[l]	4.80[l]		MeOH	Phot	RT	185
			4.80[l]	0.68[l]	0.38[a]	0.68[l]	4.80[l]		EtOH	Phot	RT	185
			4.87[l]	0.67[l]	0.41[a]	0.67[l]	4.87[l]		EtOH	Phot	295	179, 186
			4.83[l]	0.68[l]	0.22[a]	0.68[l]	4.83[l]		Pr^iOH	Phot	RT	185
			4.9[l]	0.8[l]	u	0.8[l]	4.9[l]	2.0043	Bu^tOH	Phot[f]	~RT	187
			4.91[l]	0.79[l]	0.10[a]	0.79[l]	4.91[l]		$C_5H^t_{11}OH$	Phot	RT	185
			5.04[l]	0.96[l]	0.10[a]	0.96[l]	5.04[l]		Bu_3PO_4	Phot	RT	185
			5.20[l]	0.99[l]	0.42[a]	0.99[l]	5.20[l]		Me_2CO	Phot	RT	185
			5.20[l]	1.02[l]	0.63[a]	1.02[l]	5.20[l]		Sulfolane	Phot	RT	185
			5.24[l]	1.04[l]	0.27[a]	1.04[l]	5.24[l]		THF	Phot	RT	185
			5.08[l]	0.94[l]	0.80[a]	0.94[l]	5.08[l]		THF	Phot	203	180
			5.38[l]	1.08[l]	0.38[a]	1.08[l]	5.38[l]		Dioxane	Phot	RT	185
			5.49[l]	1.15[l]	0.38[a]	1.15[l]	5.49		C_6H_6	Phot	RT	185
1023	-,2-carboxylato-	$-COO^-$	—	1.35	0.81[a]	2.51	4.90	2.0055	w/pH = 3	Radiolysis[g]	298	18

Table II.4.1 (continued)
4-HYDROXY- AND 4-ALKOXYPHENOXYL RADICALS

Neutral Species

No.	Radical	Substituent(s)	Splitting constants/Gauss a_2	a_3	a_4	a_5	a_6	g-Factor	Solvent	Method	T [K]	Ref.
1024	-,tetrafluoro-	–F	12.35[F]	3.17[F]	1.07[a]	3.17[F]	12.35[F]		Me_2CO	Phot	RT	68
			12.46[F]	3.24[F]	1.08[a]	3.24[F]	12.46[F]		THF	Phot	RT	68
			12.64[F]	3.20[F]	1.17[a]	3.20[F]	12.64[F]		CH_2Cl_2	Phot	RT	68
			12.89[F]	3.51[F]	1.11[a]	3.51[F]	12.89[F]		Dioxane	Phot	Rt	68
			13.06[F]	3.51[F]	1.27[a]	3.51[F]	13.06[F]		$CHCl_3$	Phot	RT	68
1025	-,tetrachloro-	–Cl	0.8[h]	u	0.8[a]	u	0.8[h]	2.0057	Dioxane	Flash phot	298	245
			—	—	—	—	—	2.0056[i]	Pr^iOH	Flash phot	298	245
			—	—	—	—	—	[j]	Pr^iOH	Phot	293	246
			—	—	—	—	—	[j]	EtOH	Phot	293	246
			—	—	—	—	—	[j]	MeOH	Phot	293	246
			—	—	—	—	—	2.0055	EtOH or MeOH	Flash phot	295	247

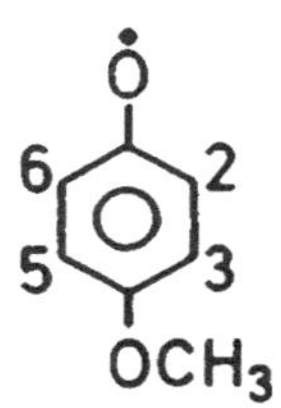

No.	Radical	Substituent(s)	a_2	a_3	a_4	a_5	a_6	g-Factor	Solvent	Method	T [K]	Ref.
1026	4-Methoxyphenoxyl radical		5.05	0.35	2.0[k]	0.35	5.05	2.00429	w/H_2SO_4	Ox CeIV	RT	256
			5.05	0.2	2.1[k]	0.2	5.05	2.00445	w/H_2SO_4	Ox CeIV	298	85, 173
			5.16	u	2.10[k]	u	5.16		w/H_2SO_4	Ox H_2O_2	298	172
			5.38	0.57	1.83[k]	0.57	5.38		C_6H_6	Phot[c]	298	176
			5.7	0.8	1.7[k]	0.8	5.7		C_6H_6	Ox PbO_2	RT	177

1027	-,2-*t*-butyl-	–CMe$_3$	—	0.6	1.9[k]	u	5.4	w/H_2SO_4	Ox CeIV	RT	173
			—	0.9	1.7[k]	u	6.2	C_6H_6	Ox PbO_2	RT	177
1028	-,3-*t*-butyl-	–CMe$_3$	3.9	—	2.1[k]	u	5.8	w/H_2SO_4	Ox CeIV	RT	173
1029	-,2-methyl-5-isopropyl-	–CHMe$_2$ –Me	7.65[l]	1.53	1.53[k]	1.53[l]	1.53	C_6H_6	Ox PbO_2	298	189
			7.65[l]	0.7	1.53[k]	1.53[l]	2.2	C_6Cl_2	Ox PbO_2	298	189
1030	-,2-isopropyl-5-methyl-	–CHMe$_2$, –Me	3.20[l]	u	1.60[k]	1.60[l]	0.80	C_6H_6	Ox PbO_2	298	189
1031	-,2,5-di-*t*-butyl-	–CMe$_3$	—	u	2.0[k]	—	4.0	w/H_2SO_4	Ox CeIV	RT	173
			0.1[2]	1.08	1.80[k]	—	4.65	C_6H_6	Ox[f]	RT	190
1032	-,2,6-dimethyl-	–Me	4.7[l]	u	2.0[k]	u	4.7[l]	w/H_2SO_4	Ox CeIV	RT	173
			5.5[l]	~0.5	1.5[k]	~0.5	5.5[l]	C_6H_6	Ox Ag_2O	RT	177
			5.46[l]	1.00	1.47[k]	1.00	5.46[l]	PhMe	Ox Ag_2O	293	191, 192
			6.15[l,d]	u	u	u	4.52[l,d]	PhMe	Ox Ag_2O	203	191, 192
1033	-,2,6-diethyl-	–Et	4.57[l]	0.93	1.52[k]	0.93	4.57[l]	PhMe	Ox Ag_2O	273	191
			5.01[l,d]	u	u	u	4.1[l,d]	PhMe	Ox Ag_2O	203	191
1034	-,2,6-diisopropyl-	–CHMe$_2$	2.70[l]	0.82	1.40[k]	0.82	2.70[l]	C_6H_6/KOH	Ox $K_3Fe(CN)_6$	RT	193
			3.05[l]	0.85	1.55[k]	0.85	3.05[l]	PhMe	Ox Ag_2O	273	191
			3.5[l,d]	u	u	u	2.9[l,d]	PhMe	Ox Ag_2O	203	191
1035	-,6-*t*-butyl-2-methyl-	–CMe$_3$, –Me	5.80[l]	1.00	1.52[k]	0.84	—	C_6H_6/KOH	Ox $K_3Fe(CN)_6$	RT	193
1036	-,6-*t*-butyl-2-isopropyl-	–CMe$_3$, –CHMe$_2$	3.20[l]	1.06	1.58[k]	0.86	—	C_6H_6/KOH	Ox $K_3Fe(CN)_6$	RT	193
1037	-,2,6-di-*t*-butyl-	–CMe$_3$	0.07[2]	1.0	1.50[k]	1.0	0.07[2]	CCl_4	Ox PbO_2	298	195
			0.06[2]	1.01	1.51[k]	1.01	0.06[2]	Heptane	Ox PbO_2	223	195
1038	-,tetramethyl-	–Me	6.17[l]	1.58[l]	0.00[k]	1.58[l]	6.17	PhMe	Ox Ag_2O	293	191

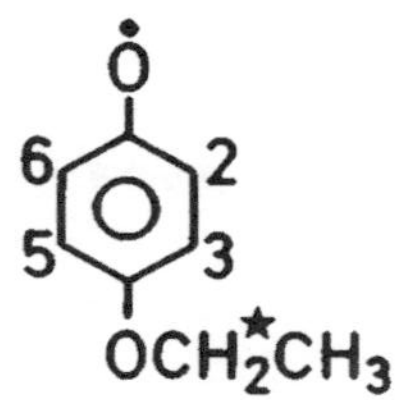

1039	4-Ethoxyphenoxyl radical										
1040	-,tetramethyl-	–Me	6.11[l]	1.55[l]	0.28*	1.55[l]	6.11[l]	PhMe	Ox Ag_2O	293	264

Table II.4.1 (continued)
4-HYDROXY- AND 4-ALKOXYPHENOXYL RADICALS

Neutral Species

No.	Radical	Substituent(s)	Splitting constants/Gauss a_2	a_3	a_4	a_5	a_6	g-Factor	Solvent	Method	T [K]	Ref.
1041	-,2,6-di-*t*-butyl-	–CMe_3	0.08[2]	0.96	1.96*	0.96	0.08[2]		Heptane	Ox PbO_2	193	195
1042	4-*t*-Butoxyphenoxyl radical											
1043	-,6-*t*-butyl-2-methyl-	–CMe_3, –Me	5.90[1]	1.08	u	0.83	—		C_6H_6/KOH	Ox $K_3Fe(CN)_6$	RT	193
1044	-,2,6-diisopropyl-	–$CHMe_2$	3.00[1]	1.00	u	1.00	3.00[1]		C_6H_6/KOH	Ox $K_3Fe(CN)_6$	RT	193
1045	-,6-*t*-butyl-2-isopropyl-	–CMe_3, –$CHMe_2$	3.10[1]	1.00	u	0.84	—		C_6H_6/KOH	Ox $K_3Fe(CN)_6$	RT	193
1046	-,2,6-di-*t*-butyl-	–CMe_3	—	1.0	u	1.0	—		C_6H_6/alk	Ox $K_3Fe(CN)_6$	RT	260

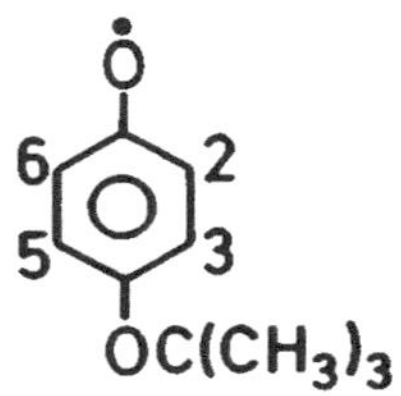

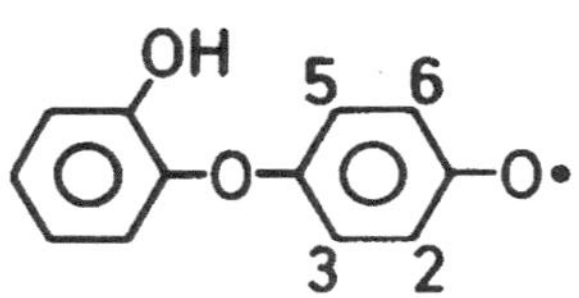

1047	4-(2-Hydroxyphenoxy)phenoxyl radical		5.25	0.65	—	0.65	5.25		w/alk	OxCeIV	RT	57
			4.75	u	—	u	4.75		w/alk	$OxK_3Fe(CN)_6$	RT	57

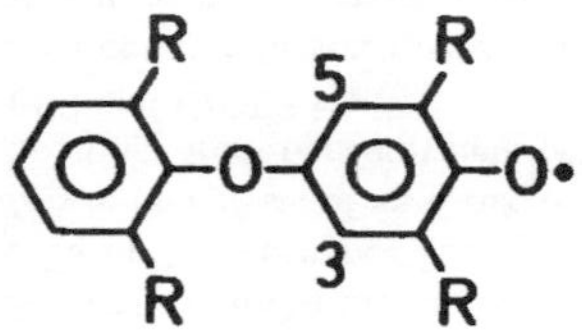

1048	[R = methyl]		5.11[l]	1.10	—	1.10	6.21[l]	2.0062	C_6H_6	$OxAg_2O$	RT	253
1049	[R = ethyl]		4.65[l]	1.15	—	1.15	4.65[l]	2.0061	C_6H_6	$OxAg_2O$	RT	253
1050	[R = isopropyl]		3.06[l]	1.05	—	1.05	3.06[l]	2.0062	C_6H_6	$OxAg_2O$	RT	253

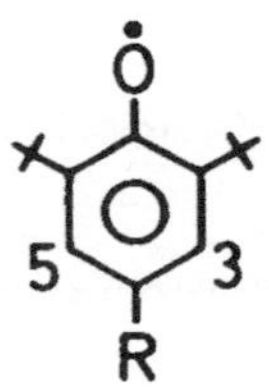

4-R-2,6-di-*t*-butylphenoxyl radical

	R =	R =										
1051	pentachlorophenoxy	$-OC_6Cl_5$	—	1.33	—	1.33	—		C_6H_6	Ox PbO_2	RT	260
1052	benzoyloxy	–OCOPh	—	1.65	—	1.65	—		C_6H_6	Ox PbO_2	RT	260
1053	acetoxy	–OCOMe	—	1.75	—	1.75	—		C_6H_6	Ox PbO_2	RT	260
1054	chloroformyloxy	–OCOCl	—	1.8	—	1.8	—		C_6H_6	Ox PbO_2	RT	260
1055	dibutylphosphinyloxy	$-OP(=O)(CH_2CH_2CH_2Me)_2$	—	1.46	5.50[m]	1.46	—		C_6H_6	Ox PbO_2	RT	260

Table II.4.1 (continued)
4-HYDROXY- AND 4-ALKOXYPHENOXYL RADICALS

Neutral Species

No.	Radical	Substituent(s)	Splitting constants/Gauss a_2	a_3	a_4	a_5	a_6	g-Factor	Solvent	Method	T [K]	Ref.
1056	diphenylphosphinyloxy	$-OP(=O)Ph_2$	—	1.55	6.66[m]	1.55	—		C_6H_6	Ox PbO_2	RT	260
1057	diethoxyphosphinyloxy	$-OP(=O)(OCH_2Me)_2$	—	1.56	7.62[m]	1.56	—		C_6H_6	Ox PbO_2	RT	260
1058	dibutoxyphosphinyloxy	$-OP(=O)(OCH_2CH_2CH_2Me)_2$	—	1.60	7.62[m]	1.60	—		C_6H_6	Ox PbO_2	RT	260
1059	diphenoxyphosphinyloxy	$-OP(=O)(OPh)_2$	—	1.68	10.1[m]	1.68	—		C_6H_6	OxPbO_2	RT	260
1060	mesyloxy	$-OSO_2Me$	—	1.7	—	1.7	—		C_6H_6	Ox PbO_2	RT	260
1061	diphenylthiophosphinato	$-SP(=O)Ph_2$	—	1.79	11.26[m]	1.79	—		C_6H_6	Ox PbO_2	RT	260

[a] Splitting from OH proton.
[b] g-Factor for Reference 85.
[d] *In situ* photolysis of an α-diketone and the phenol.
[d] Splittings from nonequivalent protons at the lowered temperature.
[e] EPR and CIDEP study.
[f] Photoreaction between benzophenone, duroquinone, and a substituted carboxylic acid.
[g] *In situ* radiolysis, aqueous solutions saturated with N_2O to convert e_{aq}^- into ·OH; see Reference 18.
[h] Splitting from chlorine (average of ^{35}Cl and ^{37}Cl). Peak-to-peak line width 0.9 G.
[i] Single sharp line of 0.58-G width.

[j] Single line of neutral radical observed with g-factor larger by 0.0008 than that of the corresponding chloranil anion radical. The peak-to-peak line width amounts to 0.63 (0.49), 0.62 (0.43), and 0.55 G (0.43) in MeOH, EtOH, and Pr^iOH, respectively, with values in brackets from the anion radical.

[k] Splitting from the protons of the methoxyl group.

[l] Radical obtained from a parent aryloxy-dienone in oxygen-free solution.

[m] Splitting from ^{31}P.

Table II.4.2
CHROMANYLOXYL RADICALS[a]

No.	Parent compound	Observed radical	R	R′	R″	R‴	a_5^R	$a_7^{R'}$	$a_8^{R''}$	$a_4^{CH_2}$	g-Factor	Ref.[b]
							Splitting constants/Gauss					
1062	α-Tocopherol[c]	α-Chromanyloxyl	Me	Me	Me	A[d]	6.07[l]	4.55[l]	0.98[l]	1.52[l]	2.0046	445, 58, 216
							5.98[l]	4.57[l]	0.94[l]	1.47[l]		448
							6.02[l]	4.58[l]	0.94[l]	1.48[l]		446
1063	2,2,5,7,8-Pentamethyl-chroman-6-ol	2,2,5,7,8-Pentamethyl-chromanyl-6-oxyl	Me	Me	Me	Me	6.07[l]	4.55[l]	0.98[l]	1.52[l]	2.0046	445
							5.94[l]	4.51[l]	0.96[l]	1.46[l]		448
1064	2-(2-Carboxylatoethyl)-2,5,7,8-tetramethylchroman-6-ol	2-(2-Carboxylatoethyl)-2,5,7,8-tetramethylchromanyl-6-oxyl	Me	Me	Me	B[c]	5.96[l]	4.50[l]	0.96[l]	1.47[l]		449
1065	5,7-Dimethyltocol	5,7-Dimethyltocoxyl	Me	Me	H	A	5.91[l]	4.64[l]	0.90	1.40[l]	2.0046	445
1066	2,2,5,7-Tetramethylchroman-6-ol	2,2,5,7-Tetramethyl-chromanyl-6-oxyl	Me	Me	H	Me	5.91[l]	4.64[l]	0.90	1.40[l]	2.0046	445

Table II.4.2 (continued)
CHROMANYLOXYL RADICALS[a]

No.	Parent compound	Observed radical	R	R′	R″	R‴	Splitting constants/Gauss a_5^R	$a_7^{R'}$	$a_8^{R''}$	$a_4^{CH_2}$	g-Factor	Ref.[b]
1067	β-Tocopherol	β-Chromanyloxyl	Me	H	Me	A	6.40[1]	4.60	0.91[1]	1.74[1]	2.0047	445
							6.20[1]	4.50	0.85[1]	1.70[1]		449
							6.25[1]	4.66	0.85[1]	1.68[1]		446
1068	2,2,5,8-Tetramethylchroman-6-ol	2,2,5,8-Tetramethyl-chromanyl-6-oxyl	Me	H	Me	Me	6.40[1]	4.60	0.91[1]	1.74[1]	2.0047	445
							6.24[1]	4.53	0.86[1]	1.71[1]		449
1069	γ-Tocopherol	γ-Chromanyloxyl	H	Me	Me	A	6.03	4.82[1]	1.12[1]	1.30[1]	2.0047	445, 216
							5.93	4.77[1]	1.17[1]	1.17[1]		449
							5.95	4.77[1]	1.16[1]	1.16[1]		446
1070	2,2,7,8-Tetramethylchroman-6-ol	2,2,7,8-Tetramethyl-chromanyl-6-oxyl	H	Me	Me	Me	6.03	4.82[1]	1.12[1]	1.30[1]	2.0047	445
							5.94	4.72[1]	1.18[1]	1.18[1]		449
1071	δ-Tocopherol	δ-Chromanyloxyl	H	H	Me	A	6.19	4.82	1.03[1]	1.45[1]	2.0049	445
							6.21	4.72	1.04[1]	1.42[1]		449
							5.92	4.60	1.11[1]	1.11[1]		446
1072	Tocol	Tocoxyl	H	H	H	A	5.92	5.13	0.79	1.38[1]	2.0049	445
							5.78	5.08	0.66	1.32[1]		449
1073	2,2-Dimethylchroman-6-ol	2,2-Dimethylchromanyl-6-oxyl	H	H	H	Me	5.92	5.13	0.79	1.38[1]	2.0049	445
							5.81	5.10	0.66	1.33[1]		449

[a] Commonly named chromanoxyl, not conforming with rules of nomenclature; also widely used in tocopheroxyl.

[b] Radical generated by treatment of phenol with 2,2-diphenyl-1-picrylhydrazyl (DPPH) in degassed solution of toluene,[445] by treatment with PbO_2 in toluene,[448,449] and by treatment with DPPH or di-*t*-butyldiperoxyoxalate in benzene.[446] All studies performed at room temperature.

[c] Vitamin E (mainly α-tocopherol).

[d] A = $(CH_2CH_2CH_2CH(Me)-)_3Me$.

[e] B = CH_2CH_2COOH.

Table II.5
PYROGALLOL-DERIVED PHENOXYL RADICALS

Neutral Species

No.	Radical	Substituent(s)	Splitting constants/Gauss $a_2 = a_6$	a_3	a_4	a_5	g-Factor	Solvent	Method	Ref.
	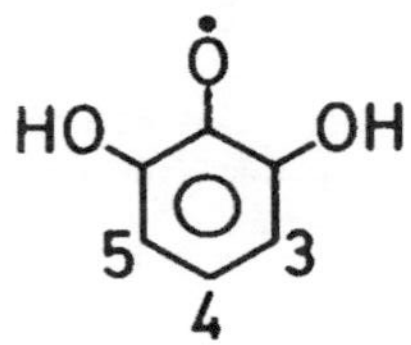									
1074	2,6-Dihydroxyphenoxyl radical		0.85[a]	1.50	7.99	1.50		w/H_2SO_4	Ox H_2O_2/TiIII	86, 83
			0.75[a]	1.5	7.6	2.0[b]		w/H_2SO_4	Ox CeIV	19
			1.07[a]	1.53	7.62	1.53	2.00426	CCl_4/dioxane[c]	Phot	170, 169
			0.17[d]	1.56	7.74	1.56		CCl_4/dioxane[c]	Phot	169
1075	-,3-carboxy-	–COOH	u[a]	—	8.1	1.3		w/H_2SO_4	Ox CeIV	83
1076	-,4-carboxy-	–COOH	u[a]	1.70	—	1.70		w/H_2SO_4	Ox CeIV	83
	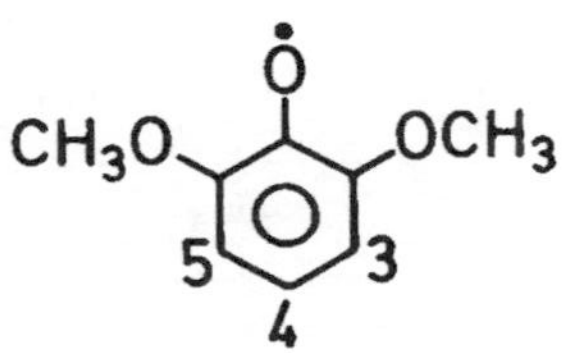									
1077	2,6-Dimethoxyphenoxyl radical		1.4[e]	1.5	7.9	1.5	2.00444	w/H_2SO_4	Ox CeIV	256, 173
			1.35[e]	1.35	8.4	1.35		C_6H_6	Ox Ag_2O/ PbO_2	177
1078	-,4-propyl-	–CH_2CH_2Me	1.35[e]	1.35	6.25[g]	1.35		w	Ox HRP[f]	212

Table II.5 (continued)
PYROGALLOL-DERIVED PHENOXYL RADICALS

Neutral Species

No.	Radical	Substituent(s)	Splitting constants/Gauss: $a_2 = a_6$	a_3	a_4	a_5	g-Factor	Solvent	Method	Ref.
1079	-,4-*t*-butyl-	–CMe_3	1.19[c]	1.19	0.30[2]	1.19		C_6H_6	Ox	213, 190
			1.18[c]	u[g]	0.28[2]	u[g]		C_6H_6	Ox	213
										213
1080	-,4-hydroxymethyl-	–CH_2OH	1.35[c]	1.35	9.87[1]	1.35		w	Ox HRP[i]	212
1081	-,4-(1-hydroxyethyl)-	–CH(OH)Me	1.35[c]	1.35	5.45[1]	1.35		w	Ox HRP[i]	212
					0.67[2]					
1082	-,4-(1-hydroxypropyl)-	–$CH(OH)CH_2Me$	1.35[c]	1.35	5.45[1]	1.35		w	Ox HRP[i]	212
			0.30[c]	0.30	<0.2[2]	0.30		w/alk	Ox air[h]	103
					3.35[1]					
1083	-,4-formyl-	–CHO	1.45[c]	1.45	0.48[1]	1.45		w	Ox HRP[i]	212
1084	-,4-acetyl-	–COMe	1.42[c]	1.42	0.30[2]	1.42		w	Ox HRP[i]	212
1085	-,4-propionyl-	–COEt	1.42[c]	1.42	0.66[2]	1.42		w	Ox HRP[i]	212
1086	-,4-(4-methoxybenzoyl)-	–C(=O)–Ph–OMe(*p*)	1.30[c]	1.78	—	1.78		C_6H_6	Ox Ag_2O	57
1087	-,4-carboxy-	–COOH	1.65[c]	1.65	—	1.65		w/H_2SO_4	Ox CeIV	173

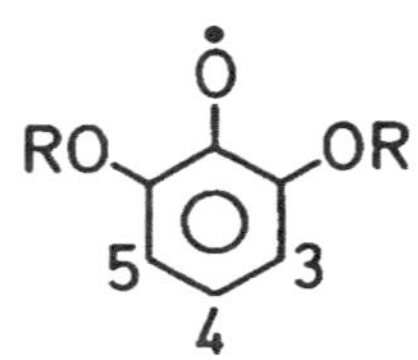

1088 2,6-Diethoxyphenoxyl radical [R = CH_2Me]

1089	-,4-*t*-butyl-	–CMe_3	1.45*	1.45	0.29[c]	1.45		C_6H_6	Ox	213
			1.42*	u[g]	0.28[c]	u[g]		C_6H_6	Ox	213

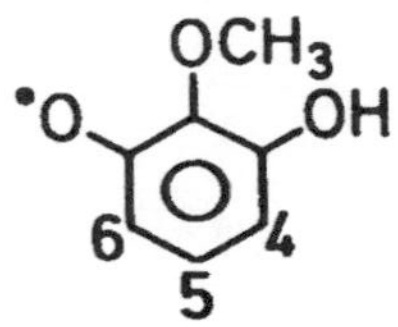

			a_2	a_4	a_5	a_6				
1090	3-Hydroxy-2-methoxy-phenoxyl radical									
1091	-,5-carboxy-	–COOH	1.65[i]	6.4	—	6.4	2.00470	w/H_2SO_4	Ox CeIV	256

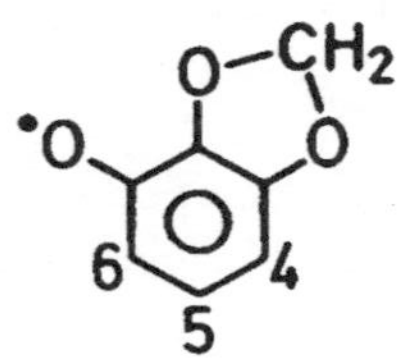

1092	2,3-Methylenedioxyphenoxyl radical		3.60[j]	6.50	0.45	2.50	2.00463	w/neut	Ox CeIV	249, 241
1093	-,4-acetyl-	–COMe	5.60[j]	—	1.70	5.05	2.00481	w	Ox CeIV	241

[a] Splittings from equivalent protons of the OH groups.
[b] This value (Reference 19) seems a misprint and should read ~1.5.
[c] CCl_4/dioxane (0.1 mol l^{-1} dioxane).
[d] Deuterium splitting, OH groups exchanged by ODs.
[e] Splitting from six equivalent methoxy group protons.
[f] Radical generated by mixing equal volumes of 0.01 *M* phenol in water and 0.03 *M* H_2O_2 containing 1 μ*M* horseradish peroxidase (HRP).
[g] C-3 and C-5 substituted with deuterium atoms.
[h] Weak signal obtained in 0.1 *M* NaOH.
[i] Splitting from three methoxyl group protons. No splitting reported from the OH proton.
[j] Splitting from two equivalent methylene protons.

Table II.6
PYROGALLOQUINONE RADICALS

Neutral Species

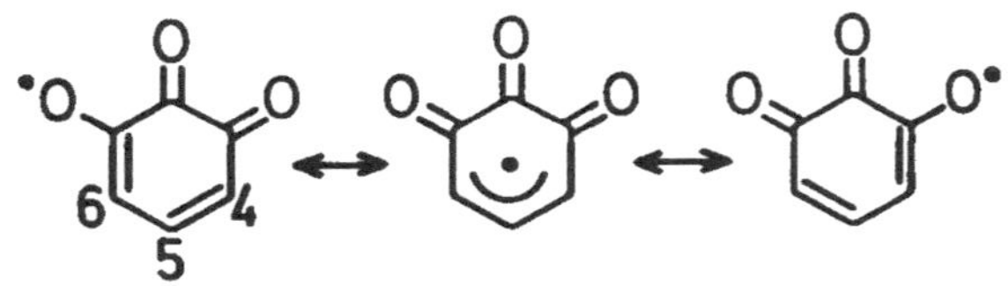

No.	Radical	Substituent(s)	Splitting constants/Gauss a_4	a_5	a_6	Solvent	Method	Ref.
1094	Pyrogalloquinone radical		12.3	3.25	12.3	w/H_2SO_4	Ox CeIV	83, 249
1095	-,4,6-di-*t*-butyl-	–CMe_3	0.03[2,a]	3.0	0.03[2,a]	w/H_2SO_4	Ox CeIV	83
1096	-,4-carboxy-	–COOH	—	3.2	12.2	w/H_2SO_4	Ox CeIV	83
1097	-,5-carboxy-	–COOH	12.3	–	12.3	w/H_2SO_4	Ox CeIV	83

[a] This splitting probably should have read 0.3 G.

Table II.7
PHENOXYL RADICALS WITH THREE KEY OXYGENS [1,2,4]

Neutral Species

No.	Radical	Substituent(s)	Splitting constants/Gauss a_2	a_3	a_4	a_5	a_6	g-Factor	Solvent	Method	Ref.
	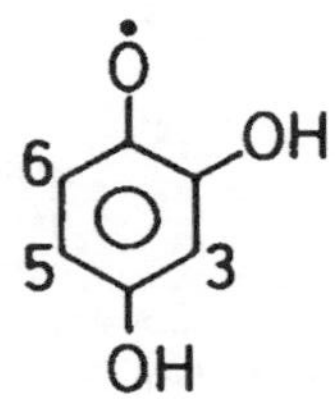										
1098	2,4-Dihydroxyphenoxyl radical		u	1.2	u	2.2	2.95	2.00435	H_2SO_4,$4H_2O$	Ox CeIV	256
	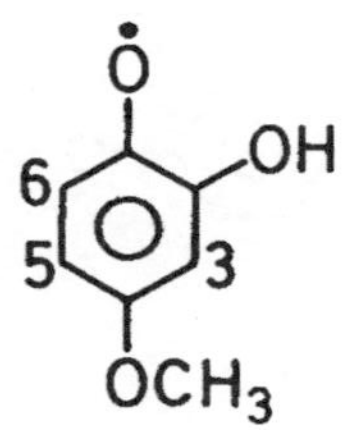										
1099	2-Hydroxy-4-methoxyphenoxyl radical										
1100	-,3,6-di-*t*-butyl-	–CMe_3	1.3[a]	u	1.3[b]	0.8	u		PhMe	Phot	131
			0[c]	u	1.3[b]	0.9	u		PhMe	Phot	131

Table II.7 (continued)
PHENOXYL RADICALS WITH THREE KEY OXYGENS [1,2,4]

Neutral Species

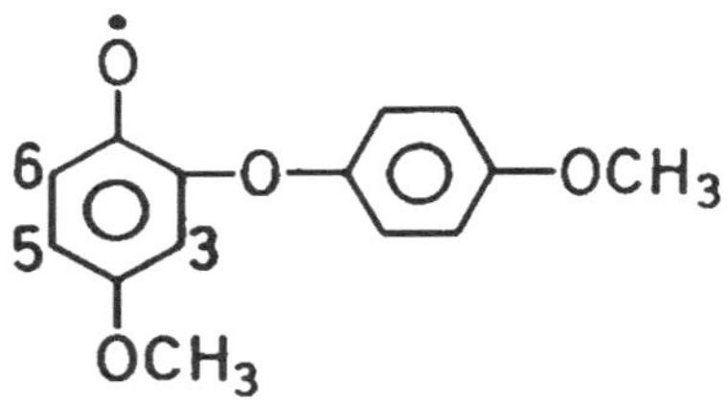

No.	Radical	Substituent(s)	Splitting constants/Gauss a_2	a_3	a_4	a_5	a_6	g-Factor	Solvent	Method	Ref.
1101	2-(4-Methoxyphenoxy)-4-methoxy-phenoxyl radical										
1102	-,5-*t*-butyl-	[R = R_6 = H, R′ = R_5 = Bu^t]	—	1.52	1.52[b]	—	4.1		CCl_4	Ox Ag_2O	291, 292
1103	-,6-*t*-butyl-	[R = R_5 = H, R′ = R_6 = Bu^t]	—	1.40[d]	1.40[b]	u[d]	—		CCl_4	Ox Ag_2O	291, 292
1104	-,6-*t*-butyl-	[R′ = R_5 = H, R = R_6 = Bu^t]	—	1.40[d]	1.40[b]	u[d]	—		CCl_4	Ox Ag_2O	291, 292
			—	1.3[d]	1.3[b]	u[d]	—		C_6H_6	Ox	177, 291

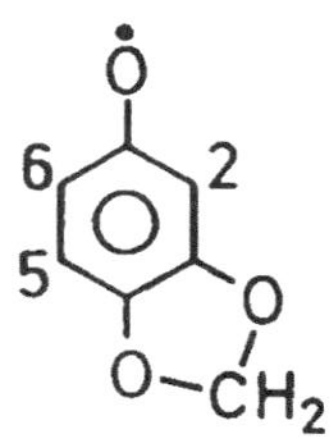

1105	3,4-Methylenedioxyphenoxyl radical	1.95	$a_H^{CH_2}$ 6.50	1.10	8.00	2.00430	H_2SO_4/w	Ox CeIV	241, 271

[a] Splitting from hydroxyl-group proton.
[b] Splitting from methoxyl-group protons.
[c] Hydroxyl-group proton exchanged with deuteron.
[d] Assignment unknown within the pair (a_3, a_5).

Table II.8
PHLOROGLUCINOL RADICALS

Anion and Neutral Species

No.	Radical	Substituent	Splitting constants/Gauss a_2	a_3	a_4	a_5	a_6	g-Factor	Solvent	Method	Ref.
	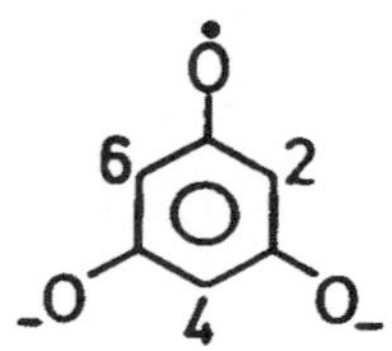										
1106	Phloroglucinol dianion radical		7.5	—	7.5	—	7.5		w/alk	Ox FeIII	27
	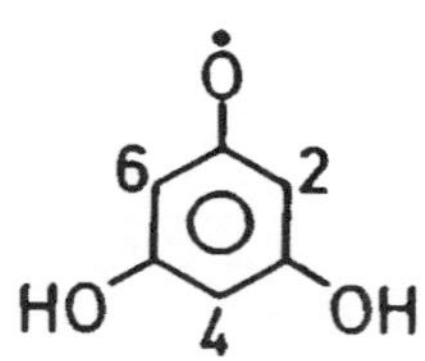										

Table II.8 (continued)
PHLOROGLUCINOL RADICALS

Anion and Neutral Species

No.	Radical	Substituent	Splitting constants/Gauss a_2	a_3	a_4	a_5	a_6	g-Factor	Solvent	Method	Ref.
1107	Phloroglucinol radical		6.0	u	11.75	u	6.0		H_2SO_4/w	Ox CeIV	84, 257
1108	-,4-methyl-	–Me	5.75	u	13.10[i]	u	5.75	2.00401	H_2SO_4/w	Ox CeIV	241
1109	-,2-acetyl-	–COMe	—	u	11.2	u	10.4	2.00421	H_2SO_4, $4H_2O$	Ox CeIV	256
	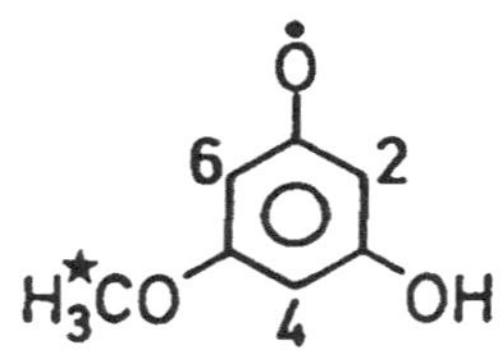										
1110	3-Hydroxy-5-methoxyphenoxyl radical		6.4	u	8.75	0.47*	8.75		H_2SO_4, $4H_2O$	Ox CeIV	257
	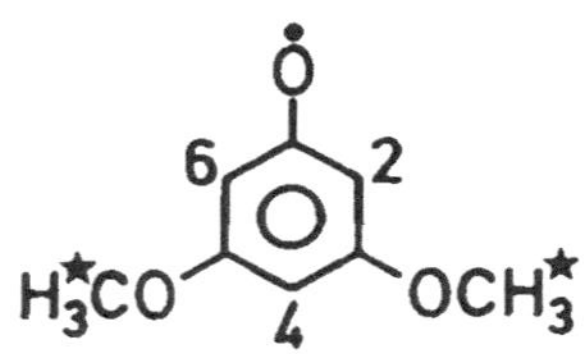										
1111	3,5-Dimethoxyphenoxyl radical		6.25	1.75*	10.75	1.75*	6.25			Ox CeIV	84
			5.75	0.435*	12.45	0.435*	5.75		H_2SO_4, $4H_2O$	Ox CeIV	257

Table II.9
PHENOXYL RADICALS WITH FOUR KEY OXYGENS [1,2,4,5]

Neutral Species

No.	Radical	Splitting constants/Gauss a_2	a_3	a_4	a_5	a_6	g-Factor	Solvent	Method	Ref.
	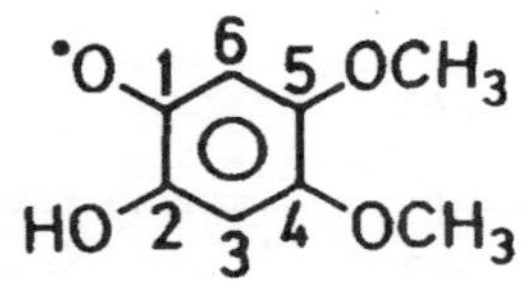									
1112	2-Hydroxy-4,5-dimethoxyphenoxyl radical	1.55[a]	—	0.13[b]	0.13[b]	—		PhMe	Re phot	131
1113	-,3,6-di-*t*-butyl-	0.26[c]	—	0.13[b]	0.13[b]	—		PhMe	Re phot	131
	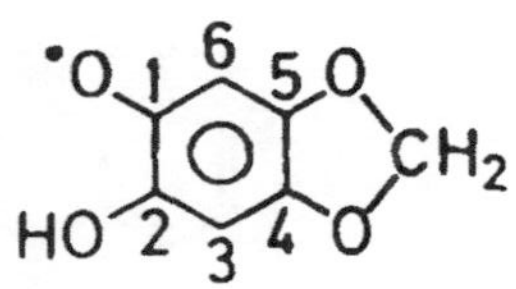			$a_H^{CH_2}$						
1114	2-Hydroxy-4,5-methylenedioxyphenoxyl radical	u	0.45	6.25		0.45	2.00433	H_2SO_4/w	Ox CeIV	241

[a] Splitting from hydroxyl-group proton.
[b] Splitting from methoxyl-group protons.
[c] Hydroxyl-group proton exchanged with deuteron.

Table II.10
1- AND 2-NAPHTHYLOXYL RADICALS

Neutral Species

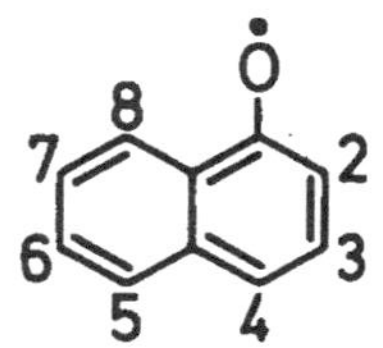

No.	Radical	Substituent(s)	a_1	a_2	a_3	a_4	a_5	a_6	a_7	a_8	Solvent	Method	Ref.
			Splitting constants/Gauss										
1115	1-Naphthyloxyl radical		—	8.25	1.75	10.75	2.5	0	2.5	0.65	w/H_2SO_4	Ox CeIV	196
1116	-,2-carboxy-	–COOH	—	—	0.9	10.4	2.9	0	2.9	0.9	w/H_2SO_4	Ox CeIV	196
1117	-,5-carboxy-	–COOH	—	8.25	2.1	10.75	—	0.6	2.5	0.9	w/H_2SO_4	Ox CeIV	196
1118	-,7-carboxy-	–COOH	—	8.5	2.0	11.0	2.5	0.5	—	0.5	w/H_2SO_4	Ox CeIV	196
1119	-,2-hydroxy-	–OH	(Data not available)										
1120	-,2-hydroxy-4-sulfonato-	–OH,–SO_3^-	—	—	2.0	—	1.25	u	1.25	0.5	w/H_2SO_4	Ox CeIV	196
1121	-,3-hydroxy-	–OH	(see Radical no. 1139)										
1122	-,4-hydroxy-	–OH*	—	6.80	0.34	1.63*	1.50	0.17	1.63	0.17	EtOH/ acid	Phot	137, 138
			—	6.92	0.35	1.70*	1.70	0.18	1.70	0.18	Pr^{i}OH	Phot	197
			—	7.18	0.27	1.55*	1.77	0.23	1.77	0.23	Pr^{i}OH	Phot	198, 346[b]
			—	7.67	0.80	1.25*	1.85	0.40	1.85	0.52	Dioxane	Phot	197
			—	3.25[c]	3.25	u	0.81	0.81	0.81	0.81	w/H_2SO_4	Ox CeIV	196
1123	-,5-hydroxy-	–OH	—	3.5	0.6	5.4	—	3.5	0.6	5.4	w/H_2SO_4	Ox CeIV	196
1124	-,5-methoxy-	–OMe	—	7.25	1.5	11.0	u	0	2.75	0	w/H_2SO_4	Ox CeIV	196
1125[a]	-,6-hydroxy-	–OH	—	8.75	2.0	10.75	2.0	—	3.0	—	w/H_2SO_4	Ox CeIV	196
1126[a]	-,7-hydroxy-	–OH	—	6.75	1.0	9.5	2.25	1.0	—	1.0	w/H_2SO_4	Ox CeIV	196

1127[a]	-,7-hydroxy-3-sulfonato-	–OH, $-SO_3^-$	—	7.0	—	9.25	2.5	1.0	—	1.0	w/H_2SO_4 Ox CeIV 196
1128	-,8-hydroxy-	–OH*	—	5.75	1.25	7.0	7.0	1.25	5.75	0.5*	w/H_2SO_4 Ox CeIV 196
1129	-,8-hydroxy-3,6-disulfonato-	–OH*, $-SO_3^-$	—	6.13	—	7.0	7.0	—	6.13	0.5*	w/H_2SO_4 Ox CeIV 196
1130	-,4-sulfonato-	$-SO_3^-$	—	8.75	2.5	—	2.5	0	2.5	0.75	w/H_2SO_4 Ox CeIV 196
1131	-,8-sulfonato-	$-SO_3^-$	—	8.75	2.0	10.75	2.0	0	2.0	—	w/H_2SO_4 Ox CeIV 196

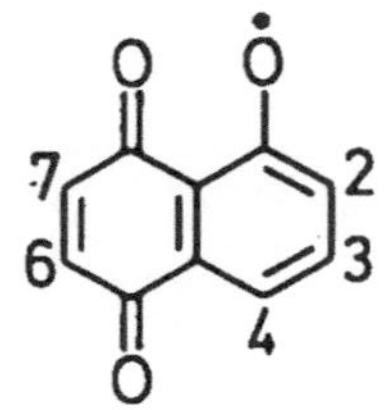

1132	5,8-Dihydro-5,8-dioxy-1-naphthyloxyl radical		—	7.0	2.2	9.5	—	0	0	—	w/pH = 0 Ox CeIV 83

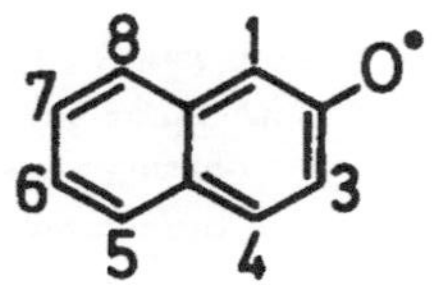

1133	2-Naphthyloxyl radical		10.75	—	1.45	0	1.45	5.4	1.2	4.3	w/H_2SO_4 Ox CeIV 196, 173
1134	-,1-carboxy-	–COOH	—	—	1.5	0	1.5	5.5	1.25	4.25	w/H_2SO_4 Ox CeIV 196
1135	-,3-carboxy-	–COOH	10.4	—	—	0.8	1.6	5.2	0.8	4.4	w/H_2SO_4 Ox CeIV 196
1136	-,1-hydroxy-	–OH	(Data not available)								
1137	-,3-hydroxy-	–OH	9.5	—	—	1.25	1.25	5.25	1.25	4.75	w/H_2SO_4 Ox CeIV 196
1138	-,3-methoxy-	–OMe	9.5	—	—	0.5	1.0	5.5	0.5	4.5	w/H_2SO_4 Ox CeIV 196
1139[d]	-,4-hydroxy-	–OH	11.9	—	1.1	—	3.6	4.7	u	3.6	w/H_2SO_4 Ox CeIV 196
1140	-,5-hydroxy-	–OH	(see Radical no. 1125)								
1141	-,6-hydroxy-	–OH	4.1	—	u	2.4	4.1	—	u	2.4	w/H_2SO_4 Ox CeIV 196
1142	-,6-methoxy-	–OMe	9.75	—	1.25	0	2.5	—	1.25	3.75	w/H_2SO_4 Ox CeIV 196

Table II.10 (continued)
1- AND 2-NAPHTHYLOXYL RADICALS

Neutral Species

No.	Radical	Substituent(s)	Splitting constants/Gauss a_1	a_2	a_3	a_4	a_5	a_6	a_7	a_8	Solvent	Method	Ref.
1143	-,7-hydroxy-	–OH	7.4	—	3.2	0.7	0.7	3.2	—	7.4	w/H_2SO_4	Ox CeIV	196
1144	-,7-methoxy-	–OMe	10.5	—	1.25	0	1.5	5.25	—	4.0	w/H_2SO_4	Ox CeIV	196
1145	-,8-hydroxy-	–OH	(see Radical no. 1126)										
1146	-,1-sulfonato-	$-SO_3^-$	—	—	1.5	0	1.5	5.5	1.25	4.5	w/H_2SO_4	Ox CeIV	196
1147	-,6-sulfonato-	$-SO_3^-$	11	—	1.5	0	1.5	—	1.5	4.25	w/H_2SO_4	Ox CeIV	196
1148	-,7-sulfonato-	$-SO_3^-$	10.75	—	1.5	0	1.25	5.5	—	4.25	w/H_2SO_4	Ox CeIV	196
1149	-,6,8-disulfonato-	$-SO_3^-$	10.65	—	1.5	0.5	2.0	—	1.5	—	w/H_2SO_4	Ox CeIV	196
1150	-,1,3,4-trichloro-	–Cl	—	—	—	—	1.0	3.5	1.0	3.5	C_6H_6	Ox PbO_2	261

[a] Structure of parent compound leads to ambiguity as to the type of radical formed, however, spectral data indicate a 1-naphthyloxyl radical.
[b] Assignment given to OH proton in Reference 346 appears incorrect.
[c] Averaging of splitting constants due to rapid proton exchange is indicated throughout with a bar in this table.
[d] Spectral data indicate a 2-naphthyloxyl type of radical; cf. footnote a.

Table II.11
ANTHRANYLOXYL AND BIANTHRONE RADICALS

Anions and Neutral Species

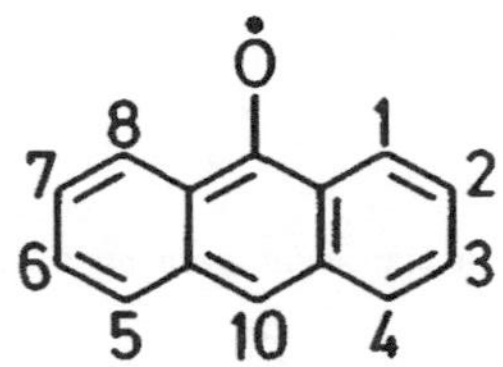

No.	Radical	Splitting constants/Gauss a_1	a_2	a_3	a_4	a_5	a_6	a_7	a_8	a_{10}	g-Factor	Solvent	Method	T [K]	Ref.
1151	9-Anthranyloxyl radical	0.875	3.54	1.06	3.05	3.05	1.06	3.54	0.875	11.86		*o*-terphenyl	Thermolysis	503	209
1152	-,10-phenyl-	0.84	3.37	1.01	3.02	3.02	1.01	3.57	0.84	1.01[o] 0.36[m] 0.36[p]		*m*-xylene	Thermolysis	293	209
1153	-,10-hydroxy-	0.44	3.28	0.44	3.28	3.28	0.44	3.28	0.44	0.88[a]		Dioxane	Phot	RT	197, 198
1154	-,10-hydroxy-2-methyl-	(Hfs data not reported)									2.00403	PhMe	CIDEP	253	258, 259
1155	-,10-(9′-anthranyl)-	0.82	3.06	1.03	3.42	3.42	1.03	3.06	0.82	0.19[b]	2.00338	*o*-terphenyl	Ox O_2	548	210
1156	-,10-[10′-hydroxyanthran-9′-yl]-	0.80	3.38	1.03	3.04	3.04	1.03	3.38	0.80	—	2.00338	Biphenyl	Phot	363	210
		0.80	3.34	0.99	3.01	3.01	0.99	3.34	0.80	—		Fluorenone	Thermolysis	423	209
		0.86	3.31	1.04	2.94	2.94	1.04	3.31	0.86	—		*o*-terphenyl	Phot	373	211
1157	-,2-methyl-10-[2′-methyl-10′-hydroxyanthran-9′-yl]-	0.75	3.75[c]	0.88	3.00	3.00	0.88	3.00	0.75	—		*o*-terphenyl	Phot	373	211
		0.79	2.83[c]	0.79	2.83	2.83	0.79	2.83	0.79	—	2.0028	MeTHF	Re $KOBu^t$	RT	289

Table II.11 (continued)
ANTHRANYLOXYL AND BIANTHRONE RADICALS

Anions and Neutral Species

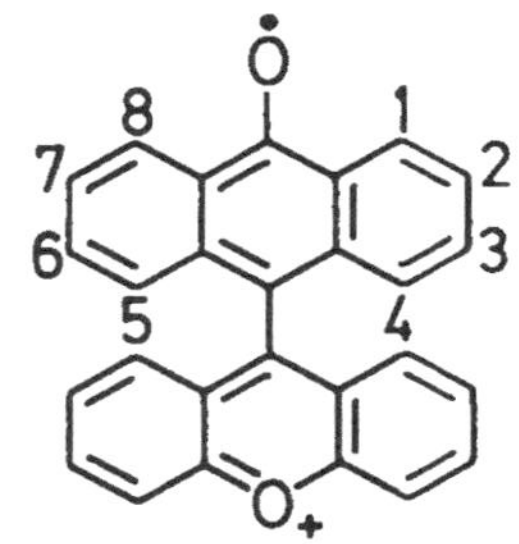

No.	Radical	Splitting constants/Gauss a_1	a_2	a_3	a_4	a_5	a_6	a_7	a_8	a_{10}	g-Factor	Solvent	Method	T [K]	Ref.
1158	Xanthylideneanthrone radical cation	0.79	3.74	0.79	3.20	3.20	0.79	3.74	0.79	—		*o*-terphenyl	Phot	373	211

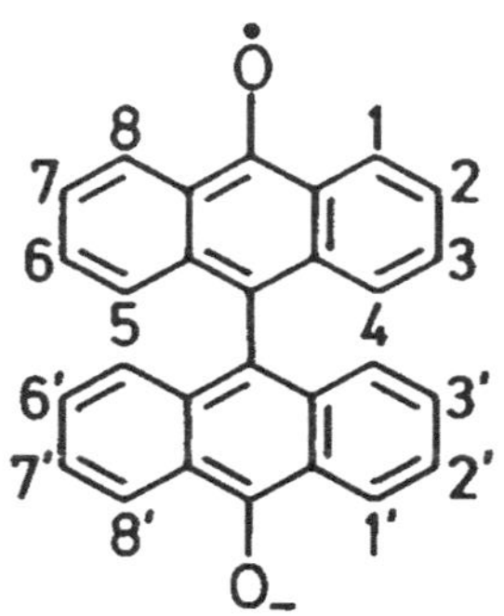

1159	Bianthrone radical anion[c]	0.39	1.28	0.40	1.42	1.42	0.40	1.28	0.39	—	o-terphenyl	Phot	373	211

[a] Splitting from the OH-group proton.
[b] Splitting from the proton at C-10′.
[c] This is the only radical in Table II.11 where coupling to protons in both halves of the molecule is observed, i.e., $a_1 = a_{1'} = a_8 = a_{8'}$, $a_2 = a_{2'} = a_7 = a_{7'}$, $a_3 = a_{3'} = a_6 = a_{6'}$, $a_4 = a_{4'} = a_5 = a_{5'}$.

III. CATIONS

Table III.1
PHENOL RADICALS WITH ONE KEY OXYGEN

Cations

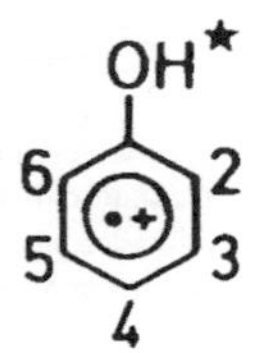

No.	Radical	Substituent(s)	Splitting constants/Gauss a_1	a_2	a_3	a_4	a_5	a_6	g-Factor	Solvent	Method	Ref.
1160	Phenol cation		—	5.3	0.8	10.7	0.8	5.3	2.00291	H_2SO_4/w	Ox CeIV	85
1161	-,2-methyl-	–Me	—	7.5[1]	1.6	9.8	0.1	4.0	2.00331	H_2SO_4/w	Ox CeIV	85
1162	-,3-methyl-	–Me	—	4.0	0.2[1]	11.0	1.2	6.4	2.00312	H_2SO_4/w	Ox CeIV	85
1163	-,4-methyl-	–Me	—	4.5	0.05	15.1[1]	0.05	4.5	2.00317	H_2SO_4/w	Ox CeIV	85
1164	-,3,5-dimethyl-	–Me	—	4.70	0.08[1]	11.85	0.08[1]	4.70	2.00284	H_2SO_4/w	Ox CeIV	241
1165	-,2,6-di-t-butyl-4-methyl-	–CMe_3, –Me	3.72*	u	0.89[a]	14.19[1]	<0.4[a]	u		CH_2Cl_2 −80°C	Ox $AlCl_3$	263, 265
1166	-,2,6-di-t-butyl-4-ethyl-	–CMe_3, –CH_2Me	3.55*	u	u	12.15[1]	u	u		CH_2Cl_2 −80°C	Ox $AlCl_3$	265
1167	-,2,6-di-t-butyl-4-isopropyl-	–CMe_3, –$CHMe_2$	3.60*	u	0.79[a]	6.29[1] 0.79[2]	u[a]	u		CH_2Cl_2 −80°C	Ox $AlCl_3$	265

Table III.1 (continued)
PHENOL RADICALS WITH ONE KEY OXYGEN

Cations

No.	Radical	Substituent(s)	Splitting constants/Gauss a_1	a_2	a_3	a_4	a_5	a_6	g-Factor	Solvent	Method	Ref.
1168	-,2-acetyl-	–COMe	—	u	0.10	10.50	2.05	7.20	2.00410	H_2SO_4/w	Ox CeIV	241
1169	-,3-acetyl-	–COMe	—	6.95	u	9.90	1.90	6.00	2.00393	H_2SO_4/w	Ox CeIV	241
1170	-,4-acetyl-	–COMe	—	6.45	1.90	0.50[b]	1.90	6.45	2.00436	H_2SO_4/w	Ox CeIV	241
1171	-,2-trifluoromethyl-	$–CF_3$	—	3.85[b]	1.25	10.55	2.05	7.35	2.00426	H_2SO_4/w	Ox CeIV	241
1172	-,3-trifluoromethyl-	$–CF_3$	—	7.10	1.50[b]	10.30	2.00	6.75	2.00425	H_2SO_4/w	Ox CeIV	241
1173	-,4-trifluoromethyl-	$–CF_3$	—	7.05	2.05	11.20[b]	2.05	7.05	2.00428	H_2SO_4/w	Ox CeIV	241
1174	-,2-amino-	$–NH_2$	—	6.75[c] 8.1[d]	2.6	1.6	6.6	0.9	2.00313	H_2SO_4/w	Ox CeIV	85
1175	-,3-amino-	$–NH_2$	—	3.05	7.0[c] 8.25[d]	8.6	2.05	10.45	2.00277	H_2SO_4/w	Ox CeIV	85
1176	-,4-amino-	$–NH_2$	—	0.4	4.0	6.6[c] 8.0[d]	4.0	0.4	2.00315	H_2SO_4/w	Ox CeIV	85
1177	-,2-nitro-	$–NO_2$	—	1.25[c]	1.15	10.40	2.10	7.35	2.00457	H_2SO_4/w	Ox CeIV	241
1178	-,3-nitro-	$–NO_2$	—	7.30	0.35[c]	9.80	2.00	6.60	2.00432	H_2SO_4/w	Ox CeIV	241
1179	-,4-nitro-	$–NO_2$	—	7.20	2.40	2.40[c]	2.40	7.20	2.00484	H_2SO_4/w	Ox CeIV	241
	-,2-hydroxy-	–OH	(see Table III.2)									
	-,3-hydroxy-	–OH	(see Table III.3)									
	-,4-hydroxy-	–OH	(see Table III.4)									
	-,2-methoxy-	–OMe	(see Table III.2)									
	-,3-methoxy-	–OMe	(see Table III.3)									
	-,4-methoxy-	–OMe	(see Table III.4)									
1180	-,2-fluoro-	–F	—	17.20[b]	1.00	10.20	2.05	5.35	2.00417	H_2SO_4/w	Ox CeIV	241

1181	-,3-fluoro-	–F	—	2.60	4.25[b]	11.05	2.05	7.90	2.00351	H_2SO_4/w	Ox CeIV	241
1182	-,4-fluoro-	–F	—	5.80	0.90	29.75[b]	0.90	5.80	2.00426	H_2SO_4/w	Ox CeIV	241
1183	-,2-chloro-	–Cl	—	1.05[e]	2.10	10.15	1.40	5.85	2.00498	H_2SO_4/w	Ox CeIV	241
1184	-,3-chloro-	–Cl	—	5.65	0.05[e]	10.85	2.10	7.25	2.00380	H_2SO_4/w	Ox CeIV	241
1185	-,4-chloro-	–Cl	—	6.05	1.35	2.20[e]	1.35	6.05	2.00566	H_2SO_4/w	Ox CeIV	241

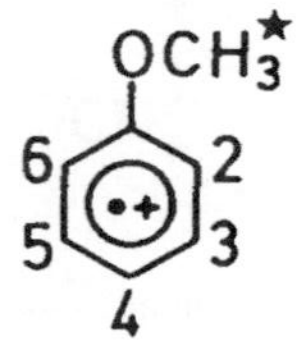

1186	Methoxybenzene cation		4.83*	4.52	1.00	9.97	0.21	5.51	2.00351	w/5°C	Radiolysis[f]	266, 265
			5.0*	4.4	1.1	10.6	0.2	5.5	2.0032	H_2SO_4/w	Ox CeIV	269
1187	-,4-methyl-	–Me	4.3*	3.5	0.8	15.25[i]	0.2	4.3	2.0032	H_2SO_4/w	Ox CeIV	269
1188	-,4-fluor-	–F	4.5*	4.5	0	31.75[b]	0	4.5	2.00385	H_2SO_4/w	Ox CeIV	269

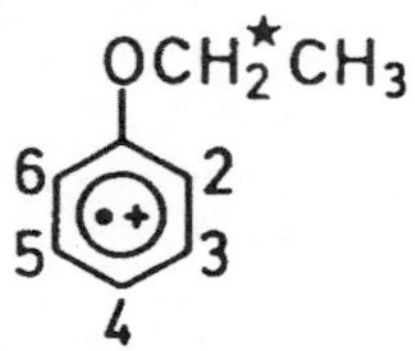

1189	Ethoxybenzene cation		5.0*	4.3	1.0	10.5	0.3	5.4	2.0033	H_2SO_4/w	Ox CeIV	269

[a] Radical stable only below 50°C, leading to nonequivalent *meta* protons. Hydroxylic hydrogen is assumed lying close to the plane of the aromatic nucleus.
[b] Splitting from fluorine nucleus (nuclei).
[c] Splitting from nitrogen nucleus.
[d] Splitting from protons of amino group.
[e] Splitting from chlorine nucleus.
[f] *In situ* radiolysis of 10^{-3} *M* Tl_2SO_4 solutions saturated with N_2O at pH = 3 to 4 and containing the anisole.

Table III.2
PHENOL RADICALS WITH TWO KEY OXYGENS [1,2]

Cations

No.	Radical	Substituent(s)	Splitting constants/Gauss a_1	a_2	a_3	a_4	a_5	a_6	g-Factor	Solvent	Method	Ref.
	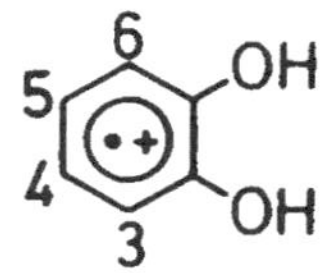											
1190	2-Hydroxyphenol cation		—[a]	—[a]	0.0	4.75	4.75	0.0		w/H_2SO_4,pH <2	Ox CeIV	84
			—[a]	—[a]	0.25	4.85	4.85	0.25	2.00355	H_2SO_4.4H_2O	Ox CeIV	85, 269
	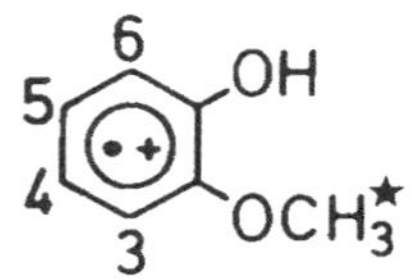											
1191	2-Methoxyphenol cation		—[a]	2.9*	1.1	6.8	2.9	1.45	2.00362	H_2SO_4.4 H_2O	Ox CeIV	85, 271
			—[a]	3.35*	0.85	5.9	3.55	0.1	2.00362	H_2SO_4.4 H_2O	Ox CeIV	256
	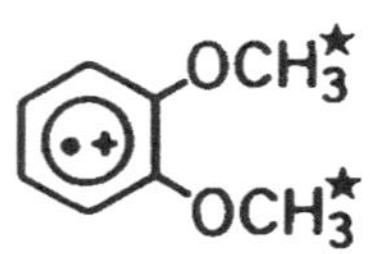											
1192	1,2-Dimethoxybenzene cation		3.25*	3.25*	0	4.9	4.9	0	2.0034	H_2SO_4.4 H_2O	Ox CeIV	269, 271
			3.33*	3.33*	0.16	4.89	4.89	0.16	2.00374	w/5°C	Radiolysis[b]	266
			3.2*	3.2*	u	4.6	4.6	u		HSO_3F/ – 70°C	Ox PbO_2	286

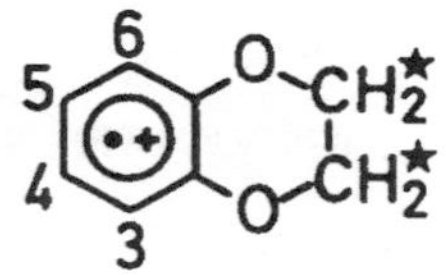

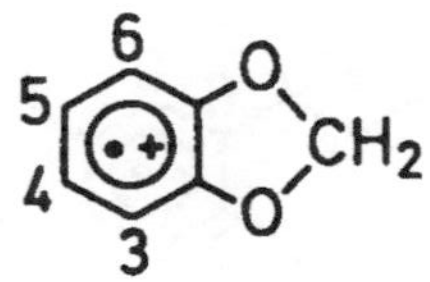

1193	1,2-Ethylenedioxybenzene cation		2.5*	2.5*	0.4	4.65	4.65	0.4	2.0034	H_2SO_4,4 H_2O	Ox CeIV	269
			$a_H^{CH_2}$									
1194	1,2-Methylenedioxybenzene cation		21.9		0.4	4.9	4.9	0.4	2.0039	H_2SO_4,4 H_2O	Ox CeIV	269, 271
1195	-,3-methyl-	–Me	21.4		0.4[1]	4.2	5.3	0.6	2.0039	H_2SO_4,4 H_2O	Ox CeIV	269
1196	-,3-isopropyl-	$–CHMe_2$	21.45		0[1]	4.15	5.2	0.6	2.0039	H_2SO_4,4 H_2O	Ox CeIV	269
1197	-,4-methyl-	–Me	19.4		0.75	7.95[1]	5.15	0.25	2.0038	H_2SO_4,4 H_2O	Ox CeIV	269
1198	-,4-(2-carboxyethyl)-	$–CH_2CH_2COOH$	20.25		0.7	6.75[1]	5.1	0.25	2.0038	H_2SO_4,4 H_2O	Ox CeIV	269
1199	-,4-allyl-	$–CH_2CH = CH_2$	19.9		0.75	7.25[1]	5.25	0.25	2.0038	H_2SO_4,4 H_2O	Ox CeIV	269
1200	-,4-(2-carboxyvinyl)-	–CH = CHCOOH	18.5		1.1	4.0[1] 1.9[2]	4.0	0.4	2.0038	H_2SO_4,4 H_2O	Ox CeIV	269
1201	-,4-carboxymethyl-	$–CH_2COOH$	20.4		0.6	4.65[1]	4.9	0.25	2.0039	H_2SO_4,4 H_2O	Ox CeIV	269
1202	-,6-isopropyl-3-methyl-	$–CHMe_2$, –Me	21.23		0.67[1]	4.61	4.61	0.27[1]	2.0039	H_2SO_4,4 H_2O	Ox CeIV	272
1203	-,4-hydroxymethyl-	$–CH_2OH$	20.0		0.75	10.5[1]	5.0	0.25	2.0038	0.5 *M* H_2SO_4	Ox CeIV	269
1204	-,4-formyl-	–CHO	22.8		0.25	u	4.05	0	2.00435	H_2SO_4,4 H_2O	Ox CeIV	269
1205	-,5-formyl-4-methyl-	–CHO, –Me	21.25		0.25	6.85[1]	0.25[1]	0.65	2.0042	H_2SO_4,4 H_2O	Ox CeIV	272
1206	-,4-carboxy-	–COOH	23.25		0.4	—	4.25	0	2.00415	H_2SO_4,4 H_2O	Ox CeIV	269
1207	-,4-cyano-	–CN	24.2		0	u	4.0	0	2.00405	H_2SO_4,4 H_2O	Ox CeIV	269
1208	-,5-chloro-4-chloromethyl-	–Cl, $–CH_2Cl$	19.5		0.75	8.9[1]	1.25[c]	0.25	2.00495	H_2SO_4,4 H_2O	Ox CeIV	269

Table III.2 (continued)
PHENOL RADICALS WITH TWO KEY OXYGENS [1,2]

Cations

No.	Radical	Substituent(s)	Splitting constants/Gauss a_1	a_2	a_3	a_4	a_5	a_6	g-Factor	Solvent	Method	Ref.
1209	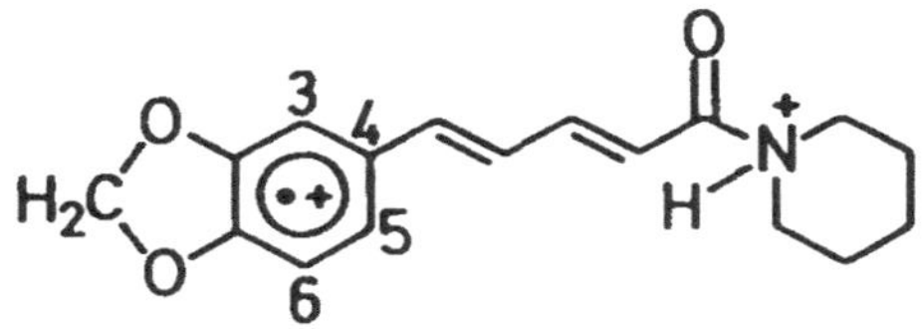 Piperine radical cation		21.25		0.5	6.0'	4.75	0		H_2SO_4.4 H_2O	Ox CeIV	269
1210	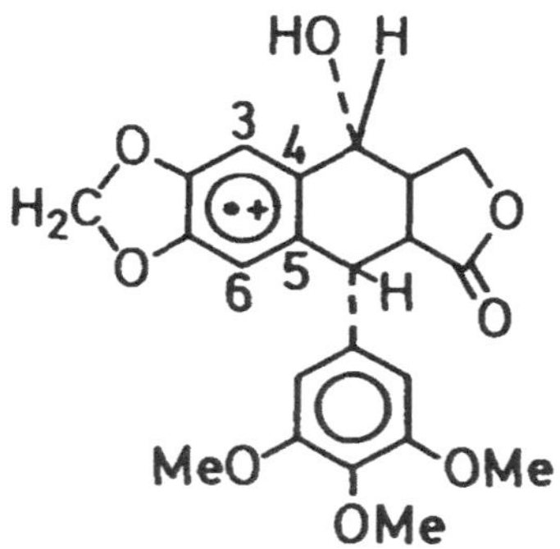 Podophyllotoxin radical cation		21.5		0.2	9.5'	5.4'	0.2		H_2SO_4.4 H_2O	Ox CeIV	269

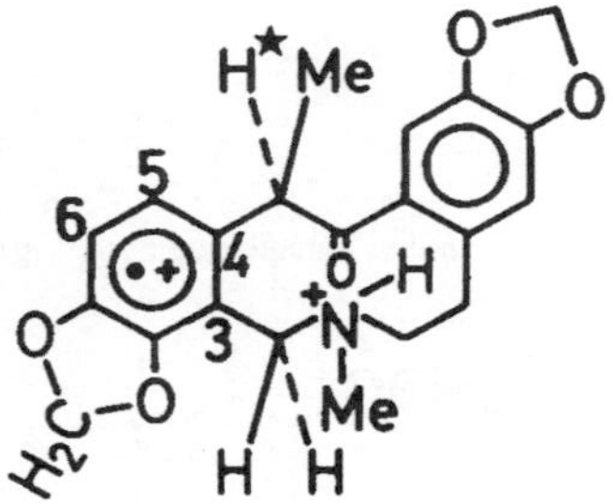

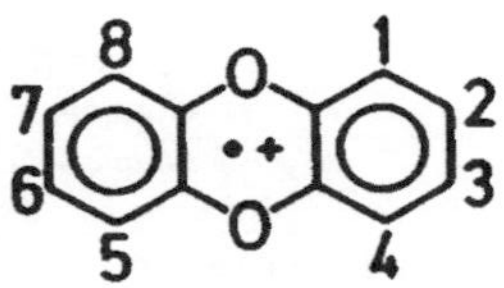

No.	Radical	Substituent	a_1	a_2	a_3	a_4	a_5	a_6	g	Solvent	Method	Ref.
1211	Corycavine radical cation		21.25		0[1]	11.5[1,2]	4.75	0		H_2SO_4,4 H_2O	Ox CeIV	269
1212	Dibenzo-*p*-dioxin cation[d]		<0.5	2.52	2.52	<0.5	<0.5	2.52	2.0034	H_2SO_4	Ox H_2O_2	282, 281
			<0.15	2.178	2.178	<0.15	<0.15	2.178	2.0038	CF_3SO_3H	Ox	282
1213	-,2,6-difluoro-	–F	0.2	6.50[e]	2.45[f]	u[g]	0.2	6.50	2.0042	H_2SO_4	Ox	285

[a] Proton exchange always fast (pH <2); no hyperfine interactions with hydroxylic protons observed.
[b] *In situ* radiolysis of 10^{-3} *M* Tl_2SO_4 solutions containing the phenol and saturated with N_2O at pH = 3 to 4.
[c] Splitting from chlorine nucleus.
[d] ^{13}C splittings reported in Reference 282.
[e] Splitting from fluorine nucleus.
[f] Splitting also observed from the proton at C-7.
[g] At C-1 and C-5 0.2 G splitting from the protons may equally well be ascribed to the protons at C-4 and C-8.

Table III.3
PHENOL RADICALS WITH TWO KEY OXYGENS [1,3]

Cations

No.	Radical	Substituent(s)	Splitting constants/Gauss a_1	a_2	a_3	a_4	a_5	a_6	g-Factor	Solvent	Method	Ref.
	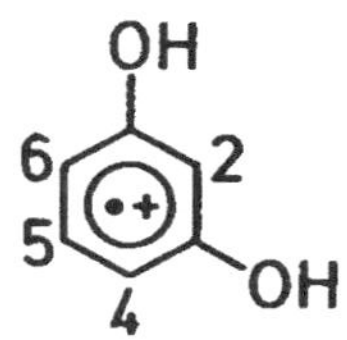											
1214	3-Hydroxyphenol cation		—	0.4	—	10.75	2.0	10.75		H_2SO_4/w	Ox CeIV	84
			—	0.4	—	10.9	2.1	10.9	2.00312	H_2SO_4,4 H_2O	Ox CeIV	85, 271
1215	-,4-methyl-	–Me	—	0.5	—	14.70[1]	1.70	8.80	2.00313	w/H_2SO_4	Ox CeIV	241
1216	-,5-methyl-	–Me	—	0.45	—	10.65	2.15[1]	10.65	2.00312	H_2SO_4,4 H_2O	Ox CeIV	256
1217	-,2-acetyl-	–COMe	u	u	1.65[a]	11.2	2.4	11.2	2.00324	H_2SO_4,4 H_2O	Ox CeIV	256
	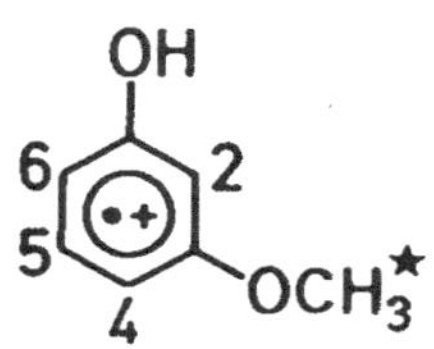											
1218	3-Methoxyphenol cation		—	0.3	2.6*	10.25	2.25	11.25	2.00315	H_2SO:.4 H_2O	Ox CeIV	85
			—	0.3	2.6*	10.85	2.15	11.25		H_2SO_4/w	Ox CeIV	271

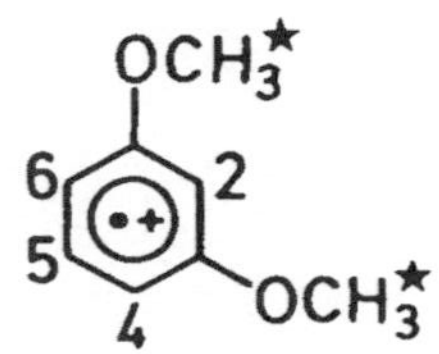

1219	1,3-Dimethoxybenzene cation	2.95*	<0.05	2.95*	10.53	2.10	10.53	2.00354	w/5°C	Radiolysis[b]	266
		3.0*	0.0	3.0*	10.75	2.1	10.75	2.0033	H_2SO_4,4 H_2O	Ox CeIV	269, 271

[a] Hydroxyl proton splitting assigned a_3 in Reference 256; no splitting reported from the hydroxyl proton at C-1.
[b] *In situ* radiolysis of 10^{-3} M Tl_2SO_4 solutions containing the phenol and saturated with N_2O at pH = 3 to 4.

Table III.4
1,4-DIHYDROXY/ALKOXYBENZENE DERIVED RADICALS

Cations

No.	Radical	Splitting constants/Gauss a_1^{OH}	a_4^{OH}	a_2	a_3	a_5	a_6	g-Factor	Solvent	Method	T [K]	Ref.
	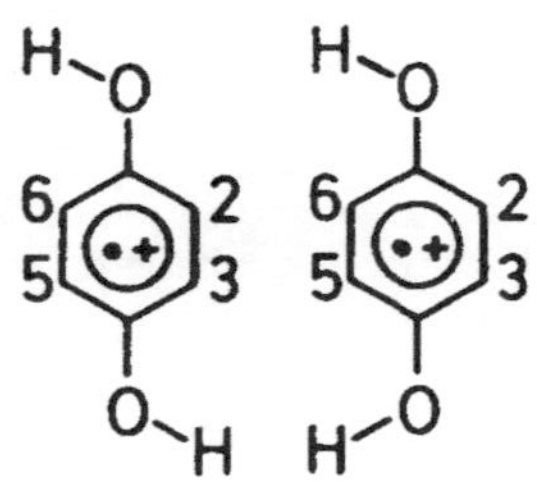 *trans* *cis*											
1220	[I] 4-Hydroxyphenol cation	3.44	3.44	2.36	2.36	2.36	2.36		H_2SO_4 abs	Ox	RT	200, 201
		3.303	3.303	2.25	2.25	2.25	2.25	2.00350	$MeNO_2$	Ox $AlCl_3$	RT	202

Table III.4 (continued)
1,4-DIHYDROXY/ALKOXYBENZENE DERIVED RADICALS

Cations

No.	Radical		Splitting constants/Gauss a_1^{OH}	a_4^{OH}	a_2	a_3	a_5	a_6	g-Factor	Solvent	Method	T [K]	Ref.
			u	u	2.25	2.25	2.25	2.25	2.00352	w/H_2SO_4	Ox CeIV	RT	85, 84
		trans[a]	3.294	3.294	2.456	2.055	2.456	2.055		$MeNO_2$	Ox $AlCl_3$	<253	203, 278
		cis[a]	3.294	3.294	2.147	2.147	2.356	2.356		$MeNO_2$	Ox $AlCl_3$	<253	203, 278
1221	-,2-methyl-		u	u	2.72^{1}	1.36	3.05	2.00	2.00347	w/H_2SO_4	Ox CeIV	296	241
1222	-,2,3-dimethyl-	*cis*[b]	3.11	3.11	1.83^{1}	1.83^{1}	2.58	2.58	2.00343	H_2SO_4	Ox	RT	204, 205
			3.138	3.138	1.828^{1}	1.828^{1}	2.583	2.583	2.00340	$MeNO_2$	Ox $AlCl_3$	258	205, 206
1223	-,2,5-dimethyl-	*trans*[b]	3.00	3.00	3.82^{1}	0.86	3.82^{1}	0.86	2.00338	H_2SO_4	Ox	RT	204, 200
			3.012	3.012	3.718^{1}	0.915	3.718^{1}	0.915	2.00340	$MeNO_2$	Ox $AlCl_3$	258	205, 206
			3.032	3.032	3.679^{1}	0.957	3.679^{1}	0.957	2.00340	$MeNO_2$	Ox $AlCl_3$	223	207
1224	-,2,5-di-isopropyl-		u	u	u	u	u	u	2.00351	$MeNO_2$	Ox $AlCl_3$	258	205
1225	-,2,5-di-*t*-butyl-		3.069	3.069	0.112^{2}	1.432	0.112^{2}	1.432	2.00345	$MeNO_2$	Ox $AlCl_3$	258	205
			3.062	3.062	0.112^{2}	1.428	0.112^{2}	1.428	2.00345	$MeNO_2$	Ox $AlCl_3$	233	207, 274
1226	-,2,6-dimethyl-		3.01	3.01	2.41^{1}	1.81	1.81	2.41^{1}	2.00337	H_2SO_4	Ox	RT	204
		trans	2.99	3.11	2.55^{1}	1.70	2.29	1.82^{1}		$MeNO_2$	Ox $AlCl_3$	213	204
		cis	2.99	3.11	2.11^{1}	1.87	2.12	2.26^{1}		$MeNO_2$	Ox $AlCl_3$	213	204
1227	-,2,6-di-*t*-butyl-		3.02	3.02	0.75^{2}	2.09	2.09	0.75^{2}		$MeNO_2$	Ox $AlCl_3$	248	208
			3.10	3.10	0.75^{2}	1.83	2.33	0.75^{2}		$MeNO_2$	Ox $AlCl_3$	203	208
1228	-,tetramethyl-		2.741	2.741	2.094^{1}	2.094^{1}	2.094^{1}	2.094^{1}		H_2SO_4	Ox		279, 270
			2.742	2.742	2.045^{1}	2.045^{1}	2.045^{1}	2.045^{1}	2.00350	$MeNO_2$	Ox $AlCl_3$	183	202
		trans[c]	2.887	2.887	2.785^{1}	1.389^{1}	2.785^{1}	1.389^{1}	2.00342	CH_2Cl_2	Ox $AlCl_3$	203	276
			2.730	2.730	2.483^{1}	1.571^{1}	2.483^{1}	1.571^{1}	2.00330	$MeNO_2$	Ox $AlCl_3$	203	276
			0.447^{D}	0.447^{D}	2.524^{1}	1.587^{1}	2.524^{1}	1.587^{1}	2.00346	$MeNO_2$	Ox D_2SO_4	213	276
		cis[c]	2.887	2.887	2.171^{1}	2.171^{1}	1.931^{1}	1.931^{1}	2.00342	CH_2Cl_2	Ox $AlCl_3$	203	276
			2.730	2.730	2.151^{1}	2.151^{1}	1.903^{1}	1.903^{1}	2.00330	$MeNO_2$	Ox $AlCl_3$	203	276
			0.447^{D}	0.447^{D}	2.168^{1}	2.168^{1}	1.895^{1}	1.895^{1}	2.00346	$MeNO_2$	Ox D_2SO_4	213	276

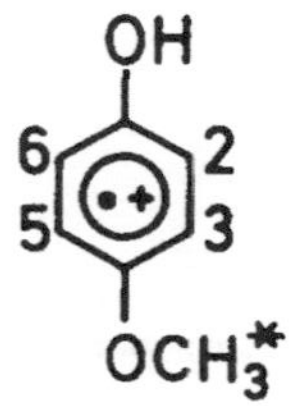

1229	4-Methoxyphenol cation		u	3.2*	3.0	1.9	1.9	2.2	2.00358	H_2SO_4,4 H_2O	Ox CeIV		85, 269
			u	3.2*	3.5	1.75	1.75	2.25	2.00318	H_2SO_4/w	Ox CeIV		256, 271
1230	-,2,6-di-*t*-butyl-		3.14	3.14*	u	1.78	2.36	u		$MeNO_2$	Ox $AlCl_3$	253	208
1231	-,tetramethyl-		2.613	3.057*	2.371^{1}	1.745^{1}	1.745^{1}	2.371^{1}		$MeNO_2$	Ox $AlCl_3$	223	277
			2.688	3.053*	2.436^{1}	1.786^{1}	1.682^{1}	2.322^{1}		80% CH_2Cl_2/ $MeNO_2$	Ox $AlCl_3$	223	277

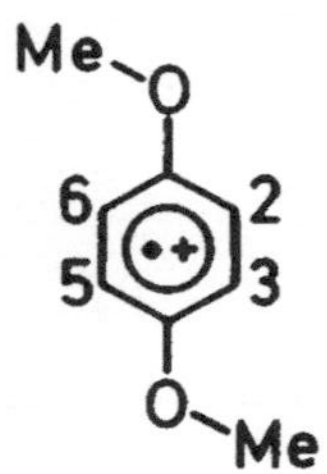 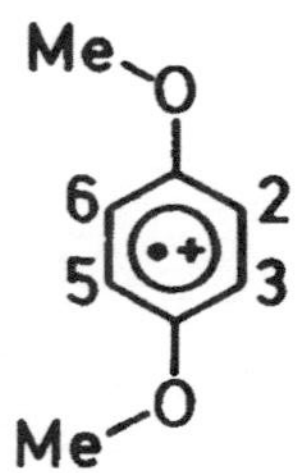

trans *cis*

			a_1^{OMe}	a_4^{OMe}									
1232	1,4-Dimethoxybenzene cation	*trans*	3.41	3.41	2.92	1.59	2.92	1.59	2.00388	w/pH = 4	Radiolysis[d]	268	266
			3.45	3.45	2.99	1.49	2.99	1.49	2.00365	H_2SO_4	Ox	RT	414, 269
			3.45	3.45	0.482^{D}	0.242^{D}	0.482^{D}	0.242^{D}		D_2SO_4	Ox	RT	414
		cis	3.24	3.24	1.88	1.88	2.61	2.61	2.00388	w/pH = 4	Radiolysis[d]	268	266
			3.33	3.33	1.81	1.81	2.68	2.68	2.00365	H_2SO_4	Ox	RT	414, 269
			3.33	3.33	0.292^{D}	0.292^{D}	0.432^{D}	0.432^{D}		D_2SO_4	Ox	RT	414
1233	-,2,3-dimethyl-	*cis*	3.07	3.07	1.48^{1}	1.48^{1}	2.90	2.90	2.00362	$MeNO_2$	Ox $AlCl_3$	248	205, 204
1234	-,2,5-dimethyl-	*trans*	3.12	3.12	4.37^{1}	0.485	4.37^{1}	0.485		H_2SO_4	Ox	RT	287, 204
			3.153	3.153	4.210^{1}	0.572	4.210^{1}	0.572	2.00356	$MeNO_2$	Ox $AlCl_3$	248	205, 204
1235	-,2,5-dipropyl-	*trans*	3.17	3.17	3.67^{1}	0.50	3.67^{1}	0.50		H_2SO_4	Ox	293	384

Table III.4 (continued)
1,4-DIHYDROXY/ALKOXYBENZENE DERIVED RADICALS

Cations

No.	Radical		Splitting constants/Gauss a_1^{OH}	a_4^{OH}	a_2	a_3	a_5	a_6	g-Factor	Solvent	Method	T [K]	Ref.
1236	-,2,5-diisopropyl-	*trans*	3.10	3.10	3.10[1]	0.50	3.10[1]	0.50		H_2SO_4	Ox	293	384
1237	-,2,5-di-*t*-butyl-	*trans*	3.242	3.242	0.103[2]	1.025	0.103[2]	1.025	2.00359	$MeNO_2$	Ox $AlCl_3$	248	205
1238[c]	-,2,5-di-*t*-butyl-	*trans*	2.87	2.87	u[2]	2.87	u[2]	2.87		H_2SO_4	Ox	293	284
			2.87	2.87	u[2]	0.48[D]	u[2]	0.48[D]		D_2SO_4	Ox	293	284
1239	-,tetramethyl-		2.955	2.955	2.119[1]	2.119[1]	2.119[1]	2.119[1]		$MeNO_2$	Ox $AlCl_3$	193	279, 288
			2.928	2.928	2.134[1]	2.134[1]	2.134[1]	2.134[1]				210	279
			2.913	2.913	2.141[1]	2.141[1]	2.141[1]	2.141[1]				228	279
			2.853	2.853	2.144[1]	2.144[1]	2.144[1]	2.144[1]				249	279
			2.814	2.814	2.148[1]	2.148[1]	2.148[1]	2.148[1]				288	279
			2.790	2.790	2.153[1]	2.153[1]	2.153[1]	2.153[1]				303	279

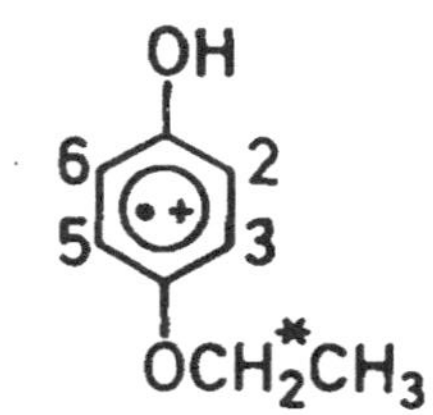

No.	Radical	a_1^{OH}	a_4^{OH}	a_2	a_3	a_5	a_6	g-Factor	Solvent	Method	T [K]	Ref.
1240	4-Ethoxyphenol cation											
1241	-,tetramethyl-	2.65[OH]	2.65*	2.068[1]	2.068[1]	2.068[1]	2.068[1]	2.00350	$MeNO_2$	Ox $AlCl_3$	253	277

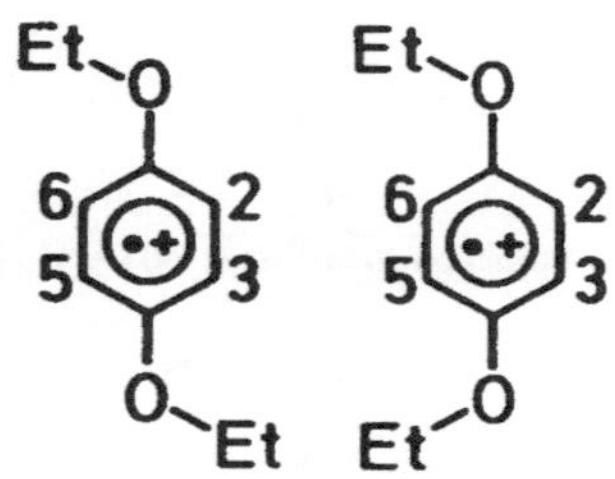

trans *cis*

			a_1^{OEt}	a_4^{OEt}									
1242	1,4-Diethoxybenzene cation	*trans*	4.103[f] 0.152[f]	4.103[f] 0.152[f]	2.925	1.614	2.925	1.614	2.0037	PhMe	Ox $AlCl_3$	190.5	413
			4.023	4.023	2.928	1.616	2.928	1.616				209	413
			4.027	4.027	2.921	1.613	2.921	1.613				213	413
			3.969	3.969	2.921	1.606	2.921	1.606				224	413
			3.951	3.951	2.911	1.602	2.911	1.602				233	413
			3.908	3.908	2.929	1.612	2.929	1.612				237	413
		cis	3.883[f] 0.152[f]	3.883[f] 0.152[f]	1.883	1.883	2.639	2.639	2.0037	PhMe	Ox $AlCl_3$	190.5	413
			3.807	3.807	1.886	1.886	2.649	2.649				209	413
			3.801	3.801	1.879	1.879	2.659	2.659				213	413
			3.786	3.786	1.875	1.875	2.640	2.640				224	413
			3.730	3.730	1.901	1.901	2.637	2.637				233	413
			3.704	3.704	1.887	1.887	2.661	2.661				237	413
		trans	3.789[f] 0.131[f]	3.789[f] 0.131[f]	2.885	1.587	2.885	1.587		H_2SO_4	Ox	293	413
		cis	3.561[f] 0.131[f]	3.561[f] 0.131[f]	1.812	1.812	2.661	2.661		H_2SO_4	Ox	293	413
1243	-,2,5-dimethyl-	*trans*	3.63	3.63	4.25[1]	0.63	4.25[1]	0.63		H_2SO_4	Ox	293	284
		trans	3.582[f] 0.136[f]	3.582[f] 0.136[f]	4.158[1]	0.581	4.158[1]	0.581	2.00363[g]	PhMe	Ox $AlCl_3$	248	205, 275
1244	-,2,5-dipropyl-	*trans*	3.46	3.46	3.46[1]	0.57	3.46[1]	0.57		H_2SO_4	Ox	293	284
1245	-,2,5-diisopropyl-	*trans*	3.50	3.50	2.83[1]	0.60	2.83[1]	0.60		H_2SO_4	Ox	293	284
1246	-,2,5-di-*t*-butyl-	*trans*	4.050[f] 0.155[f]	4.050[f] 0.155[f]	0.099[2]	1.057	0.099[2]	1.057	2.00366	PhMe	Ox $AlCl_3$	248	205, 275
1247[e]	-,2,5-di-*t*-butyl-	*trans*	3.38	3.38	u[2]	3.38	u[2]	3.38		H_2SO_4	Ox	293	284

Table III.4 (continued)
1,4-DIHYDROXY/ALKOXYBENZENE DERIVED RADICALS

Cations

No.	Radical		Splitting constants/Gauss a_1^{OH}	a_4^{OH}	a_2	a_3	a_5	a_6	g-Factor	Solvent	Method	T [K]	Ref.
1248	-,2,5-dimethylthio-	*trans*	1.967[f] 0.078[f]	1.967[f] 0.078[f]	3.457[h]	0.789	3.457[h]	0.789	2.00690	$MeNO_2$	Ox $AlCl_3$		275
1249	-,2,5-diethylthio-	*trans*	1.964	1.964	3.194[h]	0.770	3.194[h]	0.770	2.00690	$MeNO_2$	Ox $AlCl_3$		275

trans isomer

No.	Radical		a_1^{OH}	a_4^{OH}	a_2	a_3	a_5	a_6	g-Factor	Solvent	Method	T [K]	Ref.
1250	1,4-Dipropoxybenzene cation	(R = Pr)	4.114[i] 0.101[i]	4.114[i] 0.101[i]	2.965	1.645	2.965	1.645	2.0037	PhMe	Ox $AlCl_3$	215	413
1251	1,4-Dibutoxybenzene cation	(R = Bu)	4.102[i] 0.113[i]	4.102[i] 0.113[i]	3.000	1.630	3.000	1.630	2.0037	PhMe	Ox $AlCl_3$	215	413
1252	1,4-Diheptyloxybenzene cation	(R = C_7H_{15})	4.188[i] 0.122[i]	4.188[i] 0.122[i]	3.000	1.664	3.000	1.664	2.0037	PhMe	Ox $AlCl_3$	200	413
1253	1,4-Dioctyloxybenzene cation	(R = C_8H_{17})	4.135	4.135	3.031	1.675	3.031	1.675	2.0037	PhMe	Ox $AlCl_3$	210	413

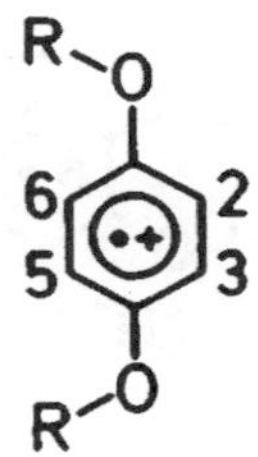

cis isomer

1250	1,4-Dipropoxybenzene cation	(R = Pr)	3.873[i] 0.101[i]	3.873[i] 0.101[i]	1.820	1.820	2.590	2.590	2.0037	PhMe	Ox $AlCl_3$	215	413
1251	1,4-Dibutoxybenzene cation	(R = Bu)	3.874[i] 0.113[i]	3.874[i] 0.113[i]	1.864	1.864	2.651	2.651	2.0037	PhMe	Ox $AlCl_3$	215	413
1252	1,4-Diheptyloxybenzene cation	(R = C_7H_{15})	3.952[i] 0.122[i]	3.952[i] 0.122[i]	1.886	1.886	2.706	2.706	2.0037	PhMe	Ox $AlCl_3$	200	413
1253	1,4-Dioctyloxybenzene cation	(R = C_8H_{17})	3.918	3.918	1.899	1.899	2.764	2.764	2.0037	PhMe	Ox $AlCl_3$	210	413

[a] *Trans* and *cis* isomers observed simultaneously. The assignment shown is in line with INDO calculations[x] (cf. b below).

[b] It has been argued that the preferred conformation for the hydroxyl proton is *anti* to a methyl substituent,[304] leading to the *cis* form for the 2,3-dimethyl compound and the *trans* form for 2,5-dialkyl compounds.

[c] Relative isomer ratio estimated to 70:30 in favor of the *trans* isomer.

[d] *In situ* radiolysis of N_2O saturated 10^{-3} *M* Tl_2SO_4 solutions containing 10^{-4} *M* of phenol.

[e] The identity of this cation observed in H_2SO_4 has been questioned. See Reference 205.

[f] Upper splitting from methylene protons of ethoxy group; lower splitting from methyl protons.

[g] g-Factor from Reference 275.

[h] Splitting from β-methylthio protons.

[i] Upper splitting from OCH_2 protons of alkoxy group; lower splitting from $OCCH_3$ protons.

Table III.5
PHENOL RADICALS WITH THREE KEY OXYGENS [1,2,3]

Cations

No.	Radical	Substituent(s)	Splitting constants/Gauss a_1	a_2	a_3	a_4	a_5	a_6	g-Factor	Solvent	Method	Ref.
	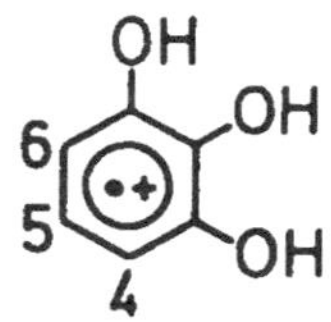											
1254	2,3-Dihydroxyphenol cation[a]											
	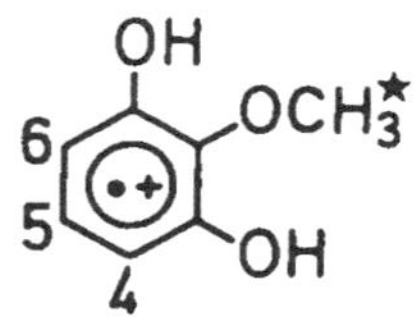											
1255	2-Methoxy-3-hydroxyphenol cation											
1256	-,5-carboxy-	–COOH	—[b]	5.15*	—[b]	0.45	—	0.45	2.00362	H_2SO_4,4 H_2O	Ox CeIV	256
	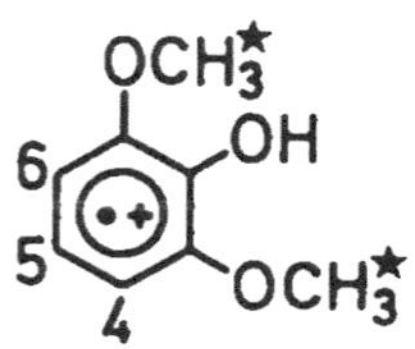											

1257	1,3-Dimethoxy-2-hydroxybenzene cation		1.8*	—[b]	1.8*	1.0	7.0	1.0	2.00360	H_2SO_4,4 H_2O	Ox CeIV	256

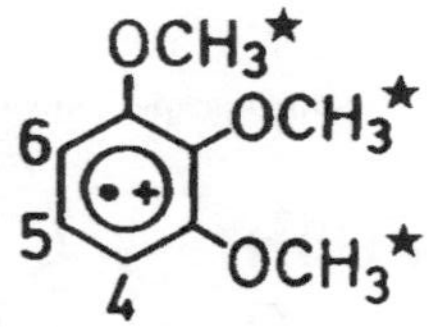

1258	1,2,3-Trimethoxybenzene cation		2.03*	5.16*	2.03*	0.47	6.26	0.47	2.00398	w/5°C	Radiolysis	266
			2.05*	5.25*	2.05*	0.45	6.25	0.45	2.0038	H_2SO_4,4 H_2O	Ox CeIV	269, 271
1259	-,5-methyl-	–Me	1.80*	5.0*	1.80*	0.05	8.95[l]	0.05	2.0037	H_2SO_4,4 H_2O	Ox CeIV	272
1260	-,5-(2-ammonioethyl)-	$–CH_2CH_2NH_3^+$	2.0*	5.15*	2.0*	0.30	5.65[l]	0.30	2.0038	H_2SO_4,4 H_2O	Ox CeIV	272
1261	-,5-carboxymethyl-	$–CH_2COOH$	2.0*	5.20*	2.0*	0.33	5.20[l]	0.33	2.0037	H_2SO_4,4 H_2O	Ox CeIV	272
1262	-,4-formyl-	–CHO	1.75*	5.70*	2.25*	0.0	5.55	0.45	2.0040	H_2SO_4,4 H_2O	Ox CeIV	272
1263	-,5-formyl-	–CHO	2.30*	5.30*	2.30*	0.80	0.10[l]	0.80	2.0043	H_2SO_4,4 H_2O	Ox CeIV	272
1264	-,5-carboxy-	–COOH	2.25*	3.23*	2.25*	0.75	—	0.75	2.0040	H_2SO_4,4 H_2O	Ox CeIV	272

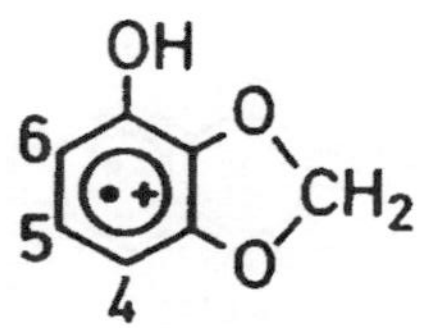

				$a_H^{CH_2}$								
1265	2,3-Methylenedioxyphenol cation		—[b]	18.80		0.55	5.85	0.95	2.00391	H_2SO_4/w	Ox CeIV	241, 249
1266	-,6-acetyl-	–COMe	—[b]	17.55		0.01	5.15	—	2.00413	H_2SO_4/w	Ox CeIV	241

Table III.5 (continued)
PHENOL RADICALS WITH THREE KEY OXYGENS [1,2,3]

Cations

No.	Radical	Substituent(s)	Splitting constants/Gauss a_1	a_2	a_3	a_4	a_5	a_6	g-Factor	Solvent	Method	Ref.
	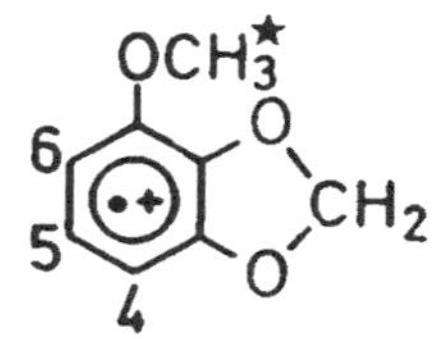											
1267	1-Methoxy-2,3-methylenedi-oxybenzene cation		1.5*	19.25		0.85	6.05	0.85	2.0040	H_2SO_4,4 H_2O	Ox CeIV	269
1268	-,5-allyl-	$-CH_2CH{=}CH_2$	1.35*	18.90		0.65	7.65^{1}	0.25	2.0044	H_2SO_4,4 H_2O	Ox CeIV	272
1269	-,5-(2-amoniopropyl)-	$-CH_2CH(Me)NH_3^+$	1.5*	18.5		0.3	$5.6^{1,c}$ $4.9^{1,c}$	0.8	2.0040	H_2SO_4,4 H_2O	Ox CeIV	269
	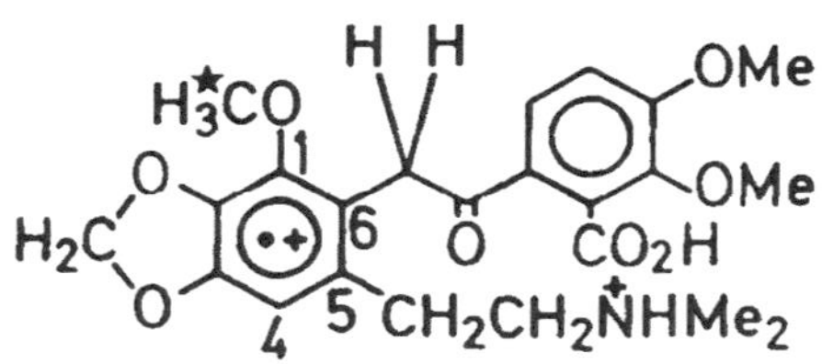											
1270	Narceine radical cation		1.3*	20		0	5.6^{1}	0.7^{1}		H_2SO_4,4 H_2O	Ox CeIV	269

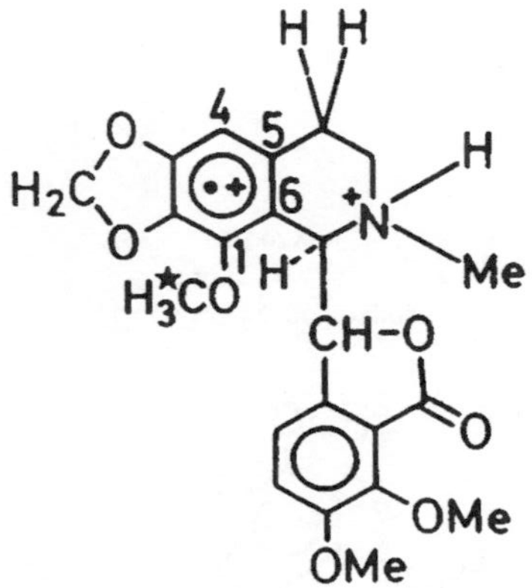

1271	Narcotine radical cation	1.4*	18.5	0.4	10.25[1,c] 11.5[1,c]	0	H_2SO_4,4 H_2O	Ox CeIV	269

[a] A spectrum of this cation has apparently not been observed, cf. References 86 and 269.
[b] Proton exchange always fast; no hyperfine interactions with hydroxylic protons observed.
[c] Methylene protons nonequivalent.

Table III.6
PHENOL RADICALS WITH THREE KEY OXYGENS [1,2,4]

Cations

No.	Radical	Substituent	Splitting constants/Gauss a_1	a_2	a_3	a_4	a_5	a_6	g-Factor	Solvent	Method	Ref.
	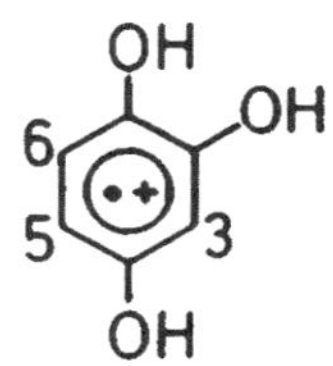											
1272	2,4-Dihydroxyphenol cation		—[a]	—[a]	0.35	—[a]	5.1	0.5	2.00353	H_2SO_4,4 H_2O	Ox CeIV	256
	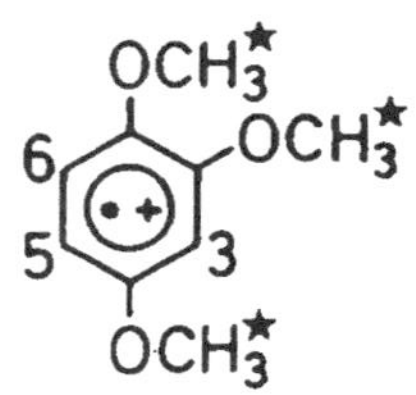											
1273	1,2,4-Trimethoxybenzene cation		(Data not reported, but see Reference 280)									
1274	-,5-carboxy-	–COOH	3.40*	1.85*	0.35	2.65*	—	0.35	2.00385	H_2SO_4,4 H_2O	Ox CeIV	272

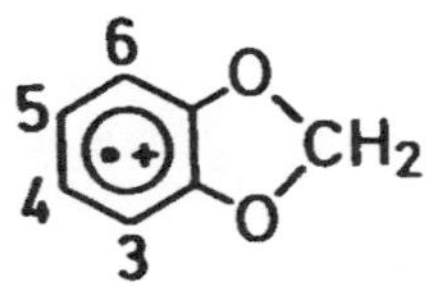

			$a_H^{CH_2}$								
1275	1,2-Methylenedioxybenzene cation										
1276	-,4-hydroxy-	–OH	13.5	0	—[a]	6.25	0.4	2.00395	0.5 *M* H_2SO_4	Ox CeIV	269, 241
1277	-,4-methoxy-	–OCH_3	14.45	0.45	2.65*	6.25	0.45	2.0040	H_2SO_4.4 H_2O	Ox CeIV	269

[a] Proton exchange always fast; no hyperfine interactions with hydroxylic protons observed.

Table III.7
PHLOROGLUCINOL RADICALS

Cations

No.	Radical	Substituent	Splitting constants/Gauss a_1	a_2	a_3	a_4	a_5	a_6	g-Factor	Solvent	Method	Ref.
	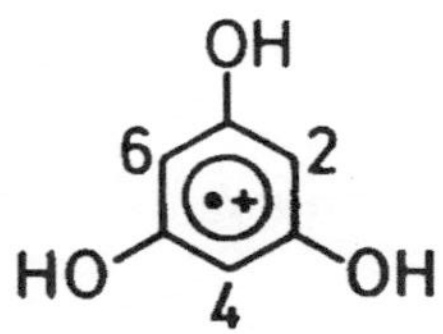											
1278	Phloroglucinol cation		u	7.25	u	7.25	u	7.25		H_2SO_4/w	Ox CeIV	84
		R_1^a	u	7.4	u	7.4	u	7.4	Δg = 4.00[b]	4.5 *M* H_2SO_4	Ox CeIV	257
		R_2^a	u	7.2[c]	u	7.2[c]	u	7.2[c]	Δg = 4.00[b]	9 *M* H_2SO_4	Ox CeIV	257

Table III.7 (continued)
PHLOROGLUCINOL RADICALS

Cations

No.	Radical	Substituent	Splitting constants/Gauss						g-Factor	Solvent	Method	Ref.
			a_1	a_2	a_3	a_4	a_5	a_6				
1279	-,2-methyl-	–Me	u	17.70[i]	u	3.10	u	3.10	2.00303	H_2SO_4/w	Ox CeIV	241
1280	-,2-acetyl-	–COMe	u	—	u	11.5	u	10.4	2.00344	H_2SO_4,4 H_2O	Ox CeIV	256
	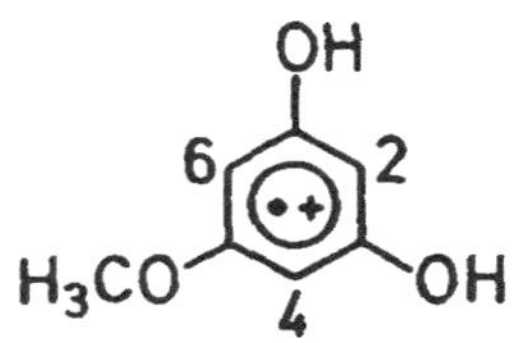											
1281	3-Hydroxy-5-methoxyphenol cation		R_1^a u	5.3	u	8.25	1.30[d]	8.25	Δg = 3.94[b]	4.5 *M* H_2SO_4	Ox CeIV	257
			R_2^a u	6.2[c]	u	7.6	1.9[d]	7.6	Δg = 3.94[b]	9 *M* H_2SO_4	Ox CeIV	257
	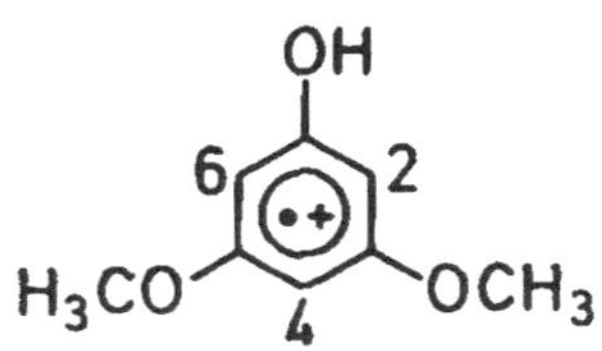											
1282	3,5-Dimethoxyphenol cation		R_1^a u	5	1.31[d]	11.7	1.31[d]	5	Δg = 3.93[b]	4.5 *M* H_2SO_4	Ox CeIV	257
			R_2^a u	5.5	1.75[d]	10.2	1.75[d]	5.5	Δg = 3.93[b]	9 *M* H_2SO_4	Ox CeIV	257

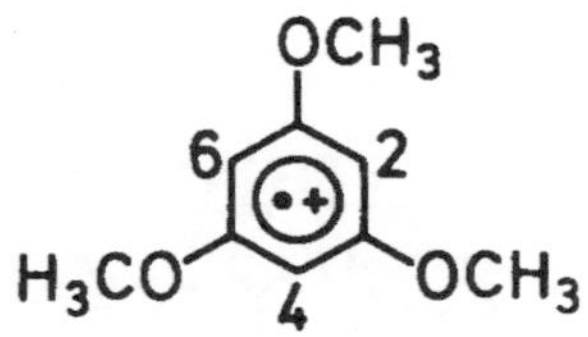

No.	Radical	Substituent	a_1	a_2	a_3	a_4	a_5	a_6	g-Factor	Solvent	Method	Ref.
1283	1,3,5-Trimethoxybenzene cation		(For data of this radical see References 257 and 280)									
1284	-,2-methyl-	–Me	0.50[d]	15.60'	0.50[d]	2.50	4.50[d]	2.50	2.00315	H_2SO_4.4 H_2O	Ox CeIV	272

[a] The radicals R_1 and R_2 represent some (unknown) mixture of interconverting types. See Reference 257 for details.
[b] The g-shift in Gauss relative to that of Fremy's salt is Δg.
[c] Values as directly measured.
[d] Splitting from methoxyl group protons.

Table III.8
PHENOL RADICALS WITH FOUR KEY OXYGENS [1,2,3,4] AND [1,2,3,5]

Cations

No.	Radical	Substituent	Splitting constants/Gauss a_1	a_2	a_3	a_4	a_5	a_6	g-Factor	Solvent	Method	Ref.
1285	1,2,3,4-Tetrahydroxybenzene cation 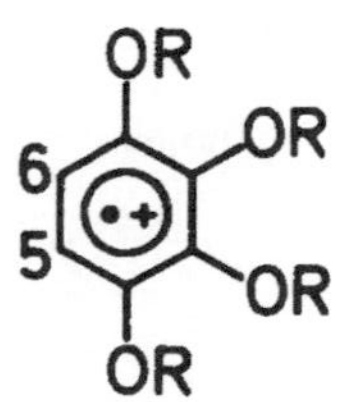	[R = H]	(Data not reported)									

Table III.8 (continued)
PHENOL RADICALS WITH FOUR KEY OXYGENS [1,2,3,4] AND [1,2,3,5]

Cations

No.	Radical	Substituent	Splitting constants/Gauss a_1	a_2	a_3	a_4	a_5	a_6	g-Factor	Solvent	Method	Ref.
1286	1,2,3,4-Tetramethoxybenzene cation 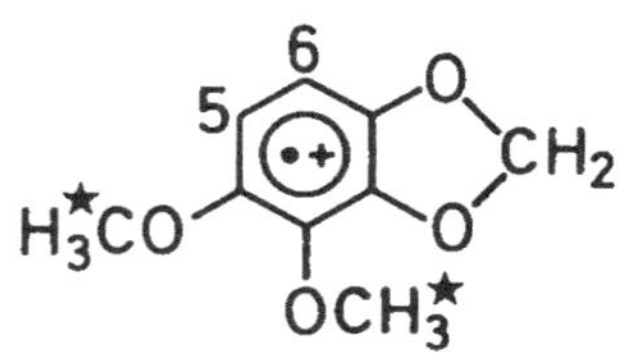	[R = CH_3^*]	2.80*	0.55*	0.55*	2.80*	3.0	3.0	2.0039	H_2SO_4.4 H_2O	Ox CeIV	272
1287	3,4-Dimethoxy-1,2-methylene-dioxybenzene cation		(Data not reported)									
				$a_H^{CH_2}$								
1288	-,5-allyl- 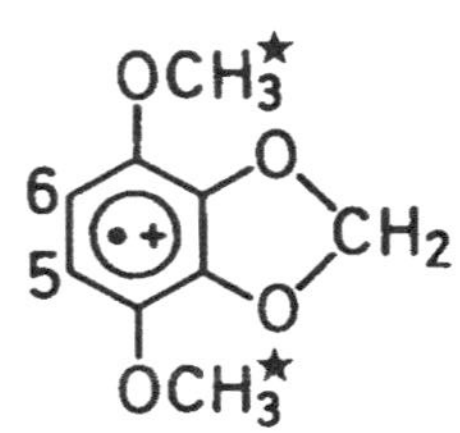	$-CH_2CH{=}CH_2$		12.15	0.0*	3.0*	6.1'	0.0	2.0038	H_2SO_4.4 H_2O	Ox CeIV	272

1289	1,4-Dimethoxy-2,3-methylene-dioxybenzene cation		(Data not reported)									
					$a_H^{CH_2}$							
1290	-,5-allyl-	$-CH_2CH{=}CH_2$	1.32*		17.42	1.32*	7.70[1]	0.60	2.0038	H_2SO_4.4 H_2O	Ox CeIV	272
	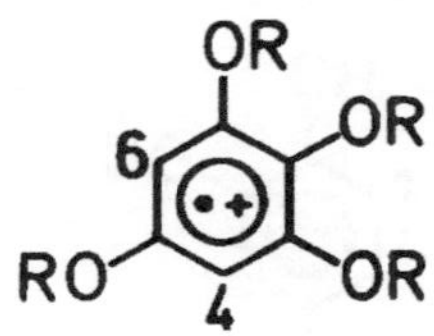											
					a_2	a_3						
1291	1,2,3,5-Tetramethoxybenzenecation	[R = CH_3]	2.20*	4.70*	2.20*	0.20	2.60*	0.20	2.00385	H_2SO_4.4 H_2O	Ox CeIV	272

Table III.9
PHENOL RADICALS WITH FOUR KEY OXYGENS [1,2,4,5]

Cations

No.	Radical	Substituent(s)	Splitting constants/Gauss a_1	a_2	a_3	a_4	a_5	a_6	g-Factor	Solvent	Method	Ref.
	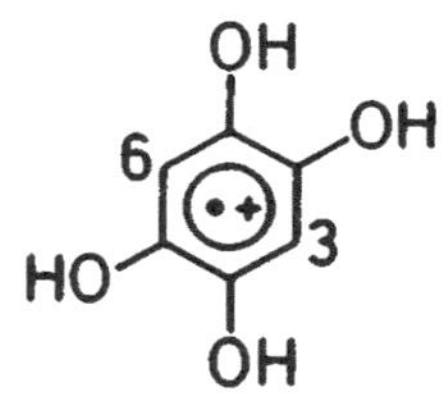											
1292	1,2,4,5-Tetrahydroxybenzene cation		1.77[a]	1.77[a]	0.98	1.77[a]	1.77[a]	0.98		H_2SO_4 conc	Re DTN	201
			1.709[a]	1.709[a]	0.946	1.709[a]	1.709[a]	0.946		$MeNO_2$/0°C	Re H_2SO_4	273
			1.726[a]	1.726[a]	0.948	1.726[a]	1.726[a]	0.948		$EtNO_2$/0°C	Re H_2SO_4	273
	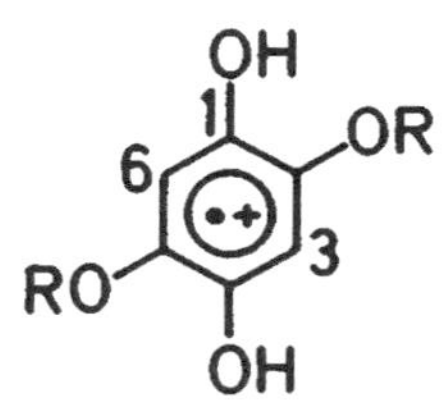											
1293	1,4-Dihydroxy-2,5-dimethoxybenzene cation	[R = CH_3]	1.842[a]	2.142*	0.898	1.842[a]	2.142*	0.898	2.00385	$MeNO_2$/ − 20°C	Ox $AlCl_3$	207

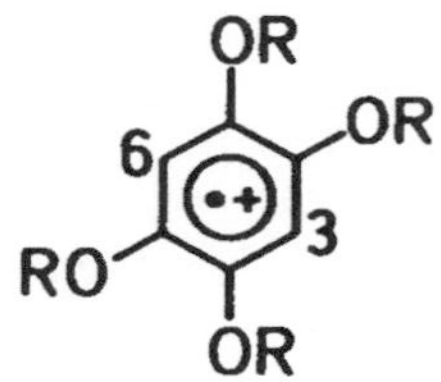

1294	1,2,4,5-Tetramethoxybenzene cation	$[R = CH_3]$	2.21*	2.21*	0.89	2.21*	2.21*	0.89		MeCN	Ox el	280
			2.35*	2.35*	0.95	2.35*	2.35*	0.95	2.0039	H_2SO_4.4 H_2O	Ox CeIV	207
			2.27*	2.27*	0.85	2.27*	2.27*	0.85		PhCN/19°C	Ox el	272
			2.22*	2.22*	0.89	2.22*	2.22*	0.89	2.00402	CH_2Cl_2/20°C	Ox[h]	274
			2.245*	2.245*	0.877	2.245*	2.245*	0.877	2.00396	$MeNO_2$/ −50°C	Ox $AlCl_3$	274
			2.242*	2.242*	0.134[c]	2.242*	2.242*	0.134[c]	2.00396	$MeNO_2$/ −50°C	Ox $AlCl_3$	207
1295	1,2,4,5-Tetraethoxybenzene cation	$[R = CH_2CH_3]$	2.656[d]	2.656[d]	0.858	2.656[d]	2.656[d]	0.858	2.00401	$MeNO_2$/ −20°C	Ox $AlCl_3$	275
			0.114[c]	0.114[c]		0.114[c]	0.114[c]					
			2.537[d]	2.537[d]	0.869	2.537[d]	2.537[d]	0.869		$MeNO_2$/20°C	Ox $AlCl_3$	275
			0.107[c]	0.107[c]		0.107[c]	0.107[c]					

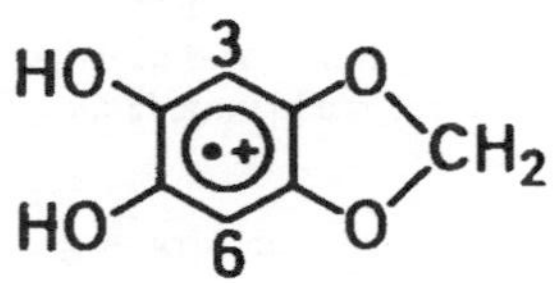

							$a_H^{CH_2}$					
1296	1,2-Dihydroxy-4,5-methylenedioxybenzene cation		—	—	1.00		11.05	1.00	2.00394	H_2SO_4/w	Ox CeIV	241

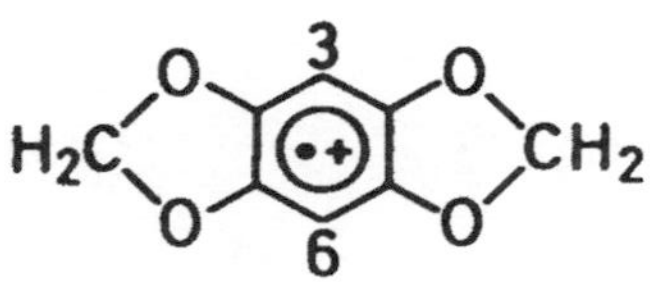

Table III.9 (continued)
PHENOL RADICALS WITH FOUR KEY OXYGENS [1,2,4,5]

Cations

No.	Radical	Substituent(s)	Splitting constants/Gauss						g-Factor	Solvent	Method	Ref.
			a_1	a_2	a_3	a_4	a_5	a_6				
			$a_H^{CH_2}$									
1297	1,2,4,5-Bis(methylenedioxy)benzene cation		11.75		1.025		11.75	1.025	2.00435	H_2SO_4,4 H_2O	Ox CeIV	272

[a] Splitting from hydroxyl proton.
[b] Parent compound oxidized with tris (*p*-bromophenyl)amminium perchlorate.
[c] Splitting from deuteron at C-3 and C-6.
[d] Splitting from two equivalent methylene protons.
[e] Splitting from three methyl protons.

Table III.10
PHENOL RADICALS WITH FIVE AND SIX KEY OXYGENS

Cations

No.	Radical	Substituent(s)	a_1	a_2	a_3	a_4	a_5	a_6	g-Factor	Solvent	Method	Ref.
			Splitting constants/Gauss									
	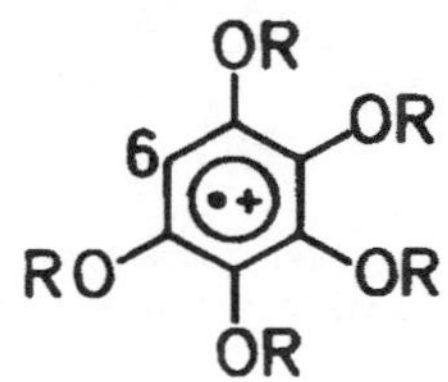											
1298	Pentamethoxybenzene cation	[R = CH_3]	2.28*	1.98*	0.30*	1.98*	2.28*	0.60		MeCN	Ox el	280
			2.25*	2.05*	0.25*	2.05*	2.25*	0.60	2.0039	H_2SO_4,4 H_2O	Ox CeIV	272, 280
	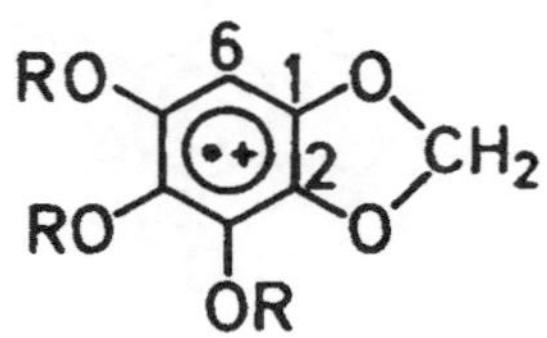											
1299	3,4,5-Trimethoxy-1,2-methylenedioxybenzene cation	[R = CH_3]		$a_H^{CH_2}$ 11.25	0.05*	2.15*	2.15*	0.75	2.0040	H_2SO_4,4 H_2O	Ox CeIV	272

Table III.10 (continued)
PHENOL RADICALS WITH FIVE AND SIX KEY OXYGENS

Cations

No.	Radical	Substituent(s)	Splitting constants/Gauss a_1	a_2	a_3	a_4	a_5	a_6	g-Factor	Solvent	Method	Ref.
	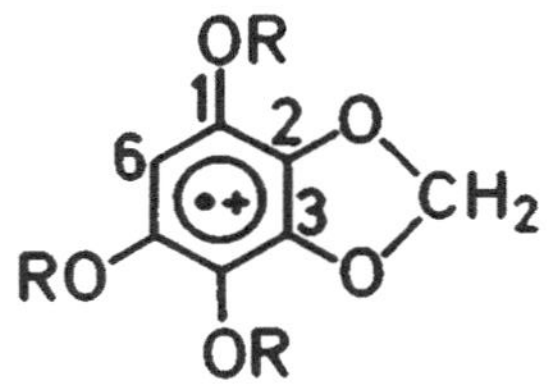			$a_H^{CH_2}$								
1300	1,4,5-Trimethoxy-2,3-methylenedioxybenzene cation	[R = CH_3]	2.30*	4.35		1.45*	2.30*	0.20	2.0040	H_2SO_4.4 H_2O	Ox CeIV	272
	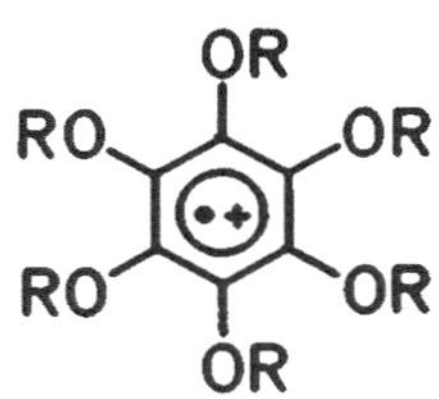											
1301	Hexamethoxybenzene cation	[R = CH_3]	1.30*	1.30*	1.30*	1.30*	1.30*	1.30*		MeCN	Ox el	280
			1.40*	1.40*	1.40*	1.40*	1.40*	1.40*		96% H_2SO_4	Ox	280

Table III.11
1,4-DIHYDROXY/ALKOXYNAPHTHALENE DERIVED RADICALS

Cations

No.	Radical	a_1^{OH}	a_4^{OH}	a_2	a_3	a_5	a_6	a_7	a_8	g-Factor	Solvent	Method	T [K]	Ref.
	Splitting constants/Gauss													
	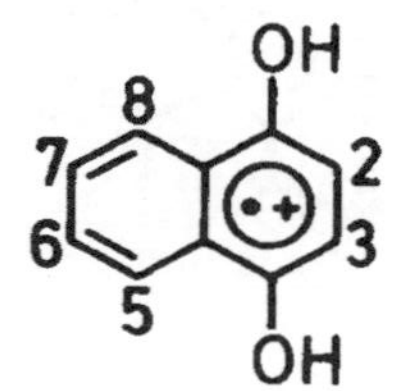													
1302	1,4-Dihydroxynaphthalene cation	2.449	2.449	3.085	3.085	0.818	1.635	1.635	0.818	2.00323	$MeNO_2/H_2SO_4$	Ox	226	202
		2.42	2.42	3.20	3.20	0.86	1.80	1.80	0.86		H_2SO_4	Re DTN		201
		2.37	2.37	3.15	3.15	0.85	1.77	1.77	0.85		H_2SO_4	Ox	RT	8
		0.38[D]	0.38[D]	3.12	3.12	0.84	1.77	1.77	0.84		D_2SO_4	Ox	RT	8
1303	-,2-methyl-	(Spectrum not analyzed)									H_2SO_4	Ox	RT	8
1304	-,6-methyl-	(Spectrum not analyzed)									H_2SO_4	Ox	RT	8
1305	-,2,3-dimethyl-	1.956	1.956	2.666[l]	2.666[l]	0.896	1.678	1.678	0.896	2.00318	$MeNO_2/H_2SO_4$	Ox	206	202
	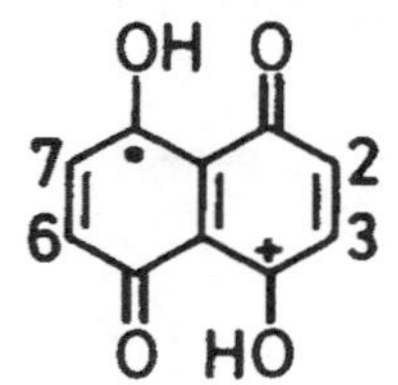													
1306	Dihydro-1,4,5,-8-naphthalenetetrone cation	3.1	3.1	0.77	0.77	—	0.77	0.77	—		H_2SO_4	Re/tetrone		164
		0.5[D]	0.5[D]	0.77	0.77	—	0.77	0.77	—		D_2SO_4	Re/tetrone		164

Table III.11 (continued)
1,4-DIHYDROXY/ALKOXYNAPHTHALENE DERIVED RADICALS

Cations

No.	Radical	Splitting constants/Gauss a_1^{OH}	a_4^{OH}	a_2	a_3	a_5	a_6	a_7	a_8	g-Factor	Solvent	Method	T [K]	Ref.
	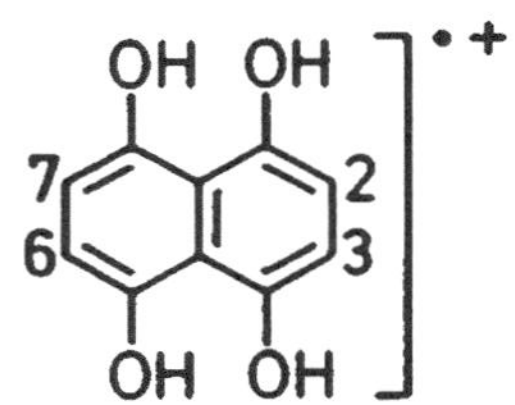													
1307	Naphthazarin cation	1.2	1.2	2.3	2.3	1.2^{OH}	2.3	2.3	1.2^{OH}		H_2SO_4	Re Zn	323	415
		0.19^{D}	0.19^{D}	2.38	2.38	0.19^{D}	2.38	2.38	0.19^{D}		D_2SO_4	Re Zn	323	415
		(0.7	1.8	2.3	2.3	0.7	2.3	2.3	1.8)[a]		H_2SO_4	Re Zn	263	415, 416
	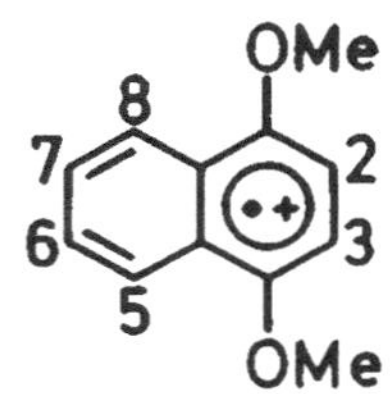													
1308	1,4-Dimethoxynaphthalene cation	2.194[b]	2.194[b]	3.354	3.354	0.701	1.497	1.497	0.701	2.00321	$MeNO_2$	Ox $AlCl_3$	200	293
		2.166[b]	2.166[b]	3.352	3.352	0.699	1.518	1.518	0.699		$MeNO_2$	Ox $AlCl_3$	265	293

[a] Data obtained from a simulation using the Bloch equation solution for a four jump model.[415] The results may need modification in that both ring and hydroxy protons show alternating linewidth effects.[416]

[b] Splitting from three methoxyl group protons.

Table III.12
9,10-DIHYDROXY/ALKOXYANTHRACENE DERIVED RADICALS

Cations

No.	Radical	a_1	a_2	a_3	a_4	a_5	a_6	a_7	a_8	a_9	a_{10}	Solvent	Method	Ref.
		Splitting constants/Gauss												
	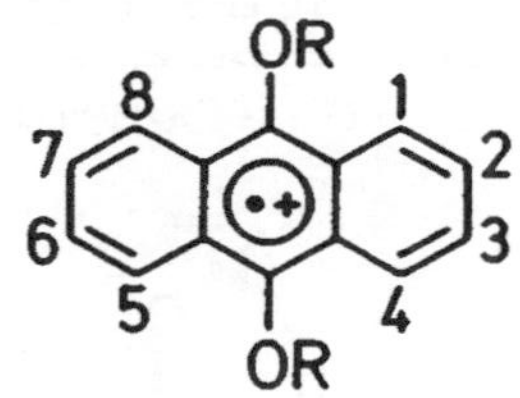													
1309	9,10-Dihydroxyanthracene cation (R = H*)	1.551	1.037	1.037	1.551	1.551	1.037	1.037	1.551	1.276*	1.276*	$MeNO_2$/−56°C	Re H_2SO_4	202[a]
		1.66	1.07	1.07	1.66	1.66	1.07	1.07	1.66	1.31*	1.31*	H_2SO_4	Re Zn	201
		1.66	1.07	1.07	1.66	1.66	1.07	1.07	1.66	0.20[D]	0.20[D]	D_2SO_4	Re Zn	201
1310	9,10-Dimethoxyanthracene cation (R = CH_3*)	1.752	1.084	1.084	1.752	1.752	1.084	1.084	1.752	1.186*	1.186*	$MeNO_2$/−55°C	Re $AlCl_3$	293[b]
1311	Dihydro-1,2,9,-10-anthracenetetrone cation[c]	(0.45(1H),[d] 1.9(3H), 0.95(3H)[e]										H_2SO_4	Re Zn	429
	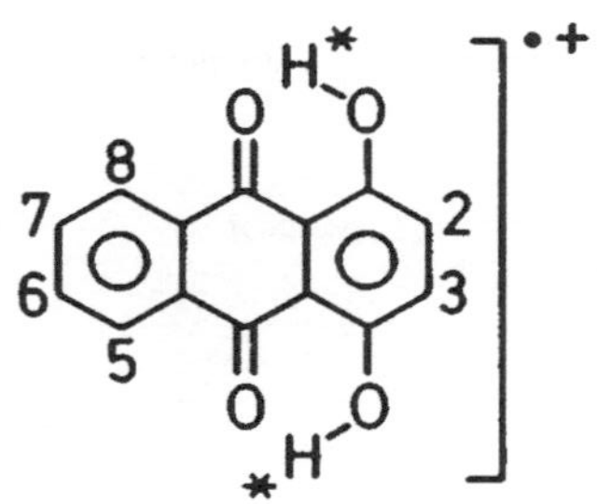													
1312	Dihydro-1,4,9,-10-anthracenetetrone cation	2.8*	1.25	1.25	2.8*	u	0.3	0.3	u	—	—	H_2SO_4	Re/tetrone	164
		0.42[D]	1.25	1.25	0.42[D]	u	u	u	u	—	—	D_2SO_4	Re/tetrone	164

Table III.12 (continued)
9,10-DIHYDROXY/ALKOXYANTHRACENE DERIVED RADICALS

Cations

No.	Radical	a_1	a_2	a_3	a_4	a_5	a_6	a_7	a_8	a_9	a_{10}	Solvent	Method	Ref.
		Splitting constants/Gauss												
1313	-,5,8-dichloro-	2.98*	1.48	1.48	2.98*	—	u	u	—	—	—	H_2SO_4	Re/tetrone	164
		0.48[D]	1.48	1.48	0.48[D]	—	u	u	—	—	—	D_2SO_4	Re/tetrone	164
1314	Dihydro-9,10,11,12-naphthacenetetrone cation	0.73	0.25	0.25	0.73	0.73	0.25	0.25	0.73	1.96*	1.96*	H_2SO_4	Re/tetrone	164
		0.73	0.25	0.25	0.73	0.73	0.25	0.25	0.73	0.31[D]	0.31[D]	D_2SO_4	Re/tetrone	164
1315	-,2,3,6,7-tetrachloro-	0.70	—	—	0.70		—	—	0.70	2.0*	2.0*	H_2SO_4	Re/tetrone	164
1316	Dihydro-1,5,9,10-anthracenetetrone cation[c]	(0.45(2H),[d] 1.75(4H), 0.9(2H)[e]										H_2SO_4	Re Zn	429
1317	Dihydro-1,8,9,10-anthracenetetrone cation[c]	(0.45(2H),[d] 1.75(4H), 0.9(2H)[e]										H_2SO_4	Re Zn	429
1318	Dihydro-2,6,9,10-anthracenetetrone cation[c]	(2.40(2H), 1.2(4H)[e]										H_2SO_4	Re Zn	429
1319	1,2,3-Trihydroxyanthraquinone[c]	(0.65(1H),[d] 2.6(2H), 1.3(3H)[e]										H_2SO_4	Re Zn	429

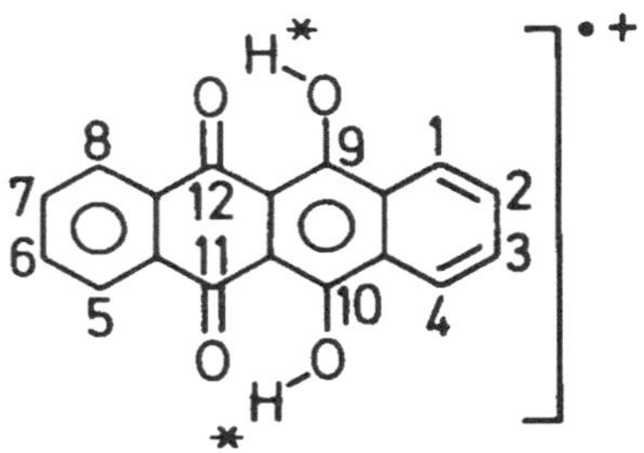

1320	1,2,5,8-Tetrahydroxyanthraquinone[c]	(0.50(3H),[d] 2.15(3H), 1.07(1H)[e]	H_2SO_4	Re Zn	429
1321	1,2,3,5,6,7-Hexahydroxyanthraquinone[c]	(1.60(2H),[d] 3.25(2H)[e]	H_2SO_4	Re Zn	429

[a] g-Factor = 2.00309.
[b] g-Factor = 2.00312.
[c] Structure of cation does not appear from Reference 429.
[d] Splitting from hydroxyl proton(s).
[e] Data seemingly obtained from low-resolution spectra. No assignment given.

IV. SEMIQUINONE-CATION COMPLEXES

Table IV.1.1
1,2-BENZOSEMIQUINONE-CATION COMPLEXES

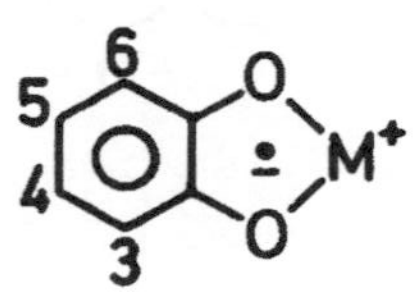

No.	Complex	Central atom		Splitting constants/Gauss a_M	$a^H_{3,6}$	$a^H_{4,5}$	$a^C_{1,2}$	$a^C_{3,6}$	$a^C_{4,5}$	g-Factor	Solvent	Method	T [K]	Ref.
C1	No. 1	IA	$^7Li(I)$	0.48	0.87	3.64	u	3.08	2.32		Bu^tOH	Ox I_2/O_2	RT	6, 54
				0.52	0.93	3.62	u	3.14	2.38		Bu^tOH/MeTHF	Ox I_2/O_2	RT	6
				0.64	1.12	3.60	u	u	u		MeTHF	Ox I_2/O_2	RT	6
C2	No. 1		$^{23}Na(I)$	0.27	0.96	3.53	u	2.93	2.20		Bu^tOH	Ox I_2/O_2	RT	6, 54
				0.23	1.00	3.52	u	u	u		Bu^tOH/MeTHF	Ox I_2/O_2	RT	6
				0.43	1.28	3.43	u	u	u		C_5H_5N	NaH[a]	298	297
				0.49	1.31	3.40	u	u	u		THF	NaH[a]	298	297
				0.50	1.29	3.60	u	u	u		MeTHF	Ox I_2/O_2	RT	6
C3	No. 1		K(I)	u	0.96	3.50	1.20	2.94	2.16		Bu^tOH	Ox I_2/O_2	RT	6
				u	0.94	3.42	1.22	2.94	2.16		Bu^tOH/MeTHF	Ox I_2/O_2	RT	6

Table IV.1.1 (continued)
1,2-BENZOSEMIQUINONE-CATION COMPLEXES

No.	Complex	Central atom	Splitting constants/Gauss a_M	$a^H_{3,6}$	$a^H_{4,5}$	$a^C_{1,2}$	$a^C_{3,6}$	$a^C_{4,5}$	g-Factor	Solvent	Method	T [K]	Ref.
			u	0.94	3.52	1.22	2.93	2.15		Bu^{t}OH/MeTHF	Ox I_2/O_2	293	6
			u	0.98	3.50	1.36	2.80	2.1		Bu^{t}OH/MeTHF	Ox I_2/O_2	277	6
			u	1.03	3.49	1.26	2.77	2.1		Bu^{t}OH/MeTHF	Ox I_2/O_2	265	6
			u	1.05	3.42	1.40	2.62	2.0		Bu^{t}OH/MeTHF	Ox I_2/O_2	236	6
			u	1.12	3.42	1.59	2.53	1.92		Bu^{t}OH/MeTHF	Ox I_2/O_2	199	6
C4	No. 1	^{85}Rb(I)	0.254	1.02	3.50	u	u	u		Bu^{t}OH	Ox I_2/O_2	RT	6
			0.254	1.02	3.50	u	u	u		Bu^{t}OH/MeTHF	Ox I_2/O_2	RT	6

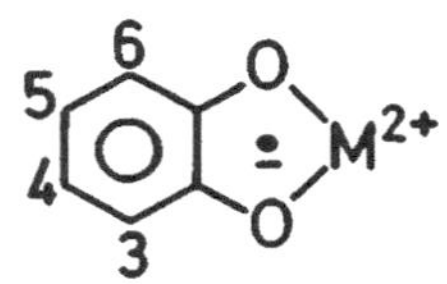

No.	Complex	Central atom	a_M	$a^H_{3,6}$	$a^H_{4,5}$	$a^C_{1,2}$	$a^C_{3,6}$	$a^C_{4,5}$	g-Factor	Solvent	Method	T [K]	Ref.
C5	No. 2	IIA Mg(II)	u	0.52	3.97	—	—	—		w	Ox alk	RT	28
			u	0.53	4.03	—	—	—	2.00423	w	Re el[b]	RT	29
			u	0.51	3.92	u	3.95	2.90	2.0042	w/alk	Phot[c]	301	30
			u	0.90	3.83	—	—	—	2.00442	DMSO	Re el[b]	RT	29
C6	No. 2	Ca(II)	u	0.65	3.85	—	—	—		w	Ox alk	RT	28
			u	0.66	3.89	—	—	—	2.00433	w	Re el[b]	RT	29
			u	0.64	3.80	u	u	u	2.0043	w/alk	Phot[c]	301	30
			u	1.03	3.76	—	—	—	2.00447	DMSO	Re el[b]	RT	29
C7	No. 2	Sr(II)	u	0.69	3.68	—	—	—		w	Ox alk	RT	28
			u	0.70	3.82	—	—	—	2.00435	w	Re el[b]	RT	29
			u	0.68	3.75	u	3.65	2.55	2.0042	w/alk	Phot[c]	301	30
			u	1.07	3.67	—	—	—	2.00430	DMSO	Re el[b]	RT	29
C8	No. 2	Ba(II)	u	0.72	3.75	—	—	—		w	Ox alk	RT	28
			u	0.73	3.85	—	—	—	2.00400	w	Re el[b]	RT	29
			u	0.71	3.71	u	3.50	2.47	2.0040	w/alk	Phot[c]	301	30
			u	1.15	3.62	—	—	—	2.00386	DMSO	Re el[b]	RT	29

C9	No. 2	IIB	^{67}Zn(II)	u	0.48	3.95	—	—	—			Ox alk	RT	28
				u	0.49	3.98	—	—	—	2.00389	w	Re el[b]	RT	29
				1.40	0.48	3.92	—	—	—	2.0039	w/alk	Phot[c]	301	30
				u	0.5	3.9	—	—	—		w/pH = 6.0	Ox HRP[d]	RT	299
				u	0.77	3.78	—	—	—	2.00412	DMSO	Re el[b]	RT	29
C10	No. 2		Cd(II)	u	0.55	3.71	—	—	—		w	Ox alk	RT	28
				u	0.53	3.93	—	—	—	2.00347	w	Re el[b]	RT	29
			^{111}Cd(II) ^{113}Cd(II)	6.79 7.11	0.50	3.81	—	—	—	2.0039	w/alk	Phot[c]	301	30

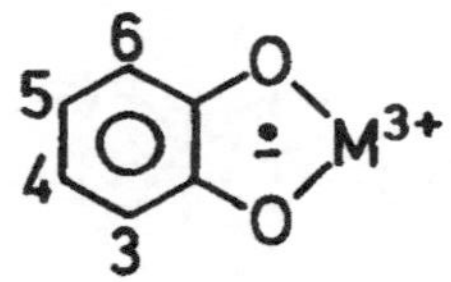

C11	No. 3	IIIA	^{27}Al(III)	1.96	u	u	—	—	—		w/alk	Ox Na_2O_2	RT	28
				2.37	~0.20	4.26	—	—	—	2.0041	w/alk	Phot[c]	301	30
C12	No. 3		^{69}Ga(III) ^{71}Ga(III)	5.05 6.42	~0.30	4.05	—	—	—	2.0040	w/alk	Phot[c]	301	30
C13	No. 3		^{115}In(III)	7.10	~0.35	3.99	—	—	—	2.0035	w/alk	Phot[c]	301	30
C14			Tl(III)	(see Tables IV.1.3 and IV.1.10)										
C15	No. 3	IIIB	^{45}Sc(III)	2.25	0.30	4.20	—	—	—	2.0038	w/alk	Phot[c]	301	30
C16	No. 3		^{89}Y(III)	0.65	0.43	3.68[e]	—	—	—		w/alk	Ox O_2	RT	28
				0.67	0.43	4.01	—	—	—	2.0038	w/alk	Phot[c]	301	30
C17	No. 3	LA	^{139}La(III)	2.07	u	u	—	—	—		DMSO	Ox alk	RT	28
				2.50	0.50	3.95	—	—	—	2.0033	w	Phot[c]	301	30

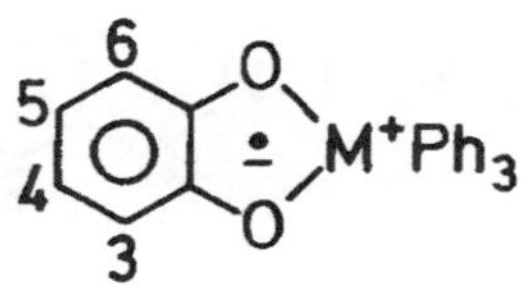

C18	No. 4	IVA	^{117}Sn(I) ^{119}Sn(I)	9.68 10.13	0.55	3.90	—	—	—	2.00395	PhMe	Ox Ph_3SnOH	298	296
C19	No. 1	VA	NH_4^+	u	0.76	3.77	—	—	—	2.00467	w/NH_3	Ox air	RT	29

Table IV.1.1 (continued)
1,2-BENZOSEMIQUINONE-CATION COMPLEXES

No.	Complex	Central atom	a_M	$a^H_{3,6}$	$a^H_{4,5}$	$a^C_{1,2}$	$a^C_{3,6}$	$a^C_{4,5}$	g-Factor	Solvent	Method	T [K]	Ref.
			Splitting constants/Gauss										
	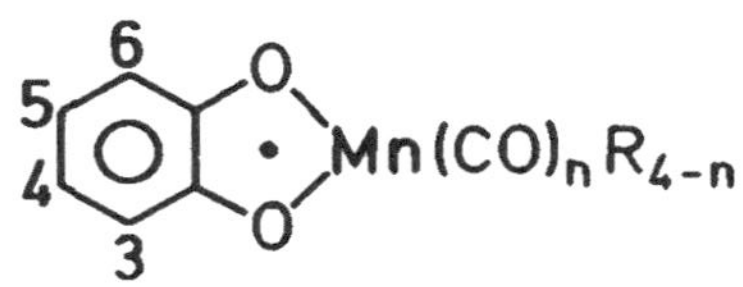												
C20	No. 5 (n = 4)	VIIB ^{55}Mn(I)	7.72	0.51	3.40	—	—	—	2.00336	PhMe	Phot/ $Mn_2(CO)_{10}$	293	374, 385
C21	No. 5 (n = 3, R = C_5H_5N)	Mn(I)	6.80 2.05[1N]	0.45	3.35	—	—	—		PhMe	L.Ex.[f]/C_5H_5N	RT	385
C22	No. 5 (n = 3, R = $P(OEt)_3$)	Mn(I)	11.2 41.7[1P]	0.68	3.12	—	—	—		PhMe	L.Ex.[f]/$P(OEt)_3$	223—243	385
C23	No. 5 (n = 2, R = $P(OEt)_3$)	Mn(I)	24.3 58.5[2P]	<0.2	1.12	—	—	—		PhMe	L.Ex.[f]/$P(OEt)_3$	243	385
	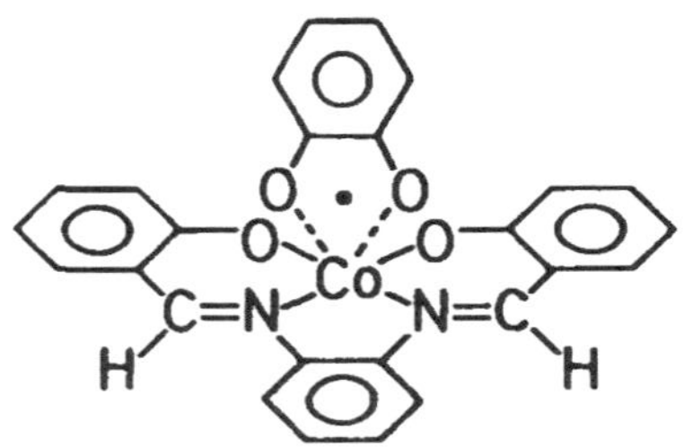												
C24	No. 6	VIIIB ^{59}Co(III)	10.85	u	3.65	—	—	—	2.00095	PhMe	Ox addition[g]	<273	405

[a] Radical obtained by treatment of phenol with NaH.
[b] Radical also generated by the flow method (3 mℓ/min).

[c] Radical generated by photolysis in flowing solutions (1 to 2 mℓ/min).
[d] Radical generated by peroxidase (horseradish, HRP). Catalyzed oxidation of catechol.
[e] Constant obtained by the editor from a simulation of spectrum in Reference 28.
[f] Ligand exchange (L.Ex.) reaction starting with the complex C20.
[g] Addition of 1,2-quinone and Co^{II} (saphen) results in the paramagnetic complex.

Table IV.1.2
4-SUBSTITUTED 1,2-BENZOSEMIQUINONE-CATION COMPLEXES

No.	Complex	Central atom	a_M	a_3	a_4	a_5	a_6	g-Factor	Solvent/pH	Method	T [K]	Ref.
			Splitting constants/Gauss									
	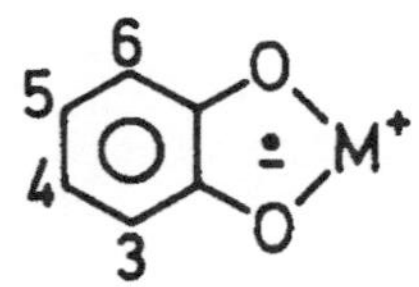											
C25	1,2-Benzosemiquinone, Na complex	IA $^{23}Na(I)$	0.43	1.28	3.43	3.43	1.28		C_5H_5N/ –	Ox NaH	298	297
C26	–,4-Me	Na(I)	0.43	0.71	4.08[1]	3.65	1.37		C_5H_5N/ –	Ox NaH	298	447
C27	–,4-CMe_3	Na(I)	0.47	0.85	0.29[2]	3.78	1.19		C_5H_5N/ –	Ox NaH	298	447
	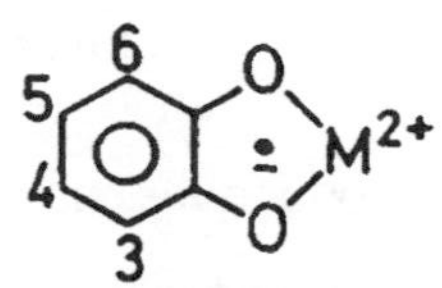											
C28	1,2-Benzosemiquinone, M(II) complex	IIB Zn(II)	—	0.5	3.9	3.9	0.5		w/6.0	Ox HRP[a]	RT	299
C29	–,4-$CH_2CH_2NH_3^+$	Zn(II)	—	0.2	3.4[1]	3.9	0.7		w/4.5	Ox HRP[a]	RT	299

Table IV.1.2 (continued)
4-SUBSTITUTED 1,2-BENZOSEMIQUINONE-CATION COMPLEXES

No.	Complex	Central atom	Splitting constants/Gauss a_M	a_3	a_4	a_5	a_6	g-Factor	Solvent/pH	Method	T [K]	Ref.
C30	$-,4-CH_2CH(NH_2)COO^-$	Zn(II)	—	<0.2	4.2[1] 2.2[1]	3.9	0.6		w/6.4	Ox HRP[a]	RT	299
C31	$-,4-CH_2CH(NH_3^+)COO^-$	IIA $^{25}Mg(II)$	—	0.28	3.08[1] 2.68[1]	3.84	0.70	2.0042	w/5.5	Phot[b]	301	344
C32		Ca(II)	—	0.43	3.14[1] 2.38[1]	3.71	0.83	2.0043	w/5.5	Phot[b]	301	344
C33		Sr(II)	—	0.48	3.08[1] 2.27[1]	3.65	0.89	2.0042	w/5.5	Phot[b]	301	344
C34		IIB $^{67}Zn(II)$	—	0.3	3.2[1] 2.6[1]	3.8	0.7		w/4.5	Ox HRP[a]	RT	299
			1.40	0.25	3.11[1] 2.65[1]	3.83	0.67	2.0039	w/5.5	Phot[b]	327	344
C35		$^{111}Cd(II)$ $^{113}Cd(II)$	6.81 7.13	0.29	3.14[1] 2.46[1]	3.74	0.69	2.0035	w/5.5	Phot[b]	303	344
C36	$-,4-COO^-$	Zn(II)	—	0.9	—	3.5	0.5		w/6.0	Ox HRP[a]	RT	299

[a] Radical generated by peroxidase (horseradish, HRP)-catalyzed oxidation of catechol.
[b] Radical generated by photolysis of 0.025-*M* deoxygenated solutions of DOPA containing 0.025 *M* metal ion,

Table IV.1.3
DIORGANOTHALLIUM(III)-1,2-BENZOSEMIQUINONE COMPLEXES

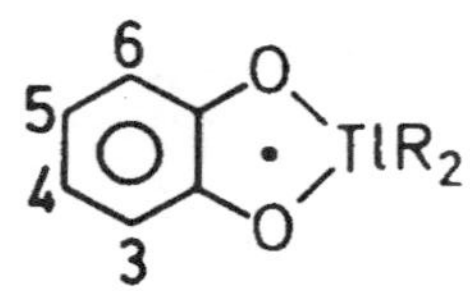

No.	Complex	Substituent(s)	R	Splitting constants/Gauss a_{Tl}	a_3	a_4	a_5	a_6	g-Factor	Solvent	Lw/G[a]	Ref.
C37	1,2-Benzosemiquinone-diaryl thallium(III)complex		Ph	10.61[b]	u	u	u	u	2.00368	MeOH		20
			Ph	10.84	u	u	u	u	2.00360	EtOH		20
			Ph	11.35	0.90	3.55	3.55	0.90	2.00379	$PrNH_2$		20
			Ph	11.72	0.88	3.56	3.56	0.88	2.00364	C_5H_5N	0.14	20
			Ph	12.61	0.89	3.48	3.48	0.89	2.00385	Pr^iNH_2		20
			Ph	13.35	u	u	u	u	2.00358	DME		20
			Ph	14.00	u	u	u	u	2.00363	THF		20
			Ph	14.05	u	u	u	u	2.00378	MeCN		20
			Ph	14.40	0.87	3.56	3.56	0.87	2.00380	Bu^tNH_2		20
			Ph	14.65	0.83	3.57	3.57	0.83	2.00360	MeTHF	0.17	20
			Ph	15.50	0.78	3.57	3.57	0.78	2.00359	Dioxane	0.17	20
			Ph	18.25	u	u	u	u	2.00340	$(CH_2Cl)_2$		20
			Ph	20.85	u	u	u	u	2.00349	PhMe		20
			Ph	21.90	u	u	u	u	2.00338	Hexane		20
			Mes[c]	23.7[d]	0.73	3.58	3.58	0.73	2.00346	MeTHF		298
C38	– ,3-methyl-	–Me	Ph	12.95	0.55[f]	2.93	3.80	0.64	2.00354	C_5H_5N	0.14	20
			Ph	15.22	0.47[f]	3.04	3.77	0.53	2.00344	MeTHF	0.17	20
			Ph	16.35	0.44[f]	3.05	3.79	0.46	2.00344	Dixoane	0.16	20
C39	– ,3-isopropyl-	$-CHMe_2$	Ph	13.88	0.49[f]	2.95	3.82	0.49	2.00355	C_5H_5N	0.16	20
			Ph	16.44	0.45[f]	3.07	3.74	0.49	2.00344	MeTHF	0.17	20
			Ph	17.18	0.45[f]	3.06	3.79	0.45	2.00346	Dioxane	0.18	20

Table IV.1.3 (continued)
DIORGANOTHALLIUM(III)-1,2-BENZOSEMIQUINONE COMPLEXES

No.	Complex	Substituent(s)	R	a_{Tl}	a_3	a_4	a_5	a_6	g-Factor	Solvent	Lw/G[a]	Ref.
				Splitting constants/Gauss								
C40	—.4-methyl-	−Me	Ph	11.58	0.31	4.38[1]	3.71	0.98	2.00364	C_5H_5N	0.14	20
			Ph	14.25	0.29	4.37[1]	3.69	0.96	2.00344	MeTHF	0.15	20
			Ph	15.18	0.23	4.46[1]	3.74	0.91	2.00346	Dioxane	0.14	20
C41	—.4-*t*-butyl-	$-CMe_3$	Ph	10.65	0.43	0.30[2]	3.78	0.79	2.00365	C_5H_5N	0.12	20
			Ph	13.55	0.32	0.32[2]	3.79	0.70	2.00342	MeTHF	0.14	20
			Ph	14.30	0.32	0.32[2]	3.82	0.65	2.00361	Dioxane	0.14	20
C42	—.4-(1,1,3,3-tetramethylbutyl)-	$-CMe_2CH_2CMe_3$	Ph	12.70	u	u	3.78	u	2.00350	C_5H_5N	1.60	20
			Ph	15.68	u	u	3.90	u	2.00335	MeTHF	1.50	20
			Ph	16.18	u	u	3.94	u	2.00337	Dioxane	1.45	20
C43	—.3,5-diisopropyl-	$-CHMe_2$	Ph	12.96	0.24[1]	2.45	3.18[1] 0.24[2]	0.24	2.00348	C_5H_5N	0.14	20
			Ph	15.63	0.20[1]	2.43	3.24[1] 0.23[2]	0.20	2.00342	MeTHF	0.16	20
			Ph	16.12	0.19[1]	2.51	3.25[1] 0.24[2]	0.19	2.00340	Dioxane	0.15	20
C44	—.3,5-di-*t*-butyl-	$-CMe_3$	Ph/Cl	11.0	u	2.9	u	u	1.99994	MeOH		21
			Ph	14.55	u	3.28	0.28[2]	0.30	2.00355	C_5H_5N	0.18	20
			Ph	17.33	u	3.41	0.28[2]	0.28	2.00336	MeTHF	0.16	20
			Ph	17.58	u	3.48	0.29[2]	0.29	2.00337	Dioxane	0.21	20
			Et	20.7	u	3.2	u	u	2.0031	Piperidine		22
			Et	21.0	u	3.2	u	u	2.0029	C_5H_5N		22
			Et	23.9	u	3.6	u	u	2.0027	THF		22
			Et	24.7	u	3.2	u	u	2.0027	MeCN		22
			Et	28.4	u	3.4	u	u	2.0026	PhMe		22
			Et	32.0	u	3.6	u	u	2.0022	Hexane		22
C45	—.3-*t*-butyl-5-triphenylmethyl-	$-CMe_3$, $-CPh_3$	Ph	15.25	u	3.12	u	0.31	2.00349	C_5H_5N	0.20	20
			Ph	18.65	u	3.36	u	0.36	2.00334	MeTHF	0.25	20
			Ph	18.70	u	3.32	u	0.36	2.00336	Dioxane	0.27	20
C46	-.3,6-di-*t*-butyl-	$-CMe_3$	Et	27.7	u	3.3	3.3	u	2.0026	THF		22

C47	-,4-carboxyvinyl-	$-CH=CHCOOH$	Ph	11.25	0.35	2.31^{1} 1.24^{2}	3.00	1.13	2.00372	C_5H_5N	0.25	24
			Me	20.8	0.43	2.30^{1} 1.21^{2}	2.90	1.21		C_5H_5N	0.25	24
C48	-,4-(2-aminoethyl)-	$-CH_2CH_2NH_2$	Mes	21.50	0.30	3.10^{1}	3.80	0.85	2.00328	C_5H_5N	0.15	24
			Ph	11.55	0.40	3.05^{1}	3.72	0.95	2.00370	C_5H_5N	0.15	24
			Et	18.10	0.50	3.00^{1}	3.70	1.00	2.00397	C_5H_5N	0.25	24
C49	-,4-(2-amino-2-methoxycarbonylethyl)-	$-CH_2CH(NH_2)COOMe$	Ph	11.90	0.50	2.95^{1} 2.45^{1}	3.65	1.00	2.00362	C_5H_5N	0.25	24
C50	-,4-(2-amino-2-carboxypropyl)-	$-CH_2C(NH_2)(Me)COOH$	Et	18.40	0.59	2.90^{1} 2.05^{3}	3.45	0.97	2.00388	C_5H_5N	0.25	24
C51	-,4-diphenylthallium(III) carboxylatomethyl-	$-CH_2COO^-\ Tl^+R_2$	Ph	11.82	0.40	3.21^{1} 0.49^{Tl}	3.67	0.88	2.00365	C_5H_5N	0.15	20
			Ph	14.22	0.32	3.10^{1} 0.60^{Tl}	3.70	0.88	2.00341	MeTHF	0.15	20
			Ph	15.35	0.31	3.10^{1} 0.53^{Tl}	3.70	0.83	2.00352	Dioxane	0.17	20
C52	-,4-(2-methylamino-1-diphenyl-thallium(III)oxidoethyl-	$O^-Tl^+R_2$ / $-CHCH_2NHMe$	Ph	11.6	0.50	3.05^{1} 0.23^{Tl}	3.65	0.90	2.00361	C_5H_5N	0.275	25
			Me	21.8	0.40	3.08^{1} 0.25^{Tl}	3.68	0.75		C_5H_5N	0.325	25
			Mes	21.8	0.40	3.23^{1} 0.25^{Tl}	3.70	0.75	2.00335	C_5H_5N	0.350	25
			Mes	22.1	0.40	3.24^{1} 0.25^{Tl}	3.68	0.75	2.00332	THF	0.312	25
			Mes	23.4	0.40	3.24^{1} 0.25^{Tl}	3.68	0.75	2.00328	MeTHF	0.312	25
			Mes	21.8	0.40	3.08^{1} 0.25^{Tl}	3.68	0.75	2.00331	Dioxane	0.325	25
C53	-,3-acetyl-6-methyl-	$-COMe, -Me$	Ph	12.0	—	5.33	1.97	1.97^{1}	2.00356	C_5H_5N	0.30	458
			Mes	22.8	—	5.25	2.15	1.7^{1}	2.00327	C_5N_5N	0.20	458
			Me	22.2	—	5.3	2.05	2.05^{1}	2.00361	C_5H_5N	0.25	458
C54	-,3-methoxycarbonyl-	$-COOMe$	Ph	12.00	u	2.77	4.84	1.90	2.00370	C_5H_5N	0.16	20
			Ph	13.75	u	2.87	4.79	1.80	2.00365	MeTHF	0.25	20
			Ph	15.40	u	2.93	4.70	1.69	2.00365	Dioxane	0.20	20

Table IV.1.3 (continued)
DIORGANOTHALLIUM(III)-1,2-BENZOSEMIQUINONE COMPLEXES

No.	Complex	Substituent(s)	R	Splitting constants/Gauss a_{Tl}	a_3	a_4'	a_5	a_6	g-Factor	Solvent	Lw/G[a]	Ref.
C55	-,3-ethoxycarbonyl-	–COOEt	Ph	12.00	u	2.85	4.75	1.84	2.00368	C_5H_5N	0.20	20
			Ph	13.96	u	2.95	4.70	1.80	2.00365	MeTHF	0.25	20
			Ph	15.68	u	3.00	4.62	1.61	2.00363	Dioxane	0.19	20
C56	-,4-formyl-	–CHO	Ph	8.40	0.81	1.98[i]	2.60	0.35	2.00439	C_5H_5N	0.17	20[i]
			Ph	11.65	0.75	1.72[i]	2.68	0.33	2.00424	MeTHF		20[i]
			Ph	13.30	0.68	1.66[i]	2.76	0.29	2.00424	Dioxane	0.17	20[i]
C57	-,4-acetyl-	–COMe	Ph	9.15	1.78	u	2.83	0.75	2.00426	C_5H_5N	0.20	458
			Mes	21.1	1.59	u	2.93	0.66	2.00397	C_5H_5N	0.28	458
			Mes	22.55	1.48	u	2.95	0.66	2.00394	MeTHF	0.29	458
			Me	19.95	1.90	u	2.75	0.85	2.00434	C_5H_5N		458
C58	-,4-acetyl-3-methyl-	–$COCH_3^*$, –Me	Ph	11.1	1.93[i]	0.40*	3.41	0.4	2.00394	C_5H_5N	0.15	458
			Mes	23.15	1.68[i]	0.35*	3.4	0.35	2.00363	C_5H_5N	0.20	458
			Me	22.05	2.05[i]	0.42*	3.32	0.42	2.00405	C_5H_5N	0.20	458
			Me	25.2	1.81[i]	0.40*	3.35	0.4	2.00387	MeTHF	0.25	458
C59	-,4-acetyl-5-methyl-	–COMe, –Me	Ph	8.6	1.79	u	3.42[i]	0.31	2.00413	C_5H_5N	0.15	458
			Mes	20.05	1.6	u	3.6[i]	0.2	2.00383	C_5H_5N	0.15	458
C60	-,4-acetyl-6-methyl-	–COMe, –Me	Ph	10.4	1.39	u	2.29	0.45[i]	2.00411	C_5H_5N	0.30	458
			Mes	22.3	1.22	u	2.41	0.35[i]	2.00383	C_5H_5N	0.15	458
			Me	21.35	1.50	u	2.24	0.51[i]	2.00422	C_5H_5N	0.32	458
			Me	24.0	1.35	u	2.45	0.40[i]	2.00408	C_5H_5N	0.25	458
C61	-,4-propionyl-6-methyl-	–COEt, –Me	Ph	10.5	1.35	u	2.35	0.45[i]		C_5H_5N		458
C62	-,4-methoxycarbonyl-	–$COOCH_3^*$	Ph	9.55	1.83	0.21*	2.84	0.82	2.00414	C_5H_5N	0.14	20[+],458
			Ph	12.50	1.72	0.20*	2.92	0.76	2.00403	MeTHF	0.14	20[+]
			Ph	14.23	1.62	0.20*	2.92	0.71	2.00400	Dioxane	0.13	20[+]
C63	-,4-(X-oxycarbonyl)-	–COOX[c]	Ph	9.46	1.81	0.20*	2.85	0.82	2.00414	C_5H_5N	0.14	20[+]
			Ph	12.40	1.70	0.20*	2.91	0.78	2.00406	MeTHF	0.16	20[+]
			Ph	14.05	1.60	0.20*	2.96	0.71	2.00403	Dioxane	0.16	20[+]
C64	-,4-isopropoxycarbonyl-	–$COOCH^*Me_2$	Ph	9.49	1.79	0.18*	2.86	0.81	2.00414	C_5H_5N	0.28	20[+]

			Ph	12.40	1.68	0.18*	2.92	0.77	2.00407	MeTHF	0.29	20
			Ph	13.98	1.59	0.18*	2.98	0.70	2.00405	Dioxane	0.28	20
C65	- ,4-diphenylthallium(III)	$-COO^-Tl^+R_2$	Ph	11.00	1.25	—	3.20	0.72	2.00405	C_5H_5N	0.20	20, 458
			Ph	12.40	1.64	—	2.96	0.72	2.00400	MeTHF	0.25	20
			Ph	14.60	1.30	—	3.20	0.62		Dioxane	0.36	20
C66	-,3-chloro-4,6-di-*t*-butyl-	$-Cl$, $-CMe_3$	Ph	13.98	—	u	3.03	u	2.00364	C_5H_5N	0.47	20
			Ph	16.35	—	u	3.16	u	2.00352	MeTHF	0.55	20
			Ph	17.16	—	u	3.18	u	2.00352	Dioxane	0.45	20
C67	-,3-nitro-4,6-di-*t*-butyl-	$-NO_2$, $-CMe_3$	Et	19.3	u	u	2.8	u	2.0027	THF		22
C68	-,tetrachloro-	$-Cl$	Et	31.0	—	—	—	—	2.0047	THF/ – 20°C		22
			Ph	9.10	—	0.44[f] 0.37[g]	0.44[f] 0.37[g]	—	2.00500	MeTHF	0.25	20
			Ph	10.32	—	0.46[f] 0.38[g]	0.46[f] 0.38[g]	—	2.00498	Dioxane	0.30	20
C69	-,tetrabromo-	$-Br$	Ph	7.38	—	—	—	—	2.00868	C_5H_5N	1.40	20
			Ph	9.90	—	—	—	—	2.00888	MeTHF	2.20	20
			Ph	11.32	—	—	—	—	2.00889	Dioxane	2.00	20
	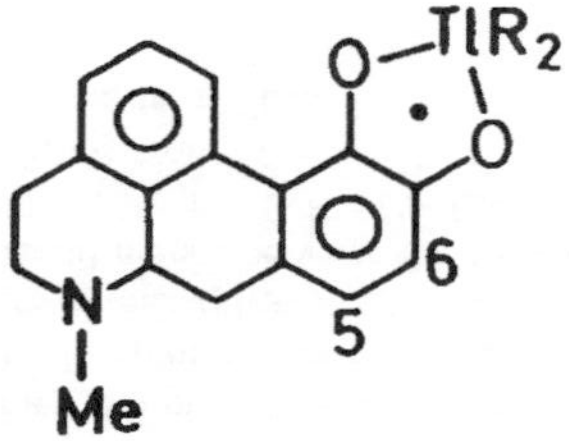											
C70	Apomorphine-diorganothallium (III)complex		Ph	13.55	—	8.60[i] 1.80[i]	3.65	1.05	2.00357	C_5H_5N	0.45	24
			Et	19.60	—	8.50[i] 1.65[i]	3.60	1.13	2.00397	C_5H_5N	0.35	24
			Bu	22.60	—	8.45[i] 1.65[i]	3.60	1.12	2.00384	C_5H_5N	0.35	24

Table IV.1.3 (continued)
DIORGANOTHALLIUM(III)-1,2-BENZOSEMIQUINONE COMPLEXES

No.	Complex	Substituent(s)	R	Splitting constants/Gauss					g-Factor	Solvent	Lw/G[a]	Ref.
				a_{Tl}	a_3	a_4	a_5	a_6				
			Mes	24.25	—	8.70[f] 1.81[f]	3.60	0.98		C_5H_5N	0.45	24

[a] Linewidth of high-field Tl lines; linewidth of low-field lines about 50 mG broader.
[b] The Tl splitting constants are average constants from ^{203}Tl- and ^{205}Tl-nuclei.
[c] Mes = mesityl = 2,4,6-$Me_3C_6H_2$.
[d] The difference between the constants of the Tl-isotopes amounts to 0.23 G in this case (see Figure 1, Reference 298).
[e] X = C_2H_5, n − C_3H_7, n − C_4H_9 or iso − C_4H_9. The * indicates splitting from COOCH proton(s).
[f] Splitting from ^{35}Cl.
[g] Splitting from ^{37}Cl.

Table IV.1.4
3,5-DI-*t*-BUTYL-1,2-BENZOSEMIQUINONE-CATION COMPLEXES

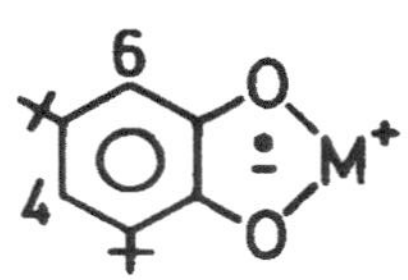

No.	Complex	Central atom	L	Splitting constants/Gauss			g-Factor	Solvent	Method	T [K]	Ref.
				a_M	a_4	a_5					
C71	No. 1 IA	Li(I)	—	u	3.32	u		Ethers[a]	Re Li		13[b]

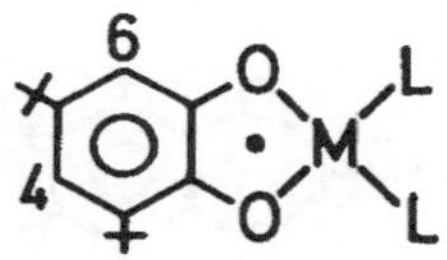

C72	No. 2	IB	^{63,65}Cu(I)	Ph_3P	10.8 16.9^{2P}	3.4	u	2.0046	THF	NaSQ CuCl/2L^c		384
C73	No. 2		Cu(I)	$(EtO)_3P$	10.8 21.6^{2P}	3.3	u	2.0046	THF	NaSQ CuCl/2L		384
C74	No. 2		Cu(I)	Ph_3P/Ph_3As	10.1 17.2^{1P} 17.2^{1As}	3.3	u	2.0047	THF	NaSQ CuCl/L,L′		384
C75	No. 2		Cu(I)	Ph_3As	10.3 16.9^{2As}	3.3	u	2.0048	THF	NaSQ CuCl/2L		384
C76	No. 2		Cu(I)	Et_3As	10.9 18.1^{2As}	3.2	u	2.0050	THF	NaSQ CuCl/2L		384
C77	No. 2		Cu(I)	1,5–Cod	7.0	3.4	u	2.0035	THF	NaSQ CuCl/L		384
C78	No. 2		Cu(I)	PhC ≡ CPh	6.3	3.6	u	2.0033	THF	NaSQ CuCl/L		384
C79	No. 2		Cu(I)	PhC ≡ CH	6.4	3.6	u	2.0032	THF	NaSQ CuCl/L		384

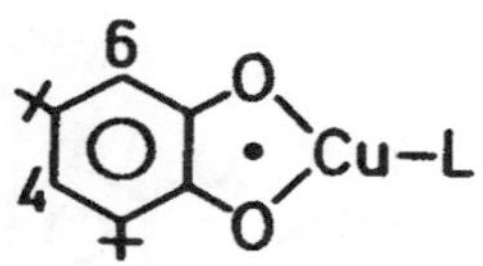

C80	No. 3		^{63,65}Cu(I)	Ph_3P	3.5	3.5	u	2.0036	THF	NaSQ CuCl/L		384
C81	No. 3		Cu(I)	CO	3.7	3.7	u	2.0026	Hexane	NaSQ CuCl/L		384
					4.8	3.6	u	2.0033	PhMe	NaSQ CuCl/L		384
					5.3	3.6	u	2.0028	Diglyme	NaSQ CuCl/L		384
					6.0	3.4	u	2.0032	THF	NaSQ CuCl/L		384
					6.2	3.6	u	2.0034	$Me_2C_6H_3NH_2$	NaSQ CuCl/L		384
					7.1	3.2	u	2.0038	DMSO	NaSQ CuCl/L		384
					9.5	3.1	u	2.0040	Et_3N	NaSQ CuCl/L		384
C82	No. 2		^{107,109}Ag(I)	Ph_3P	1.7 6.7^{2P}	3.4	u	2.0046	THF	NaSQ $AgOCOCF_3$/2L	RT	302

Table IV.1.4 (continued)
3,5-DI-*t*-BUTYL-1,2-BENZOSEMIQUINONE-CATION COMPLEXES

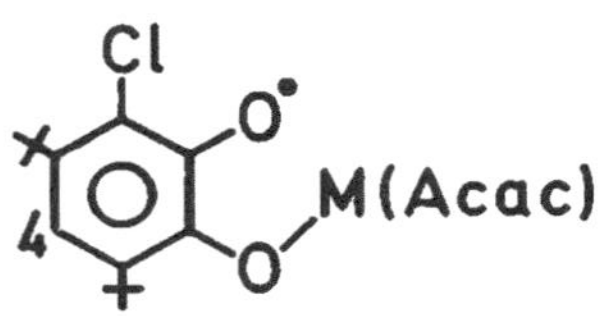

No.	Complex		Central atom	L	Splitting constants/Gauss a_M	a_4	a_5	g-Factor	Solvent	Method	T [K]	Ref.
C83	No. 4	IIA	Be(II)	—	u	3.75	u		$PhNO_2$	Q $Be(Acac)_2^d$	193	314
C84	No. 4		Mg(II)	—	u	2.4[c]	u		$PhNO_2$	Q $Mg(Acac)_2$	193	314
C85	No. 4		Ca(II)	—	u	3.0[c]	u		$PhNO_2$	Q $Ca(Acac)_2$	193	314

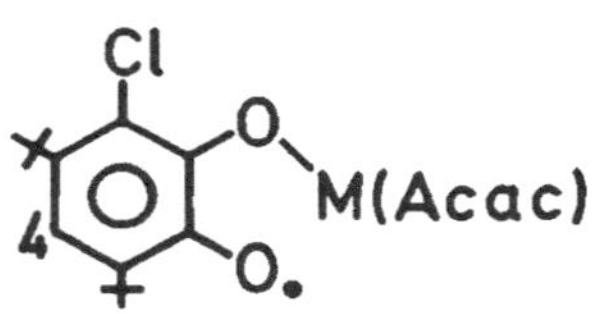

No.	Complex		Central atom	L	a_M	a_4	a_5	g-Factor	Solvent	Method	T [K]	Ref.
C86	No. 5		Sr(II)	—	u	2.0	u		$PhNO_2$	Q $Sr(Acac)_2$	193	314
C87	No. 5		Ba(II)	—	u	2.2	u		$PhNO_2$	Q $Ba(Acac)_2$	193	314
C88	No. 4	IIB	Zn(II)	—	u	3.5	u		$PhNO_2$	Q $Zn(Acac)_2$	193	314
C89	No. 4		Cd(II)	—	u	3.1	u		$PhNO_2$	Q $Cd(Acac)_2$	193	314

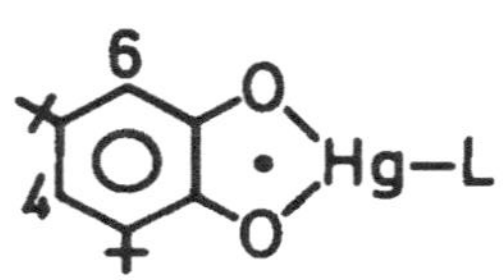

C90	No. 6		^{199}Hg(II)	Bz	22.7[f]	2.70	u	1.9994	THF	Phot Bz_2Hg	213	303
C91	No. 6		Hg(II)	Ph	u	2.69	u	1.9993	THF	Phot Ph_2Hg	213	303
C92	No. 6		Hg(II)	Et	6.7[f]	2.8	u	1.9996	PhMe	NaSQ EtHgCl	193	304
C93	No. 6		Hg(II)	$GeEt_3$	11.7[g]	3.5	u	1.9989	PhMe	Q $(Et_3Ge)_2Hg$	233	304
					~3.5							
C94	No. 1		^{203,205}Tl(I)	—	62.0[h]	3.2	u	1.9973	THF	Re Tl	RT	328

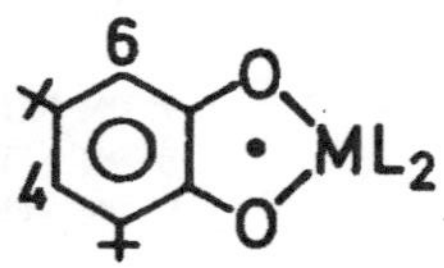

C95	No. 7	IIIA	^{27}Al(III)	Et	2.2	4.4	u	2.0030	PhMe	Q $AlEt_3$	213—233	404
C96	No. 7		Al(III)	Cl	2.2	4.4	u		THF	Q $AlCl_3$	RT	305
C97	No. 7	IIIA	^{27}Al(III)	Br	2.1	4.2	u		THF	Q $AlBr_3$	RT	305
C98	No. 7		Al(III)	I	2.6	5.2	u		THF	Q AlI_3	RT	305
C99	No. 7		^{69,71}Ga(III)	Et	15.1[i]	4.1	u	2.0033	PhMe	Q $GaEt_3$	213—233	404
					19.3							
C100	No. 7		Ga(III)	Cl	8.2[i]	3.5	~0.3[2]		Me_2CO	Q $GaCl_3$	RT	305
					10.4							
C101	No. 7		Ga(III)	Br	8.0[i]	3.6	~0.3[2]		Me_2CO	Q $GaBr_3$	RT	305
					10.4							
C102	No. 7		Ga(III)	I	8.2[i]	3.5	~0.3[2]		Me_2CO	Q GaI_3	RT	305
					10.6							
C103	No. 7		^{115}In(III)	Et	8.8[j]	—	—	2.0021	PhMe	Q $InEt_3$	RT	404
C104	No. 7		In(III)	Cl	7.7[j]	3.9	u		THF	Q $InCl_3$	RT	306
C105	No. 7		^{203,205}Tl(III)	Et	28.4	3.4	u	2.0026	PhMe	Q $TlEt_3$	RT	404
					29.9	3.6	u	2.0027	THF	NaSQ/Et_2TlCl	RT	22
C106	No. 7	IIIB	^{45}Sc(III)	Cl	2.2	4.5	u		THF	Q $ScCl_3$	293	307
C107	No. 7		Y(III)	Cl	u	3.7	u		THF	Q YCl_3	RT	306
C108	No. 7		^{175}Lu(III)	Cl	5.5	5.5	u		THF	Q $LuCl_3$	RT	306
C109	No. 7	LA	^{139}La(III)	Cl	3.2	3.2	u		EtOH	Q $LaCl_3$	RT	306

Table IV.1.4 (continued)
3,5-DI-*t*-BUTYL-1,2-BENZOSEMIQUINONE-CATION COMPLEXES

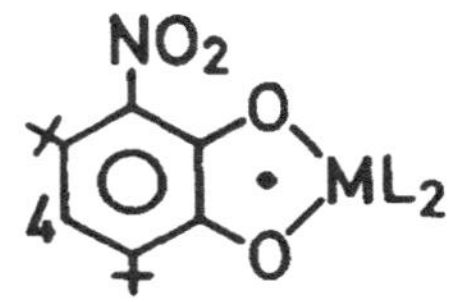

No.	Complex		Central atom	L	Splitting constants/Gauss a_M	a_4	a_5	g-Factor	Solvent	Method	T [K]	Ref.
C110	No. 8	IIIA	$^{27}Al(III)$	Cl	2.3	2.3	u		THF	Q $AlCl_3$	RT	305
C111	No. 8		Al(III)	Br	2.4	2.4	u		THF	Q $AlBr_3$	RT	305
C112	No. 8		Al(III)	I	2.5	2.5	u		THF	Q AlI_3	RT	305
C113	No. 8		$^{69,71}Ga(III)$	Cl	8.0[i]/10.0	3.2	u		Me_2CO	Q $GaCl_3$	RT	305
C114	No. 8		Ga(III)	Br	8.0[i]/10.1	3.0	u		Me_2CO	Q $GaBr_3$	RT	305
C115	No. 8		Ga(III)	I	8.2[i]/10.5	3.3	u		Me_2CO	Q GaI_3	RT	305
C116	No. 8		$^{115}In(III)$	Cl	6.5[j]	3.4	u		THF	Q $InCl_3$	RT	306
C117	No. 8	IIIB	$^{45}Sc(III)$	Cl	2.3	2.3	u		THF	Q $ScCl_3$	293	307
C118	No. 8		Y(III)	Cl	u	3.2	u		THF	Q YCl_3	RT	306
C119	No. 8		$^{175}Lu(III)$	Cl	5.1	5.1	u		THF	Q $LuCl_3$	RT	306
C120	No. 8	LA	$^{139}La(III)$	Cl	3.0	3.0	u		EtOH	Q $LaCl_3$	RT	306

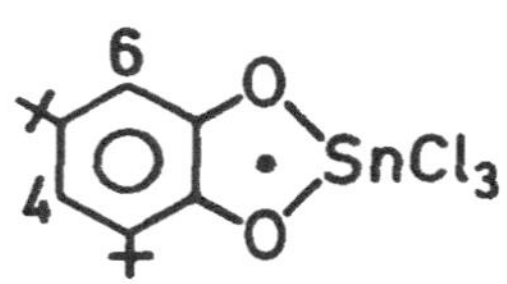

No.	Complex		Central atom	L	a_M	a_4	a_5	g-Factor	Solvent	Method	T [K]	Ref.
C121	No. 9	IVA	Sn(IV)	—	9.5[k]	4.2	u		THF	Re $SnCl_4$	RT	306

C122 No. 9 (6–NO_3)	Sn(IV)	—	8.4[k]	3.8	u		THF	Re $SnCl_4$	RT	306

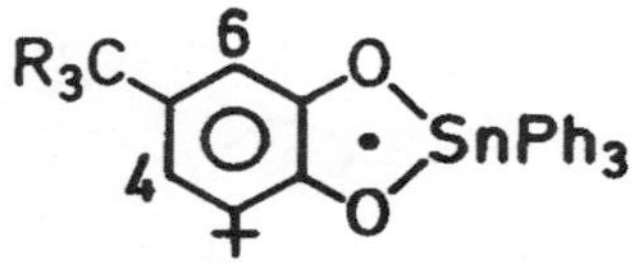

C123 No. 10 (R = Me)	Sn(IV)	—	9.94[l] / 10.40	2.85	0.35[2] 0.55[m]	2.00380	PhMe	Re[UV] Ph_3SnOH	298	296
C124 No. 10 (R = Ph)	Sn(IV)	—	9.70[l] / 10.15	2.83	u 0.43[m]	2.00380	PhMe	Re[UV] Ph_3SnOH	298	296

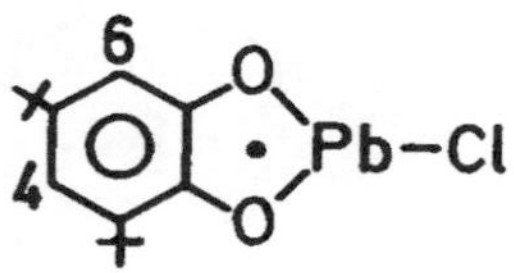

C125 No. 11	^{207}Pb(II)	—	55	3.3	u	1.9997	THF	—	RT	308[n]

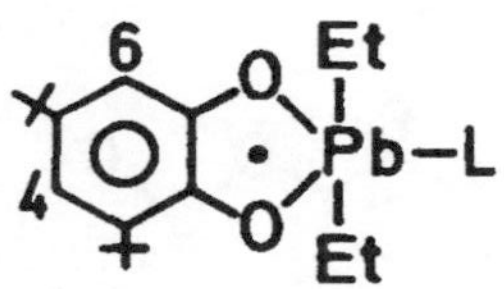

C126 No. 12	^{207}Pb(IV)	Et	9.3	3.3	u	2.0022	THF	NaSQ ET_3PbCl	RT	308
C127 No. 12	Pb(IV)	OAc	9.0	3.2	u	2.0041	PhMe	—	RT	308[n]
C128 No. 12	Pb(IV)		u	3.4	u	2.0034	THF	—	RT	308

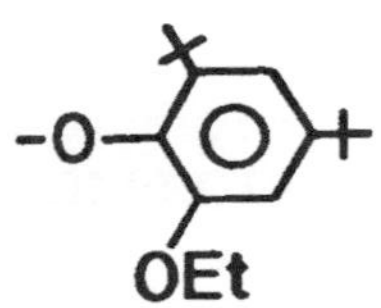

Table IV.1.4 (continued)
3,5-DI-*t*-BUTYL-1,2-BENZOSEMIQUINONE-CATION COMPLEXES

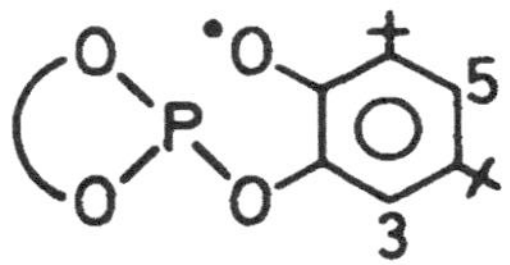

No.	Complex	Central atom	L	Splitting constants/Gauss a_M	a_4	a_5	g-Factor	Solvent	Method	T [K]	Ref.
C129	No. 13[o] VA	$^{31}P(III)$	—	27	u	u		PhMe	—	233	309
C130	No. 13[o]	$^{75}As(III)$	—	24[f]	u	u		DEE	Q As_4	293	309
				20[f]	u	u		DEE	Q As_4	293	309
C131	No. 13	$^{209}Bi(III)$	—	46.4[q]	u	u		—[q]	KSQ + QK_2/ $BiCl_3$	—[q]	309

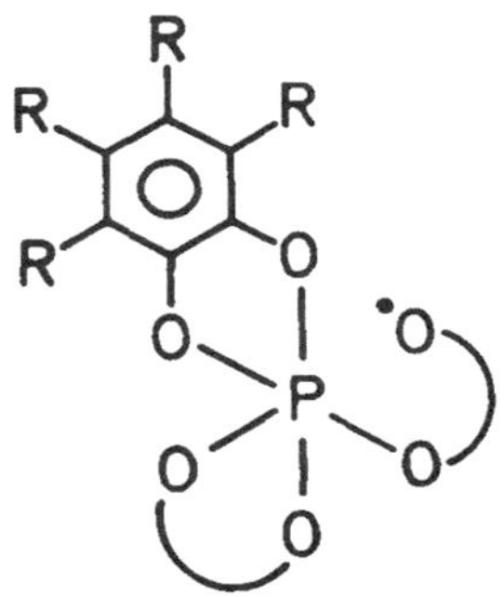

No.	Complex	Central atom	L	a_M	a_4	a_5	g-Factor	Solvent	Method	T [K]	Ref.
C132	No. 14 (R = H)	$^{31}P(V)$	—	4.35	4.35	u		PhMe	Phot Q/SpP[r]	173	309
C133	No. 14 (R = Cl)	P(V)	—	4.35	4.35	u		PhMe	Phot Q/SpP[r]	173	390

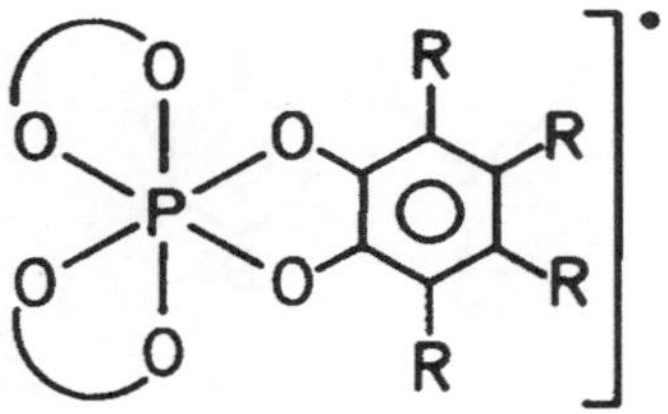

C134 No. 15 (R = H)		^{31}P(V)	—	4.35	1.45	u	PhMe	Phot Q/SpP^{r}	233	390
C135 No. 15 (R = Cl)		P(V)	—	4.35	2.17	u	PhMe	Phot Q/SpP^{r}	223	390

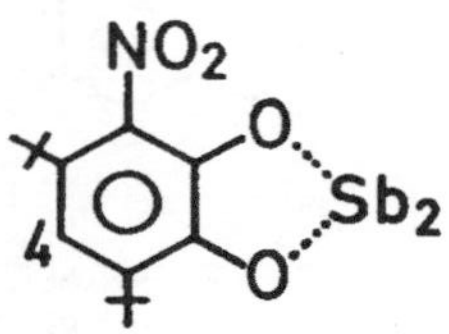

C136 No. 16^{s}		^{121,123}Sb(V)	—	19.7^{t} 10.9	u	u	Diglyme	Q $SbCl_5$	RT	392
	VB	(V, W, see References 394 and 461)								
	VIB	(Mo, W, see Reference 394)								
	VIIB	(Re, see after VIIIB)								

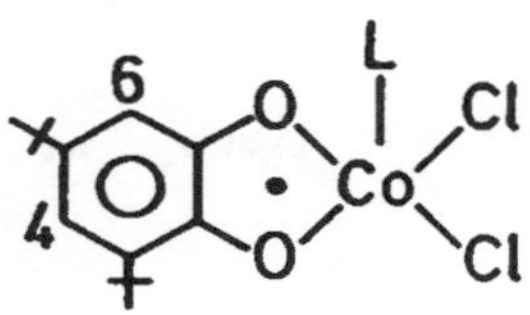

C137 No. 17	VIIIB	^{59}Co(III)	$(C_2H_2)_3(NH_4)_4$	9.76	3.5	u	MeOH	Ox $Pb(OAc)_4$	298	326

Table IV.1.4 (continued)
3,5-DI-*t*-BUTYL-1,2-BENZOSEMIQUINONE-CATION COMPLEXES

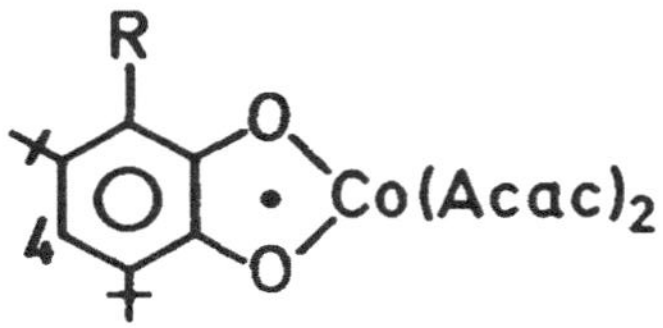

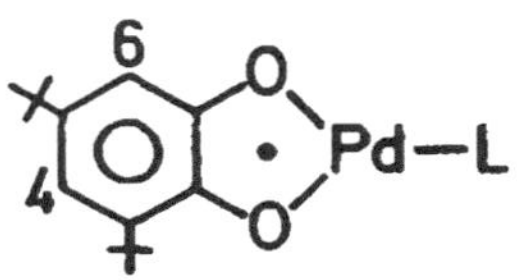

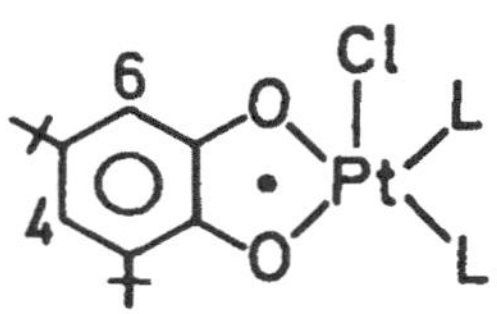

No.	Complex	Central atom	L	Splitting constants/Gauss a_M	a_4	a_5	g-Factor	Solvent	Method	T [K]	Ref.
C138	No. 18	^{59}Co(III)	(R = H)	11.1	3.45	u		—	Q Co(Acac)$_2$		301
C139	No. 18	Co(III)	(R = Cl)	11.1	3.5	u		—	Q Co(Acac)$_2$		301
C140	No. 19	VIIIB ^{105}Pd(II)	C_3H_5	2.4	3.2	u	2.0026	THF	NaSQ/ ((C$_3$H$_5$)PdCl)$_2$	RT	393
				—	—	—	1.993(g∥)[a] 2.008(g⊥)	THF		77	393

C141 No. 20		^{195}Pt(II)	Ph_3P	13.0 3.3^{2P}	3.3	u	1.9990	THF	$(Ph_3P)_2PtCat^{v}$/ $CuCl_2$	RT	375

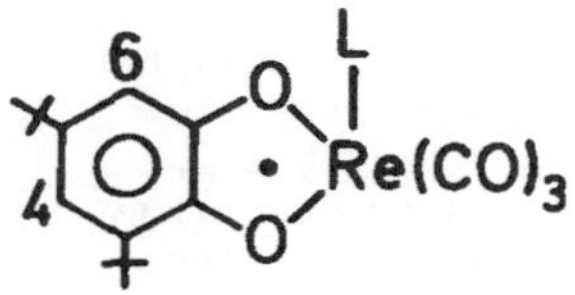

				a_M	a_L	a_4					
C142 No. 21	VIIB	^{185}Re(I)	CO	28.2	—	3.46	2.0032	C_6H_6	Phot Q/$Re(CO)_{10}$	RT	311
C143 No. 21		Re(I)	Ph_3P	38.0	24.7^{1P}	—	2.0012	C_6H_6	L.Ex.^{x}C142/L	RT	311
C144 No. 21		Re(I)	Bu_3P	40.1	26.9^{1P}	—	2.0023	C_6H_6	L.Ex. C142/L	RT	311
C145 No. 21		Re(I)	$(C_6H_4Me)_3P$	38.4	24.7^{1P}	—	2.0013	C_6H_5	L.Ex. C142/L	RT	311
C146 No. 21		Re(I)	$(OPh)_3P$	32.8	23.8^{1P}	—	2.0019	C_6H_6	L.Ex. C142/L	RT	311
C147 No. 21		Re(I)	$Ph_2PCH_2P\ Ph_2$	38.7	24.0^{1P}	—	2.0010	C_6H_6	L.Ex. C142/L	RT	311
C148 No. 21		Re(I)	$Ph_2P(CH_2)_2P\ Ph_2$	37.5	25.4^{1P}	—	2.0023	C_6H_6	L.Ex. C142/L	RT	311
C149 No. 21		Re(I)	$Ph_2P(CH_2)_3P\ Ph_2$	37.9	26.3^{1P}	—	2.0016	C_6H_6	L.Ex. C142/L	RT	311
C150 No. 21		Re(I)	$Ph_2P(CH_2)_4P\ Ph_2$	37.9	25.8^{1P}	—	2.0015	C_6H_6	L.Ex. C142/L	RT	311
C151 No. 21		Re(I)	$(C_6H_4NMe_2)_3P$	38.0	24.7^{1P}	—	2.0015	C_6H_6	L.Ex. C142/L	RT	311
C152 No. 21		Re(I)	$Ph_2P(CH_2)_2AsPh_2$	37.5	25.1^{1P}	—	2.0016	C_6H_6	L.Ex. C142/L	RT	311
C153 No. 21		Re(I)	Ph_3As	36	36^{1As}	—	2.0017	C_6H_6	L.Ex. C142/L	RT	311
C154 No. 21		Re(I)	Ph_3Sb	36	$\frac{93^{t}}{50}$	—	2.0024	C_6H_6	L.Ex. C142/L	>RT	311
C155 No. 21		Re(I)	Et_3N	24.6	u	—	2.0022	C_6H_6	L.Ex. C142/L	RT	311
C156 No. 21		Re(I)	$O = P(NMe_2)_3$	9.5	2.2^{1P}	—	2.0047	C_6H_6	L.Ex. C142/L	RT	311

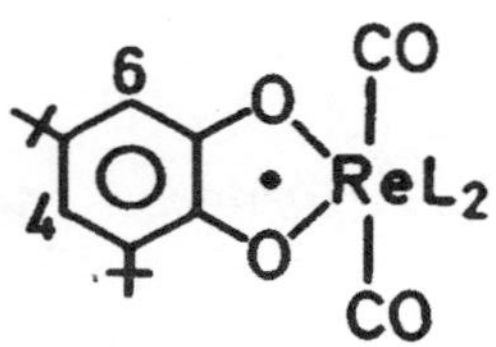

C157 No. 22		Re(I)	$(OMe)_3P$	49.9	30.8^{2P}	—	2.0009	C_6H_6	L.Ex C142/2L	RT	311

Table IV.1.4 (continued)
3,5-DI-*t*-BUTYL-1,2-BENZOSEMIQUINONE-CATION COMPLEXES

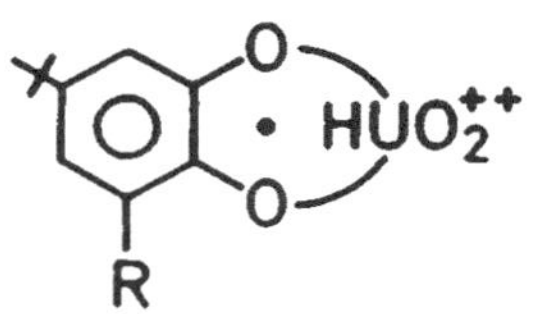

No.	Complex		Central atom	L: a_H^{OH}	a_3^R	Splitting constants/Gauss: a_4	a_5	a_6	g-Factor	Solvent	Method	T [K]	Ref.
C158	No. 23 (R = H)	AC	U(VI)	0.73[y]	0.74	3.63	0.37[z]	0.74	1.9933	THF(?)	Phot Q/$UO_2^{+\cdot}$	193	376
C159	No. 23 (R = Bu^t)		U(VI)	u	u	3.61	0.36[z]	0.72	1.9923	THF(?)	Phot Q/$UO_2^{+\cdot}$	193	376

[a] THF, dioxane, or ether not specified.
[b] Spectral data of other alkali metal complexes (low resolution) are listed in Table I.1 under Reference 13.
[c] Complex prepared by exchange reaction of CuCl with sodium- or thallium-3,5-di-*t*-butylbenzosemiquinolate (SQ) in presence of neutral ligands (L).
[d] Complex prepared by reacting the quinone with metal acetylacetonate (Acac) in vacuum.
[e] Value indicates rapid equilibrium between structures no. 4 and 5.
[f] Splitting from ^{201}Hg not resolved.
[g] Number above line from ^{199}Hg, below from ^{201}Hg. Hg splittings in Figure 2 of Reference 304 are interchanged.
[h] Splittings from ^{203}Tl and ^{205}Tl superimposed.
[i] Number above line from ^{69}Ga, below from ^{71}Ga.
[j] $^{bf,a}In^{113}$ = $^{bf,a}In^{115}$, I = 9/2.
[k] Superimposed splitting from ^{115}Sn, ^{117}Sn, and ^{119}Sn.
[l] Number above line from ^{117}Sn, below from ^{119}Sn.
[m] Splitting from proton at C-6.

[n] For preparation of the complexes C125, C127, and C128, see Reference 308 for details.

[o]  represents 3,5-di-*t*-butylcatechol ligands in this table.

[p] Splittings from two different rotamers occurring simultaneously.
[q] For solvent/temperature dependence of a_{Bi209} see Reference 309.
[r] Complex obtained by photolytic reaction between pertinent quinone and bis(3,5-di-*t*-butyl-1,2-phenylenedioxy)spirophosphorane (SpP).
[s] Hfs of spectrum caused by interaction of unpaired electron with two antimony nuclei. Absence of proton splitting indicates existence of weak complex with quinonoid derivative.
[t] Number above line from ^{121}Sb, below from ^{123}Sb.
[u] The complex C140 in THF solution at 77K anisotropic with g-tensor of axial symmetry.
[v] Cat = fully reductive form of pertinent quinone (cf. no. C244, Table IV.1.5).
[x] Complex prepared by ligand exchange (L.Ex.) starting from complex C142.
[y] Extra proton splitting assumed to derive from a proton placed between the two carbonyl oxygens.

Table IV.1.5
3,6-DI-*t*-BUTYL-1,2-BENZOSEMIQUINONE-CATION COMPLEXES

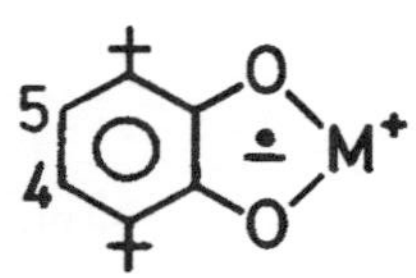

No.	Complex	Central atom	L	Splitting constants/Gauss a_M	a_4	a_5	g-Factor	Solvent	Method	T [K]	Ref.
C160	No. 1	IA ^{7}Li(I)	—	0.61	3.50	3.50		DME	Re Li	333	313
				0.62	3.50	3.50				273	
				0.63	3.50	3.50				213	
				0.61	3.50	3.50		THF	Re Li	333	313
				0.66	3.50	3.50				273	
				0.70	3.50	3.50				213	
				0.66	3.52	3.52		DEE	Re Li	333	313

Table IV.1.5 (continued)
3,6-DI-*t*-BUTYL-1,2-BENZOSEMIQUINONE-CATION COMPLEXES

No.	Complex	Central atom	L	Splitting constants/Gauss a_{M1}	a_4	a_5	g-Factor	Solvent	Method	T [K]	Ref.
				0.71	3.52	3.52				273	
				0.71	3.52	3.52				213	
C161	No. 1	^{23}Na(I)	—	0.47	3.40	3.40		DME	Re Na	333	313
				0.53	3.40	3.40				273	
				0.61	3.40	3.40				213	
				0.47	3.40	3.40		THF	Re Na	333	313
				0.54	3.40	3.40				273	
				0.61	3.40	3.40				213	
				0.57	3.40	3.40		DEE	Re Na	333	313
				0.64	3.40	3.40				273	
				0.72	3.40	3.40				213	
				0.38	3.45	3.45		Pr^{i}OH/DEE	Re Na	333	313
				0.42	3.45	3.45		(α = 0.05)		273	
				0.45	3.45	3.45				213	
				0.29	3.25	3.25		Pr^{i}OH/DEE	Re Na	333	313
				0.32	3.25	3.25		(α = 0.13)		273	
				0.37	3.25	3.25				213	

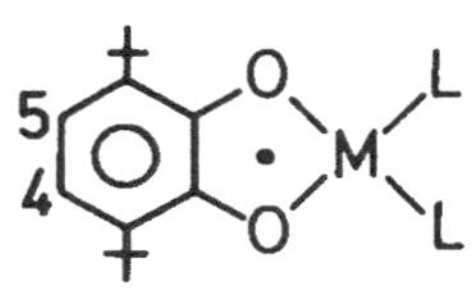

No.	Complex	Central atom	L	a_{M1}	a_4	a_5	g-Factor	Solvent	Method	T [K]	Ref.
C162	No. 2a IB	^{63,65}Cu(I)	Ph_3P	11.0[a] 11.8 17.6[2P]	3.1	3.1	2.0041	THF	NaSQ CuCl/2L		384
C163	No. 2a	Cu(I)	$(EtO)_3P$	11.3[a] 12.1	3.2	3.2	2.0044	THF	NaSQ CuCl/2L		384

C164	No. 2a		Cu(I)	Ph_3As	22.7^{2P} 11.0^{b}	3.3	3.3	2.0048	THF	NaSQ CuCl/2L		384
C165	No. 2a		Cu(I)	Et_3As	18.5^{2As} 11.6^{b}	3.3	3.3	2.0049	THF	NaSQ CuCl/2L		384
C166	No. 2a		Cu(I)	1,5–Cod	19.1^{2As} $\frac{6.8^{a}}{7.3}$	3.5	3.5	2.0037	THF	NaSQ CuCl/L		384
C167	No. 2a		Cu(I)	PhC≡CPh	6.5^{b}	3.3	3.3	2.0030	THF	NaSQ CuCl/L		384

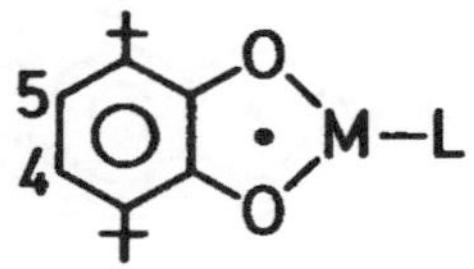

C168	No. 3		Cu(I)	Ph_3P	3.4^{b}	3.4	3.4	2.0037	THF	NaSQ CuCl/L		384
C169	No. 3		Cu(I)	CO	5.4^{b}	3.5	3.5	2.0034	THF	NaSQ CuCl/L		384
C170	No. 2a		$^{107,109}Ag(I)$	Ph_3P	1.7^{b} 5.6^{2P}	3.2	3.2	2.0030	THF	NaSQ $AgOCOCF_3$/2L	RT	302

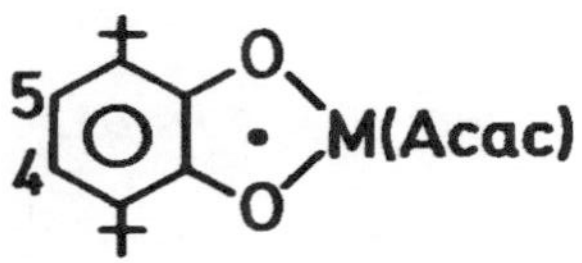

C171	No. 4	IIA	Be(II)	—	u	3.5	3.5		EtOH abs	Q $Be(Acac)_2$	233	314
C172	No. 4		Mg(II)	—	u	3.5	3.5		PhMe	Q $Mg(Acac)_2$	>RT	314
					u	(4.3	$2.7)^{c}$		PhMe	Q $Mg(Acac)_2$	253	314
C173	No. 4		Ca(II)	—	u	3.5	3.5		PhMe	Q $Ca(Acac)_2$	>RT	314
					u	(3.85	$3.15)^{c}$		PhMe	Q $Ca(Acac)_2$	253	314
C174	No. 4		Ba(II)	—	u	3.5	3.5		PhMe	Q $Ba(Acac)_2$	>RT	314
					u	(3.85	$3.15)^{c}$		PhMe	Q $Ba(Acac)_2$	253	314
C175	No. 1	IIIA	$^{113,115}In(I)$	—	3.8^{b}	3.8	3.8		THF	NaSQ $InCl_3$	273	315
				—	3.1^{b}	3.8	3.8		THF	NaSQ $InCl_3$	233	315
C176	No. 1		$^{203,205}Tl(I)$	—	58.2^{b}	3.3	3.3	1.9975	THF	Re Tl	298	328

Table IV.1.5 (continued)
3,6-DI-*t*-BUTYL-1,2-BENZOSEMIQUINONE-CATION COMPLEXES

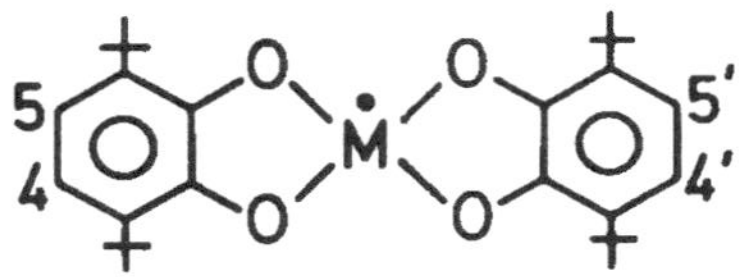

No.	Complex		Central atom	L	a_M	a_4	a_5	g-Factor	Solvent	Method	T [K]	Ref.
C177	No. 5	IIIA	$^{10}B(III)$	—	0.83	4.5	4.5		THF	NaSQ BCl_3	173	396
			$^{11}B(III)$		2.5							
C178	No. 5		$^{27}Al(III)$	—	2.58	4.0	4.0		THF	NaSQ $AlCl_3$	173	396
						$a_4 = a_{4'}$	$a_5 = a_{5'}$					
C179	No. 5		B(III)	—	—	2.25	2.25		THF	NaSQ BCl_3	>293	396
C180	No. 5		Al(III)	—	—	2.0	2.0		THF	NaSQ $AlCl_3$	233	396

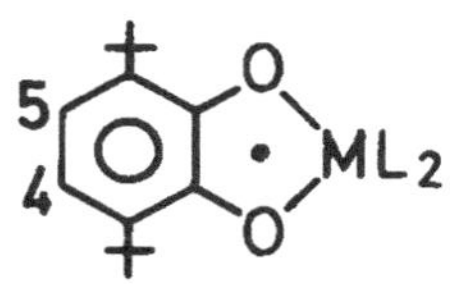

No.	Complex		Central atom	L	a_M	a_4	a_5	g-Factor	Solvent	Method	T [K]	Ref.
C181	No. 2b	IIIA	$^{27}Al(III)$	Et	2.2	4.4	4.4	2.0035	PhMe	Q $AlEt_3$	213—233	404
C182			Al(III)	Cl	2.3	3.8	3.8		THF	Q $AlCl_3$	213	307
C183			Al(III)	Br	2.4	4.0	4.0		THF	Q $AlBr_3$	213	307
C184			Al(III)	I	2.3	4.0	4.0		THF	Q AlI_3	213	307

C185	No. 2b		^{69,71}Ga(III)	Et	16.0[d]/20.2	3.6	3.6	2.0033	PhMe	Q $GaEt_3$	213—233	404
C186			Ga(III)	Cl	6.9[d]/8.8	3.9	3.9		THF	Q $GaCl_3$	230	307
C187			Ga(III)	Br	6.4[d]/8.1	4.0	4.0		THF	Q $GaBr_3$	230	307
C188	No. 2b		^{113,115}In(III)	Et	8.9[b]	3.8	3.8	2.0030	PhMe	Q $InEt_3$	RT	404
C189			In(III)	Cl	7.3[b]	3.6	3.6		THF	Q $InCl_3$	293	307
C190			In(III)	Br	6.4[b]	3.7	3.7		THF	Q $InBr_3$	293	307
C191			In(III)	I	5.0[b]	3.8	3.8		THF	Q InI_3	293	307
C192	No. 2b		^{203,205}Tl(III)	Et	32.2[b]	3.2	3.2	2.0036	PhMe	Q $TlEt_3$	RT	404
					27.7[b]	3.3	3.3	2.0026	THF		RT	23

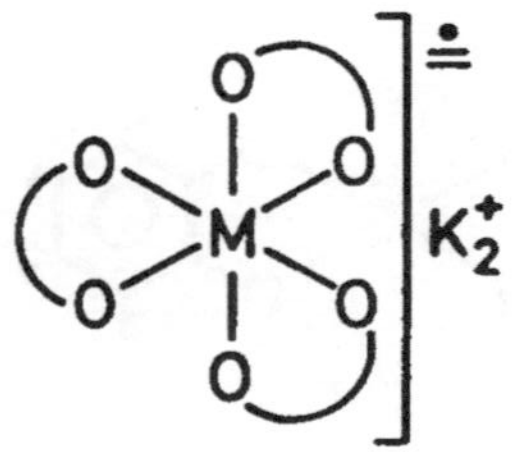

C193	No. 6[e]		^{27}Al(III)	—	2.25	3.75[f]	3.75[f]		PhMe	Re K	RT	315
C194	No. 6		^{69,71}Ga(III)	—	4.2[d]/5.35	3.75[f]	3.75[f]		THF	Re K	273	315
					4.7[d]/6.0	3.75[f]	3.75[f]		THF	Re K	213	315
C195	No. 6		^{113,115}In(III)	—	7.5[b]	1.25[g]	1.25[g]		THF	Re K	233	315
C196	No. 2b	IIIB	^{45}Sc(III)	Cl	2.5	3.8	3.8		EtOH	Q $ScCl_3$	293	307
C197	No. 2b		Y(III)	Cl	u	3.7	3.7		THF	Q YCl_3	293	307
C198	No. 2b		^{175}Lu(III)	Cl	5.3	5.3	5.3		THF	Q $LuCl_3$	293	307
C199	No. 2b	LA	^{139}La(III)	Cl	3.1	3.7	3.7		EtOH	Q $LaCl_3$	293	307
C200	No. 2b		La(III)	Br	3.4	3.4	3.4		EtOH	Q $LaBr_3$	293	307

Table IV.1.5 (continued)
3,6-DI-*t*-BUTYL-1,2-BENZOSEMIQUINONE-CATION COMPLEXES

No.	Complex	Central atom	L	Splitting constants/Gauss a_M	a_4	a_5	g-Factor	Solvent	Method	T [K]	Ref.
	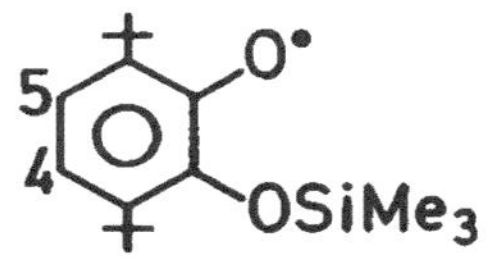										
C201	No. 7 IVA	Si(IV)	—	—	8.4	0.62		PhMe	Phot PhOH/Bu^{t}O$_2$	273	331
	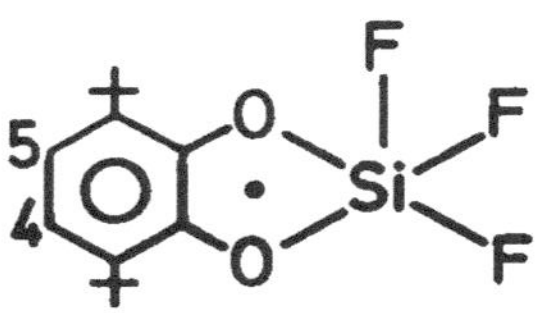			a^F							
C202	No. 8	Si(IV)	—	0.35^{1F} 10.1^{2F}	4.2	3.6		PhMe	Q SiF_4	313	316
	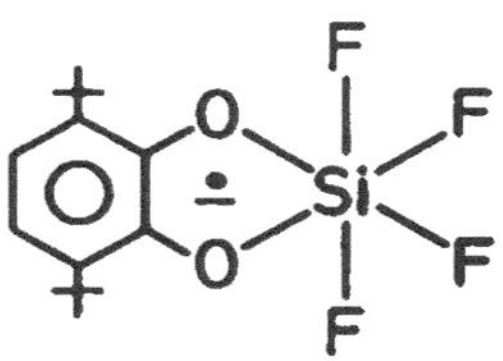										
C203	No. 9	Si(IV)	—	0.58^{2F} 10.5^{2F}	3.9	3.9		PhMe	Q SiF_4/HF	233	316

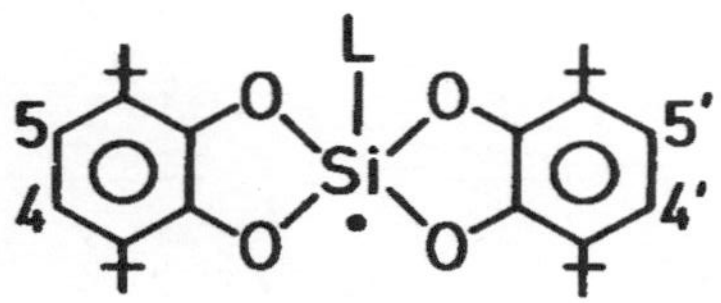

				$a_4 = a_{4'}$	$a_5 = a_{5'}$				
C204 No. 10	Si(IV)	F	5^{19F}	2.13	2.13	Pentane	$SpSi^h/F_2$ or HF	>313	330, 332
C205 No. 10	Si(IV)	Cl,Br,I	—	2.13	2.13	Pentane	$SpSi/Cl_2$, Br_2, I_2	>313	330
C206 No. 10	Si(IV)	Acac	—	2.40	1.65	Pentane	SpSi/AcacH	>293	330
			—	2.20	2.20	PhMe	SpSi/SQ	233	397, 330

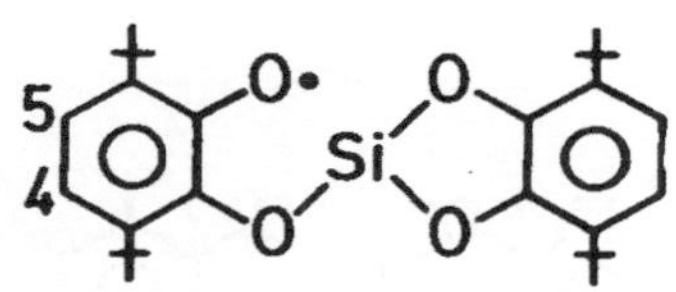

			a_M	a_4	a_5				
C207 No. 11	Si(IV)	Cl,Br,I	—	6.37	2.37	Pentane	$SpSi/Cl_2$,Br_2,I_2	173	330, 332
C208 No. 11	Si(IV)	Acac	—	4.8	3.3	Pentane	SpSi/AcacH	>293	330

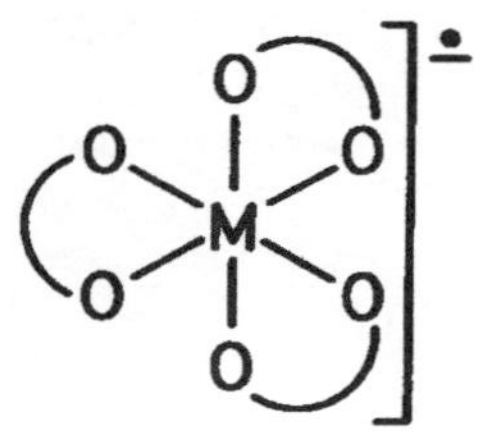

Table IV.1.5 (continued)
3,6-DI-*t*-BUTYL-1,2-BENZOSEMIQUINONE-CATION COMPLEXES

No.	Complex		Central atom	L	Splitting constants/Gauss a_M	a_4	a_5	g-Factor	Solvent	Method	T [K]	Ref.
C209	[I] No.12[c]	IVA	Si(IV)	—	—	1.25[g]	1.25[g]		THF	Re K or el	233	324
					—	3.75[f]	3.75[f]		THF	Re K or el	183	324
C210	No. 12		^{117,119}Sn(IV)	—	u	1.3[g]	1.3[g]		THF/THBCw[i]	Re K	373	319
					4.0[b]	3.9[f]	3.9[f]		DEE	Re K	183—313	319

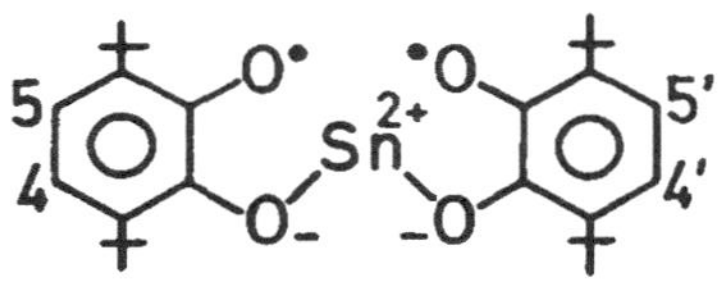

No.	Complex		Central atom	L	a_M	a_4	a_5	g-Factor	Solvent	Method	T [K]	Ref.
C211	No. 13		^{117,119}Sn(II)	—	5.0[b]	4.1	4.1[j]		PhMe:CCl_4 (3:1)	Re Sn(*in* Hg)	RT	319

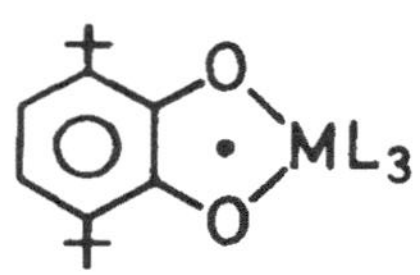

No.	Complex		Central atom	L	a_M	a_4	a_5	g-Factor	Solvent	Method	T [K]	Ref.
C212	No. 14		^{117,119}Sn(IV)	Me	12.8[k] / 13.3	3.6	3.6		PhMe	NaSQ $ClSnMe_3$	223	398
				Me	13.0[k] / 13.5	3.6	3.6		$SnMe_4$	Q $SnMe_4$	223	398
C213	No. 14		Sn(IV)	2Me, 1Cl	21.5[k] / 22.5	5.2	2.6		PhMe	NaSQ Cl_2SnMe_2	223	398
				2Me, 1Cl	—	3.6	3.6		PhMe	NaSQ Cl_2SnMe_2	RT	398
				2Me, 1Cl	21.7[k] / 22.9	5.2	2.6		PhMe	Q $ClSnMe_3$	223	398

			2Me, ICl	—	3.6	3.6		PhMe	Q ClSnMe$_3$	RT	398
C214 No. 14		Sn(IV)	Ph	11.3[k] / 12.0	3.7	3.7	2.0030	PhMe	Q (C$_5$H$_5$)SnPh$_3$	263	318
C215 No. 14	IVA	^{207}Ph(IV)	Me	6.3	3.4	3.4	2.0044	PhMe	Q (C$_5$H$_5$)PhMe$_3$	258	318
C216 No. 14		Pb(IV)	Et	u	3.2	3.2	2.0044	THF	NaSQ Et$_3$PbCl	—	308
C217 No. 14		Pb(IV)	Ph	u	3.0	3.0	2.0038	PhMe	Q (C$_5$H$_5$)PbPh$_3$	233	318
C218 No. 14		Pb(IV)	C$_5$H$_5$,2Ph	2.5	4.3[l]	2.7[l]	2.0035	PhMe	Q (C$_5$H$_5$)$_2$PbPh$_2$	331	318
				3.0	4.3[l]	2.7[l]		PhMe	Q (C$_5$H$_5$)$_2$PbPh$_2$	337	318

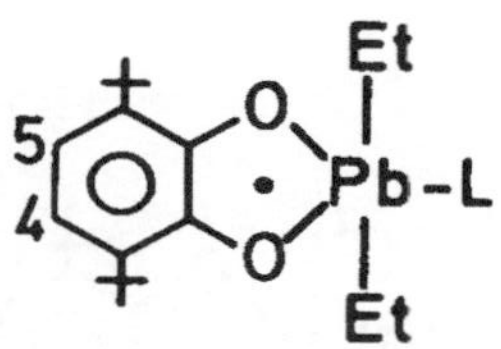

C219 No. 15		^{207}Pb(IV)	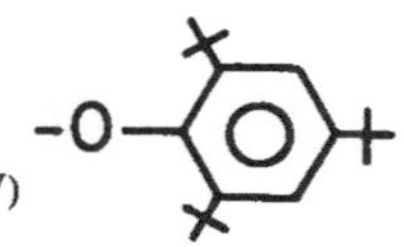	u	4.3[l]	2.5[l]	2.0044	PhMe	Q Et$_4$Pb	—	308[m]
C220 No. 15		Pb(IV)	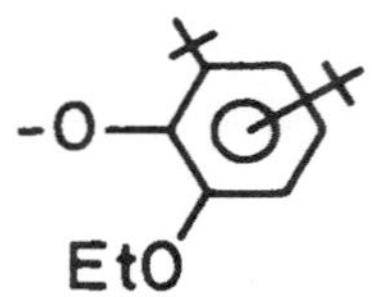	9.3	4.1[l]	2.7[l]	2.0023	THF	Q Et$_4$Pb	—	308[m]

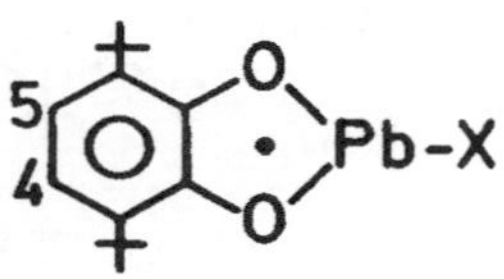

C221 No. 16		^{207}Pb(II)	X (?)	54.2	3.5	3.5	1.9999	THF	Q Et$_4$Pb	—	308[m]

Table IV.1.5 (continued)
3,6-DI-*t*-BUTYL-1,2-BENZOSEMIQUINONE-CATION COMPLEXES

No.	Complex		Central atom	L	Splitting constants/Gauss a_M	a_4	a_5	g-Factor	Solvent	Method	T [K]	Ref.
	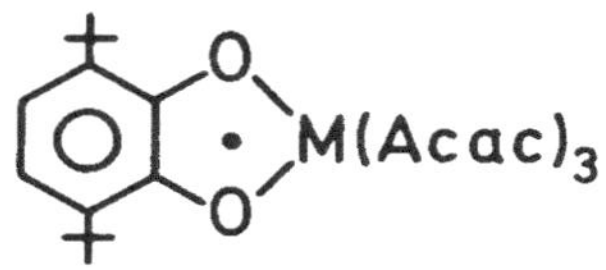											
C222	No. 17	IVB	$^{91}Zr(IV)$	—	2.0	3.72	3.72		PhMe		>RT	301
					1.78	3.85	3.85		PhMe		>RT	301
					—	4.3	3.3		PhMe		<233	301
C223			Hf	(Complex spectrum not interpreted, see Reference 301.)								
	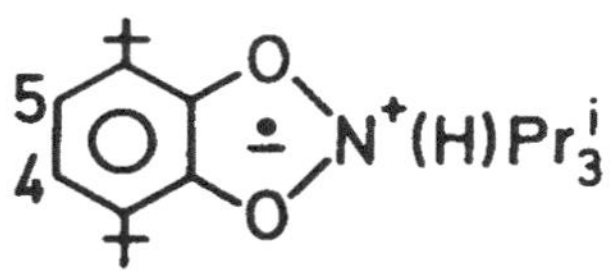											
C224	No. 18	VA	N	—	u	3.35	3.35		Pr^i_3N		253	320
				—	u	3.7	3.0				193	320
	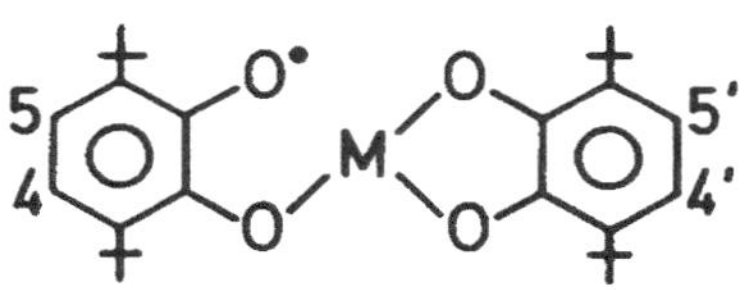											
C225	No. 19		$^{31}P(III)$	—	58	13	u		PhMe	Q P_4		309
C226	No. 19		$^{75}As(III)$	—	19.5	1.9	1.9		DEE	Q As_4	428	309, 310
				—	—	7.65	0		DEE	Q As_4	293	310

C227 No. 19		^{121}Sb(III)	—	47.5	u	u				309
		^{123}Sb(III)		25.5						
C228 No. 19		^{209}Bi(III)	—	32.5	u	u	DEE	QK$_2$ BiCl$_3$	273	309

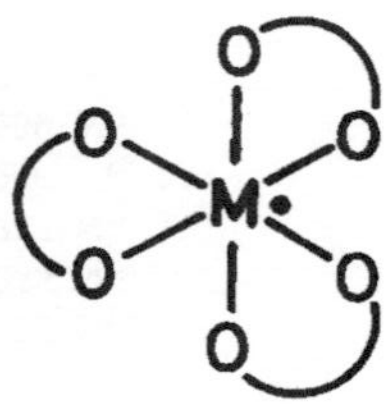

C229 No. 20[c]		^{31}P(V)	—	4.0	1.45[g]	1.45[g]	PhMe	Phot Q/SpP[n]	273	322
			—	4.0	4.35[f]	4.35[f]	PhMe	Phot Q/SpP	193	322
			—	4.35	1.45[g]	1.45[g]	DEE	Re PbO$_2$	>293	391
			—	4.35	4.35[f]	4.35[f]	DEE	Re PbO$_2$	<173	391
C230 No. 20		^{75}As(V)	—	0.4	1.45[g]	1.45[g]	DEE	Q As$_4$	>293	310
					4.35[f]	4.35[f]	DEE	Q As$_4$	193	310

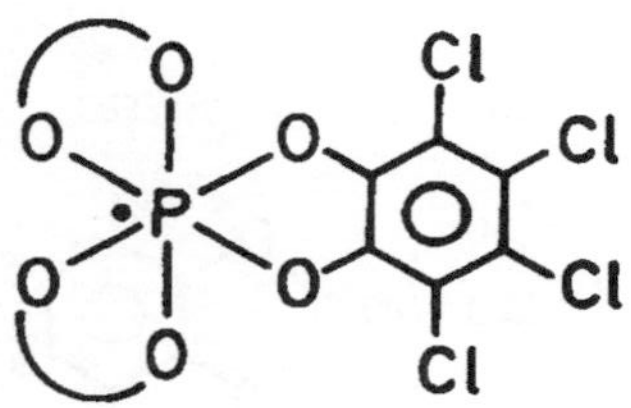

C231 No. 21[c]	VA	^{31}P(V)	—	4.35	2.18[2H]	2.18[2H]	PhMe	Phot Q/SpP	253	322
C232 No. 22 (Co = VO)	VB	^{51}V(V)	—	12	4.35	4.35	PhMe	Q VO(Acac)$_2$	293	301

Table IV.1.5 (continued)
3,6-DI-*t*-BUTYL-1,2-BENZOSEMIQUINONE-CATION COMPLEXES

No.	Complex	Central atom	L	Splitting constants/Gauss a_M	a_4	a_5	g-Factor	Solvent	Method	T [K]	Ref.
	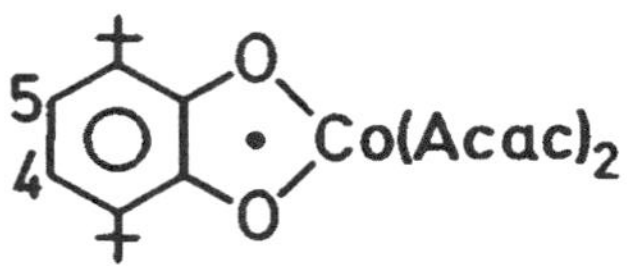										
C233	No. 22 VIIIB	^{59}Co(III)	—	11.6	3.45	3.45		PhMe	Q VO(Acac)$_2$		301
	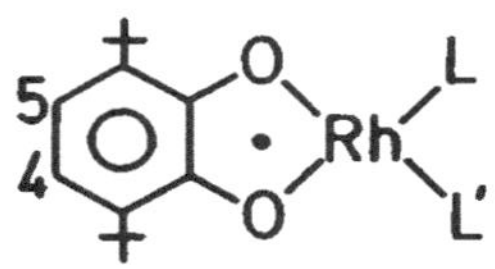										
			L/L′								
C234	No. 23	^{103}Rh(I)	CO/CO	—	3.2	3.2	2.0020	PhMe	TISQ/ ((CO)$_2$RhCl)$_2$	RT	386
C235	No. 23	Rh(I)	CO/P(OPh)$_3$	3.2^{1P}	3.2	3.2	2.0022	PhMe	L.Ex./L′	RT	386
C236	No.23	RH(I)	P(OPh$_3$)/P(OPH$_3$)	3.2^{2P}	3.2	3.2	2.0039	PhMe	L.Ex./L,L′	RT	386
C237	No. 23	Rh(I)	CO/Ph$_3$P	3.2^{1P}	3.2	3.2	2.0045	PhMe	L.Ex./L′	RT	386
C238	No. 23	Rh(I)	CO/Ph$_3$As	3.4^{1As}	3.4	3.4	3.0058	PhMe	L.Ex./L′	RT	386
C239	No. 23	Rh(I)	1,5-Cod/-	—	3.2	3.2	2.0039	PhMe	L.Ex./L	RT	386
C240	No. 23	Rh(I)	1,5-Cod/Et$_3$As	4.2 40.0^{1As}	3.2	3.2	2.0017	PhMe	L.Ex./L,L′	RT	386

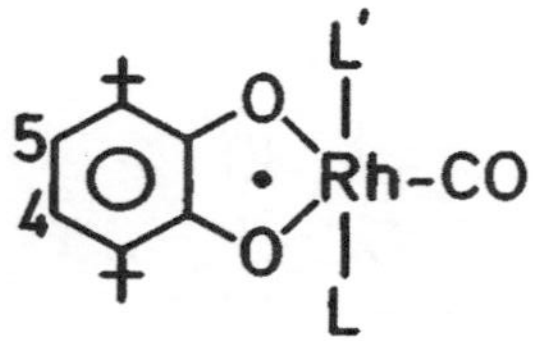

C241 No. 24	$^{103}Rh(I)$	Ph_3P/Ph_3P	1.6 17.0^{2P}	3.2	3.2	2.0011	PhMe	L.Ex./L,L′	RT	386
C242 No. 24	Rh(I)	Et_3As/Et_3As	17.1^{2P}	3.2	3.2	2.0025	PhMe	L.Ex./L,L′	RT	386

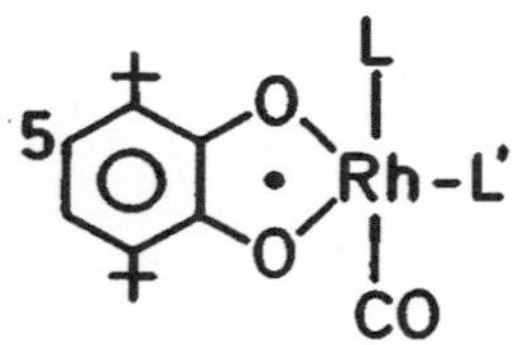

C243 No. 25	$^{103}Rh(I)$	Et_3As/Et_3As	2.6 8.6^{1As} 2.6^{1As}	3.2	3.2	2.0006	PhMe	L.Ex./L,L′	RT	386

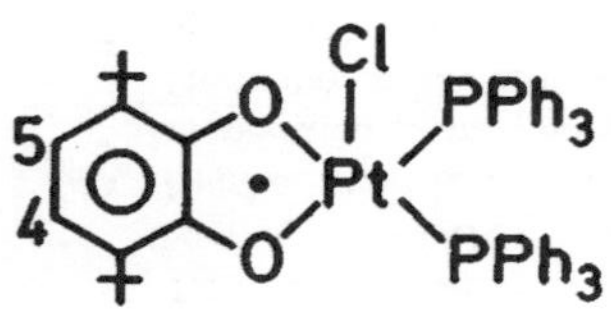

C244 No. 26	$^{195}Pt(II)$	—	12.8 3.2^{2P}	3.2	3.2	1.9991	THF	$(Ph_3P)PtCat^{o}$/ $CuCl_2$	RT	375

[a] Number above line from ^{63}Cu, below from ^{65}Cu.
[b] Splittings from the isotopes not resolved.
[c] Migration frequency of cation between O-atoms fallen off at the lowered temperature leading to nonequivalent ring proton splittings.
[d] Number above line from ^{69}Ga, below from ^{71}Ga.

Table IV.1.5 (continued)
3,6-DI-*t*-BUTYL-1,2-BENZOSEMIQUINONE-CATION COMPLEXES

[c]  Represents 3,6-di-*t*-butylcatechol in this table.
[f] Triplet observed from unpaired electron interacting with two equivalent protons of one semiquinone ligand.
[g] Septuplet observed from unpaired electron interacting with six "equivalent" protons of three aromatic ligands.
[h] SpSi = bis(3,6-di-*t*-butyl-1,2-phenylenedioxy)spirosilane.
[i] THBCw = crown ether (tetrahydrobenzo-18-crown-6).
[j] $a_4 = a_{4'} = 4.1$ G; $a_5 \approx a_{5'}$, unresolved.
[k] Number above line from ^{117}Sn, below from ^{119}Sn.
[ℓ] Introduction of one cyclopentadienyl ligand (C_5H_5) or other bulky ligand makes the two proton couplings appear nonequivalent at all accessible temperatures.
[m] Preparation of the complexes C219 and C220 (see Reference 308 for details).
[n] SpP = spirophosphorane.
[o] Cat = fully reductive form of pertinent quinone. Complex has tetrahedral pyramid geometry with Cl being at the apex.

Table IV.1.6
TETRAHALOGENO-1,2-BENZOSEMIQUINONE-CATION COMPLEXES

No.	Complex	Central atom	L/L′	Splitting constants/Gauss a_M	$a^C_{1,2}$	$a^C_{4,5}$	$a^C_{3,6}$	g-Factor	Solvent	Method	Ref.

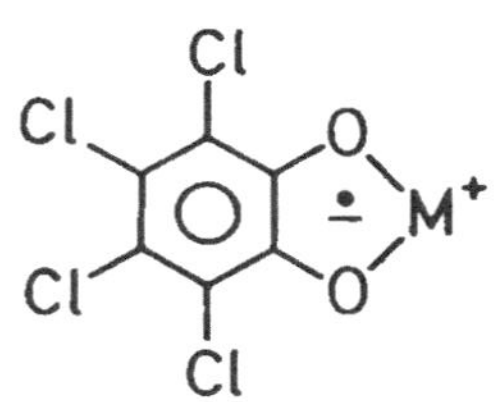

C245	No. 1	IA	^{7}Li(I)	—	—[a]	u	2.92	4.30	2.00500	THF	Re Li	221
				—	—	u	2.85	4.02	2.00507	DME	Re Li	221
C246	No. 1		^{23}Na(I)	—	—[a]	1.78	2.67	3.57	2.00515	THF	Re Na	221
				—	—	1.65	2.65	3.53	2.00523	DME	Re Na	221
C247	No. 1		K(I)	—	u	2.02	2.63	3.34	2.00522	THF	Re K	221
					u	2.04	2.61	3.31	2.00525	DME	Re K	221
C248	No. 1		^{85,87}Rb(I)	—	—[b]	u	u	u	2.00496	THF	Re Rb	221
				—	—	u	2.66	3.26	2.00509	DME	Re Rb	221
C249	No. 1		^{133}Cs(I)	—	—[b]	u	u	u	2.00481	THF	Re Cs	221
				—	—	2.23	2.70	3.21	2.00453	DME	Re Cs	221
89	No. 1 (free ion)			—	—	1.42	2.59	2.59	2.00527	THF/DMF	Re el	221

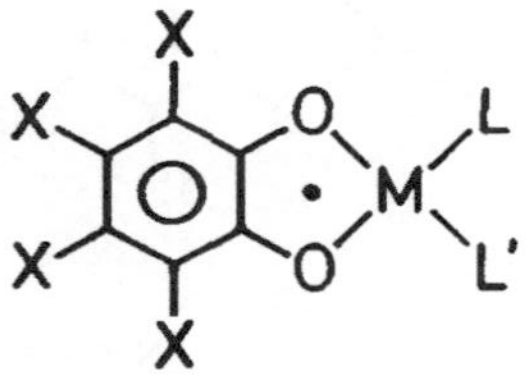

					a_M	a_L	a_4^{Cl}	a_5^{Cl}				
C250	No. 2 (X = Cl)	IB	^{63}Cu(I)	Ph_3P/Ph_3P	11.7	19.4[2P]	—	—	2.0067	THF	NaSQ CuCl/ 2L	302
C251	No. 2 (X = Br)		Cu(I)	Ph_3P/Ph_3P	11.9	19.5[2P]	—	—	2.0106	THF	NaSQ CuCl/ 2L	302

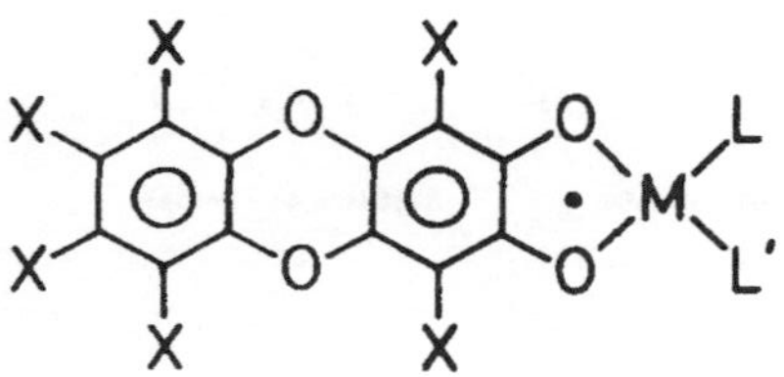

C252	No. 3	IB	^{63}Cu(I)	Ph_3P/Ph_3P	10.2[c]	15.7[2P]	—	—	2.0054	THF	NaSQ CuCl/	384

Table IV.1.6 (continued)
TETRAHALOGENO-1,2-BENZOSEMIQUINONE-CATION COMPLEXES

No.	Complex		Central atom	L/L′	Splitting constants/Gauss a_M	$a^C_{1,2}$	$a^C_{4,5}$	$a^C_{3,6}$	g-Factor	Solvent	Method	Ref.
C252	No. 3 (X = Cl)	IB	^{63}Cu(I)	Ph_3P/Ph_3P	10.2^c 10.9	15.7^{2P}	—	—	2.0054	THF	NaSQ CuCl/ 2L	384
C253	No. 3		Cu(I)	$(EtO)_3P/(EtO)_3P$	9.9^c 10.5	20.9^{2P}	—	—	2.0052	THF	NaSQ CuCl/ 2L	384
C254	No. 3		Cu(I)	$(EtO)_3P/Ph_3P$	9.8	14.8^{1P} 19.5^{1P}	—	—	2.0052	THF	NaSQ CuCl/ L,L′	384
C255	No. 3		Cu(I)	Ph_3As/Ph_3As	10.0	15.1^{2As}	—	—	2.0056	THF	NaSQ CuCl/ 2L	384
C256	No. 3 (X = Br)		Cu(I)	Ph_3P/Ph_3P	10.1	15.2^{2P}	—	—	2.0061	THF	NaSQ CuCl/ 2L	302
C257	No. 3 (X = Cl)		^{107,109}Ag(I)	Ph_3P/Ph_3P	1.3	4.5^{2P}	—	—	2.0037	THF	NaSQ AgO-$COCF_3$/2L	302
C258	No. 4	IIIA	^{27}Al(III)	Cl/Cl	2.2	—	—	—		THF	Q $AlCl_3$	300
C259	No. 4		Al(III)	Br/Br	2.5	—	—	—		THF	Q $AlBr_3$	300
C260	No. 4		Al(III)	I/I	2.4	—	—	—		THF	Q AlI_3	300
C261	No. 4		^{69}Ga(III)	Cl/Cl	7.0 8.9^c	—	—	—		THF	Q $GaCl_3$	300
C262	No. 4		Ga(III)	Br/Br	6.7 8.6^c	—	—	—		THF	Q $GABr_3$	300

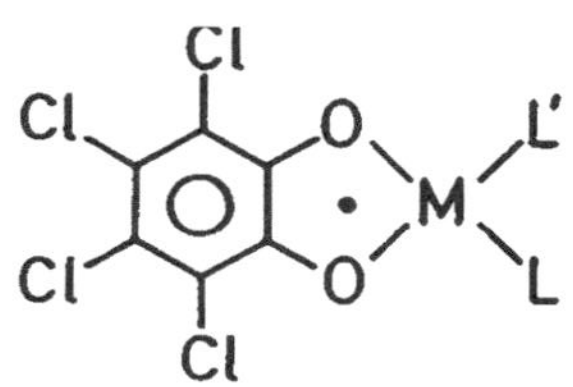

C263	No. 4		Ga(III)	I/I	6.8							
					9.0[c]	—	—	—		THF	Q GaI_3	300
C264	No. 4		^{115}In(III)	Cl/Cl	6.9	—	—	—		THF	Q $InCl_3$	300
C265	No. 4		In(III)	Br/Br	5.8	—	—	—		THF	Q $InBr_3$	300
C266	No. 4		In(III)	I/I	4.5	—	—	—		THF	Q InI_3	300
C267	No. 3		In(III)	Cl/Cl	5.0/5.5[d]	—	—	—		THF	Q $InCl_3$	323
C268	No. 1		^{203,205}Tl(I)	—	73.2	—	—	—	~1.998	THF	Re Tl	328

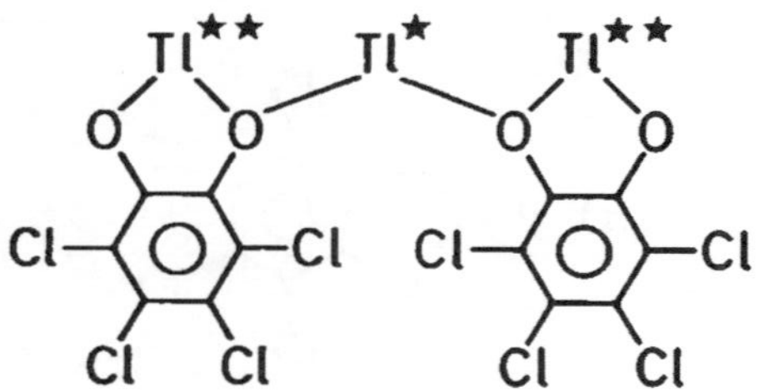

C269	No. 5[c]		^{203,205}Tl(I)	—	92.0**	—	—	—	2.0030	THF	Re Tl	328
					10.0*							
			Tl(III)	(see Table IV.1.3. no.C68).								
C270	No. 4	IIIB	^{45}Sc(III)	Cl/Cl	2.2	—	—	—		THF	Q $ScCl_3$	307
C271	No. 4		^{175}Lu(III)	Cl/Cl	4.7	—	—	—		THF	Q $LuCl_3$	300
C272	No. 4	LA	^{139}La(III)	Cl/Cl	2.6	—	—	—		THF	Q $LaCl_3$	300

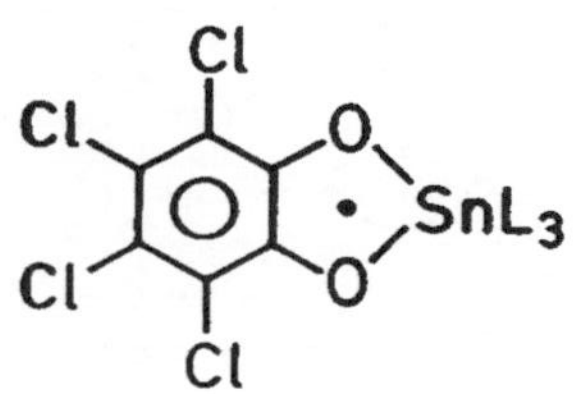

C273	No. 6	IVA	^{117,119}Sn (IV)	Ph/ –	7.72[i] / 8.07	—	0.62	0.62	2.00548	PhMe	Q Ph_6Sn_2	374
C274	No. 6		Sn(IV)	Cl/ –	11.4	—	—	—		DEE	Q $SnCl_4$	300

Table IV.1.6 (continued)
TETRAHALOGENO-1,2-BENZOSEMIQUINONE-CATION COMPLEXES

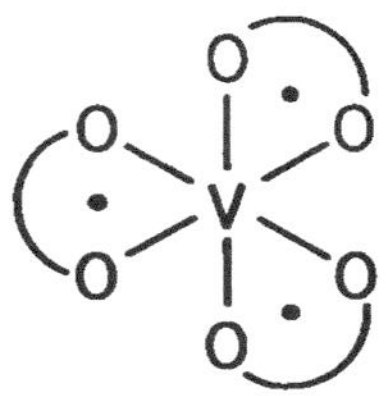

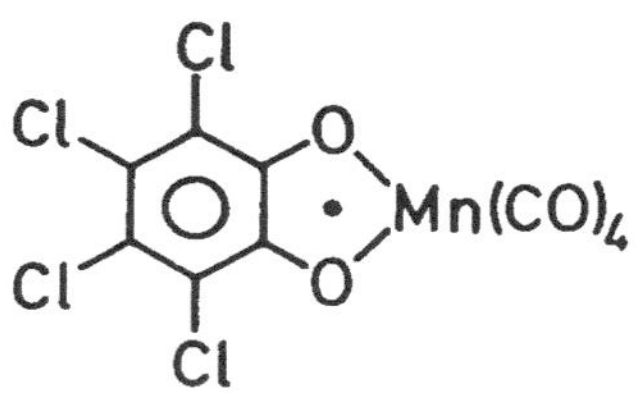

No.	Complex	Central atom		L/L′	Splitting constants/Gauss a_M	$a^C_{1,2}$	$a^C_{4,5}$	$a^C_{3,6}$	g-Factor	Solvent	Method	Ref.
C275	No. 7	VB	$^{51}V(III)$[g]	—	4.1	—	—	—	2.0079	Ether	—	406
C276	$Cr(O_2C_6Cl_4)_3^{\bullet -}$	VIB	$^{53}Cr(I)$[h]	—	27.5	—	—	—	1.9687	Ch_2Cl_2	273 K and 100 K	407
C277	No. 8	VIIB	$^{55}Mn(I)$	—	9.13	—	0.50	0.50	2.00434	PhMe	Phot $Mn_2(CO)_{10}$	374

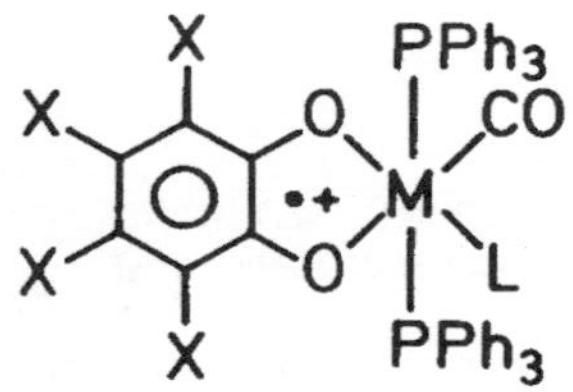

C278	No. 9 (X = Cl)	VIIIB	Ru(II)	CO/ –	—	24.5[2P]	—	—	2.004	CH_2Cl_2	Re el	334
C279	No. 9 (X = Br)		Ru(II)	CO/ –	—	24.7[2P]	—	—	2.007	CH_2Cl_2	Re el	334
C280	No. 4		^{59}Co(III)	Acac/Acac	10.6	—	—	—				301
C281	No. 9 (X = Cl)		Rh(III)	Cl/ –	—	19.3[2P]	—	—	2.003	CH_2Cl_2	Re el	334

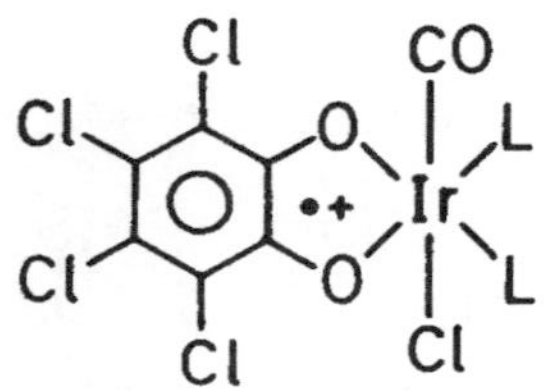

C282	No. 10		Ir(III)	Ph_3P/ –	—	<8[I]	—	—	2.002	CH_2Cl_2	Re el	334
C283	No. 10		Ir(III)	$MePh_2P$	—	<8[I]	—	—	2.002	CH_2Cl_2	Re el	387

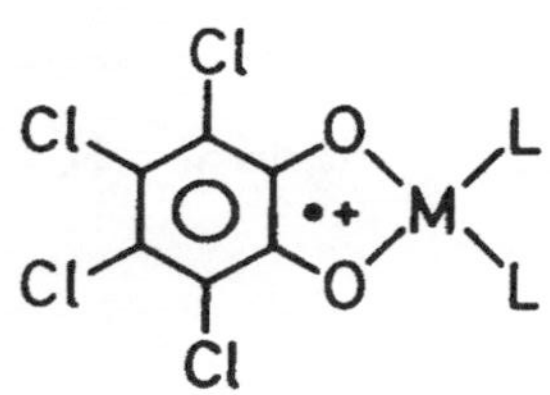

Table IV.1.6 (continued)
TETRAHALOGENO-1,2-BENZOSEMIQUINONE-CATION COMPLEXES

No.	Complex	Central atom	L/L′	a_M	$a^C_{1,2}$	$a^C_{4,5}$	$a^C_{3,6}$	g-Factor	Solvent	Method	Ref.
				Splitting constants/Gauss							
C284	No. 11	Pd(II)	Ph_3P	—	<6[i]	—	—	2.006	CH_2Cl_2	Re el	334
C285	No. 11	Pt(II)	Ph_3P	—	<8[i]	—	—	2.004	CH_2Cl_2	Re el	334
C286	No. 12	VIIIB $^{195}Pt(II)$	Ph_3P	12.5	2.6^{2P}	—	—	2.0004	CH_2Cl_2	$(Ph_3P)_2$ PtCat/ $FeCl_3$	375

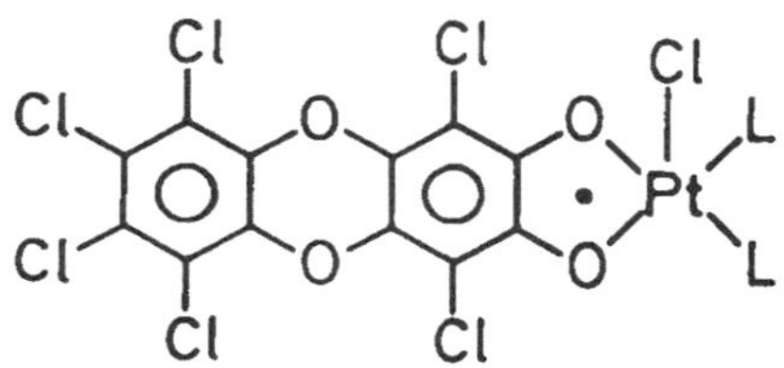

[a] Alkali metal splitting recorded as a function of temperature; see Figure 62.
[b] Alkali metal splitting recorded as a function of temperature; see Figure 63.
[c] Splitting above line from ^{63}Cu, below from ^{65}Cu, or from ^{69}Ga above, ^{71}Ga below.
[d] Two ten-component spectra observed, seemingly from two conformers.
[e] Proposed structure based on the observation of a Tl doublet (10.0 G) and a Tl triplet (92.0 G).
[f] Splitting above line from ^{117}Sn, below from ^{119}Sn.
[g] V(III) spin couples to two semiquinone ligands only; the third remains coordinated as a paramagnetic radical anion.  Represents tetrachloro-1,2-benzosemiquinone ligand in this table.
[h] The low g-factor and the large Cr splitting make $Cr(O_2C_6Cl_4)_3^{-}$ best described as a Cr(I) complex with an unexpectedly localized electronic structure (see Reference 407).
[i] Peak-to-peak linewidth. No phosphorus hfs observed.
[j] Cf. no. C244, Table IV.1.5

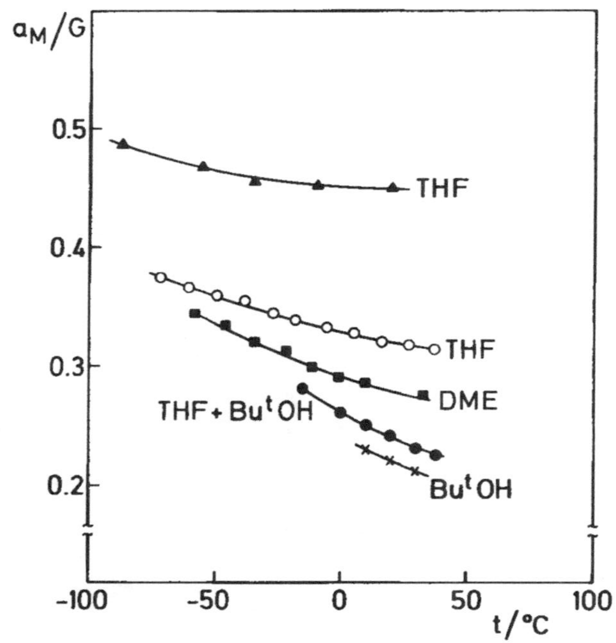

FIGURE 62. Temperature and solvent dependence of Li (▲) and Na (○,■,●,x) metal splitting constants of tetrachloro-1,2-benzosemiquinone (nos. C245 and C246). (From Pasimeni, L., Brustolon, M., and Corvaja, C., *J. Magn. Resonance*, 21, 259, 1976. With permission.)

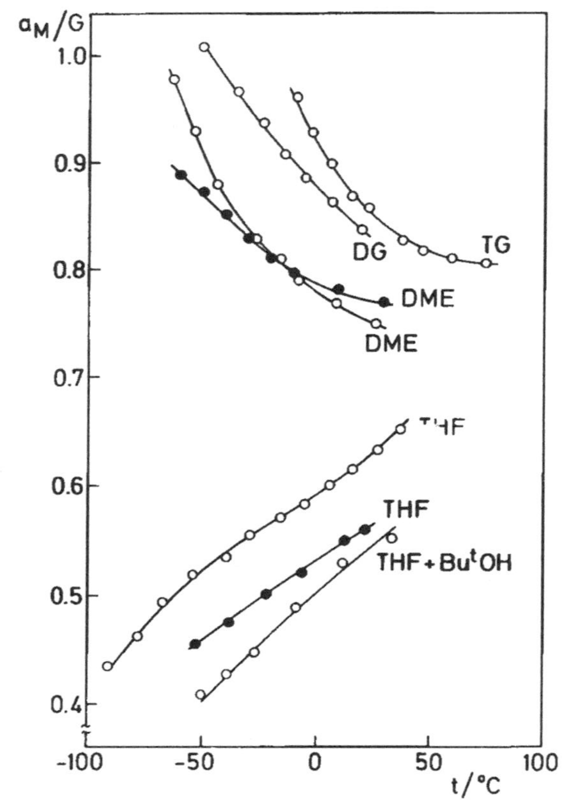

FIGURE 63. Temperature and solvent dependence of Rb (●) and Cs (○) splitting constants of tetrachloro-1,2-benzosemiquinone (nos. C248 and C249). DG = diglyme, TG = triglyme. (From Pasimeni, L., Brustolon, M., and Corvaja, C., *J. Magn. Resonance*, 21, 259, 1976. With permission.)

Table IV.1.7
1,4-BENZOSEMIQUINONE-CATION COMPLEXES

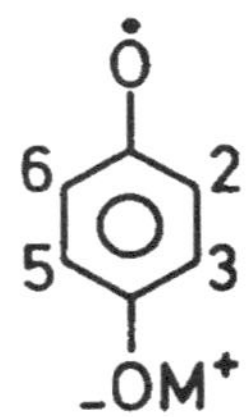

No.	Complex	Central atom		Splitting constants/Gauss a_M	a_2	a_3	a_5	a_6	g-Factor	Solvent	Method	T [K]	Ref.
C287	No. 1	IA	^{7}Li(I)	0.14	3.06	1.68	1.68	3.06		$C_5H_{11}{}^tOH$	—[a]	298	335, 336
				u	2.75	2.00	2.00	2.75		Bu^tOH	Q/HQ LiOBu	RT	54, 336
C288	No. 1		^{23}Na(I)	0.28	4.84[b]		4.84[b]			$C_5H_{11}{}^tOH$	—[a]	298	335, 336
				0.23	4.84		4.84			Bu^tOH	—	298	336
				0.1	4.90		4.90			DME	—	333	336
				u	2.61	2.30	2.30	2.61		DME	—	213	336
				1.07	2.66	2.05	2.05	2.66		DME	Re Na	283	339
				1.09	2.74	2.06	2.06	2.74		DME	Re Na	298	339
				0.29	3.68[c]	1.74	1.74	3.68[c]		HMPA	Re Na NaI[d]	RT	75
C289	No. 1		^{39}K(I)	u	2.885	1.935	1.935	2.885	2.00543	DME	Re K	198	340, 378
				u	2.93	1.95	1.95	2.93		DME	—[a]	183—223	336
				u	2.89[e]	1.95	1.95	2.89	ΔH = 0.11[f]	DME	Re K	190	341
				u	3.17[e]	1.69	1.69	3.17	ΔH = 0.25[f]	DME	Re K	190	341
				u	2.66	2.16	2.16	2.66		DME	Re K	213	339, 349
				0.30	2.30	1.74	1.74	2.30		HMPA	Re K/KI[d]	RT	75
				0.45	u	u	u	u		HMPA	Re K/KI[d]	318	75

C290	No. 1		Rb(I)	u	2.84	2.04	2.04	2.84		DME	—[a]	213—256	336
				u	2.81[e]	2.03	2.03	2.81	ΔH = 0.17[f]	DME	Re Rb	200	341
				u	3.19[e]	1.67	1.67	3.19	ΔH = 0.22[f]	DME	Re Rb	200	341
C291	No. 1		^{133}Cs(I)	~0.28	2.43[e]	2.43	2.43	2.43	ΔH = −0.12[f]	DME	Re Cs	235—293	341
				0.0	3.19[e]	1.66	1.66	3.19	ΔH = 0.23[f]	DME	Re Cs	200	341

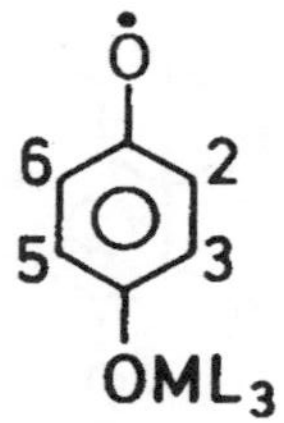

C292	No. 2 (L = Ph)	IVA	Ge(IV)	u	5.85	0.88	0.88	5.85		Pr^iOH	Phot Ph_3GeH	RT	346
C293	No. 2 (L = Bu)		Sn(IV)	u	4.90	u	u	4.90		Pr^iOH	Phot Bu_6Sn_2	≥353	346
				u	4.55	0.28	0.28	4.55		THF:HMPA (10:1)	Phot Bu_6Sn_2	>353	346
		VIIB	Mn	(see Reference 374 for a preliminary result; cf. Reference 382)									

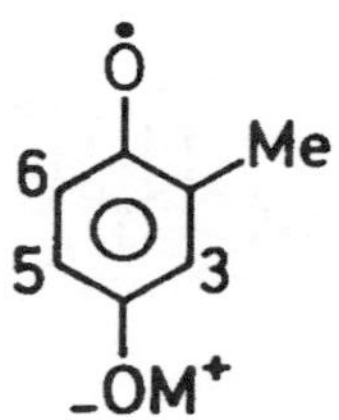

C294	No. 3	IA	^{7}Li(I)	u	2.81[f]	1.05	1.75	3.16		$C_5H_{11}{}^tOH$	—[d]	298	336
C295	No. 3		^{23}Na(I)	0.36	u	u	u	u		$C_5H_{11}{}^tOH$	—[d]	298	336
C296	No. 3		^{39}K(I)	0.33	u	u	u	u		HMPA	Re K/KI[d]	RT	37

Table IV.1.7 (continued)
1,4-BENZOSEMIQUINONE-CATION COMPLEXES

No.	Complex	Central atom	a_M	a_2	a_3	a_5	a_6	g-Factor	Solvent	Method	T [K]	Ref.
	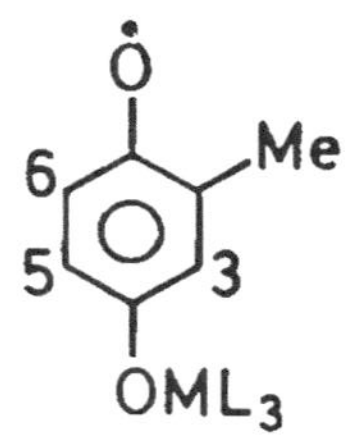											
C297	No. 4 (L = Ph)	IVA Si(IV)	u	6.20[1]	1.11	1.55	6.20		Pr^iOH	Phot Ph_3SiH	RT	346
C298	No. 4 (L = Ph)	Ge(IV)	u	5.79[1]	0.68	1.20	5.79		Pr^iOH	Phot Ph_3GeH	RT	346
C299	No. 4 (L = Bu)	Sn(IV)	u	4.20[1]	u	0.48	4.73		THF:HMPA (10:1)	Phot Bu_6Sn_2	RT	346
	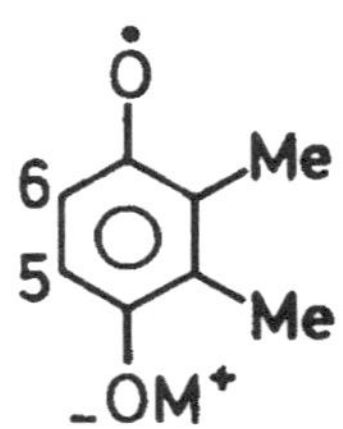											
C300	No. 5	IA K(I)	u	2.20[1]	1.22[1]	2.25	3.23	ΔH = 0.14[1]	DME	Re K	187	50, 345
			u	1.71[1]	1.71[1]	2.74	2.74		DME	Re K	411	50
C301	No. 5	$^{133}Cs(I)$	0.10	u	u	u	u	ΔH = −0.05[1]	DME	Re Cs	188	50
			0.26	1.71[1]	1.71[1]	2.74	2.74		DME	Re Cs	383	50

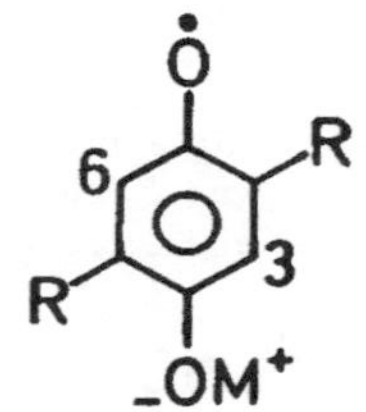

C302	No. 6 (R = Me)	IA	[7]Li(I)	0.34	u	u	u	u		$C_5H_{11}OH$	Re Li	298	50
C303	No. 6 (R = Me)		Na(I)	u	2.74[l]	1.40	1.54[l]	2.60		DME	Re Na	188	50, 345
C304	No. 6 (R = Me)		K(I)	u	2.63[l]	1.55	1.65[l]	2.52	ΔH = 0.14[l]	DME	Re K	175	50, 345
C305	No. 6 (R = Bu[t])		Li(I)	u	u	1.35	u	2.90		Bu[t]OH	Q/QH LiOBu	298	54
				u	u	1.05	u	3.31		THF	Re Li	300	368
C306	No. 6 (R = Cl)		Li(I)	u	u	1.30	u	2.75		Bu[t]OH	Q/QH LiOBu	298	54

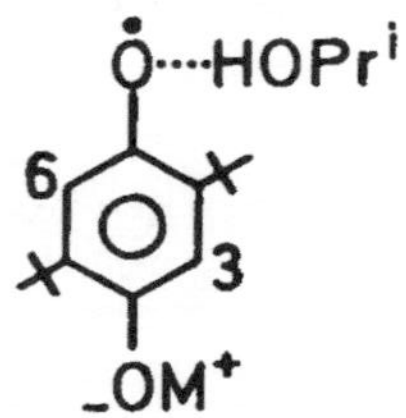

C307	No. 7		Na(I)	u	u	1.850	u	2.694		Pr[i]OH	Re Na	313	364
				u	u	1.892	u	2.650		—	—	293	364
				u	u	1.937	u	2.603		—	—	273	364
				u	u	1.970	u	2.567		—	—	253	364
				u	u	2.013	u	2.552		—	—	233	364
				u	u	2.061	u	2.473		—	—	213	364
C308	No. 6 (R = Bu[t])	IA	[23]Na(I)	u	u	1.62	u	2.92		DME	Re Na	—	368
				u	u	1.53	u	2.95		DME	Re Na	290	342, 348
				u	u	1.439	u	3.099		THF	Re Na	313	364, 351

Table IV.1.7 (continued)
1,4-BENZOSEMIQUINONE-CATION COMPLEXES

No.	Complex	Central atom	Splitting constants/Gauss a_M	a_2	a_3	a_5	a_6	g-Factor	Solvent	Method	T [K]	Ref.
			u	u	1.471	u	3.073		THF	Re Na	293	364, 368
			u	u	1.503	u	3.046		THF	Re Na	273	364
			u	u	1.534	u	3.020		THF	Re Na	253	364
			u	u	1.566	u	2.993		THF	Re Na	233	364
			u	u	1.598	u	2.967		THF	Re Na	213	364
			0.063[g]	u	u	u	u		THF	Re Na	295	347
			u	u	1.32	u	3.20		MeTHF	Re Na	—	368
C309	No. 6 (R = Bu^t)	K(I)	u	u	1.83	u	2.75		DME	Re K	—	368
			u	u	1.86	u	2.74		DME	Re K	200	342
			u	u	1.77	u	2.85		THF	Re K	—	368
			u	u	1.61	u	2.99		MeTHF	Re K	—	368
C310	No. 6 (R = Bu^t)	Rb(I)	u	u	1.80	u	2.80		THF	Re Rb	—	368
			u	u	1.71	u	2.89		MeTHF	Re Rb	—	368
C311	No. 6 (R = Bu^t)	Cs(I)	u	u	1.79	u	2.85		MeTHF	Re Cs	—	368

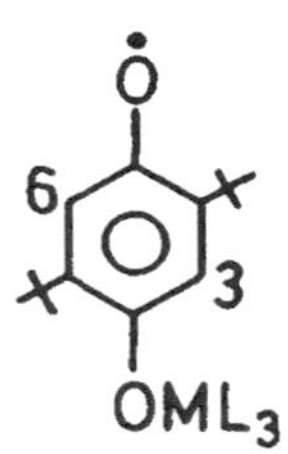

No.	Complex	Central atom	a_M	a_2	a_3	a_5	a_6	g-Factor	Solvent	Method	T [K]	Ref.
C312	No. 8 (L = Ph)	IVA Si(IV)	u	u	1.12	u	4.68		Pr^iOH	Phot Q/Ph_3SiH	RT	346
C313	No. 8 (L = Ph)	Ge(IV)	u	u	0.85	u	4.68		Pr^iOH	Phot Q/Ph_3GeH	RT	346
C314	No. 8 (L = Bu)	Sn(IV)	u	u	u	u	3.55		Pr^iOH	Phot Q/Bu_6Sn_2	RT	346

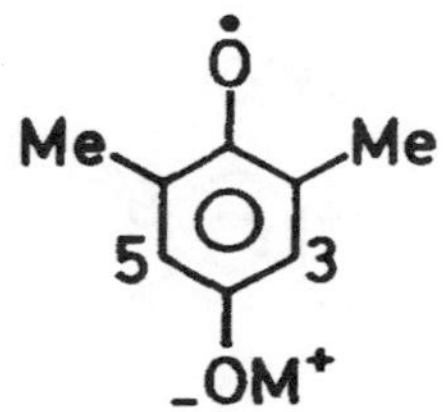

C315	No. 9	IA	^{7}Li(I)	0.21	2.91[l]	1.13	1.13	2.91[l]	$C_5H'_{11}OH$	—[a]	298	335, 336
				u	2.84[l]	1.22	1.22	2.84[l]	$C_5H'_{11}OH$	Re el	RT	31
C315	No. 9	IA	^{7}Li(I)	u	2.63[l]	1.40	1.40	2.63[l]	$C_5H'_{11}OH$	Re Li	246	50
				u	2.91[l]	1.18	1.18	2.91[l]	DME	Re Li	328	50
				0.20	2.71[l]	1.46	1.46	2.71[l]	DME	Re Li	185	50
C316	No. 9		^{23}Na(I)	0.35	2.77[l]	1.26	1.26	2.77[l]	$C_5H'_{11}OH$	—[a]	333	336, 335
				0.31	2.74[l]	1.29	1.29	2.74[l]	$C_5H'_{11}OH$	Re Na	293	50, 335
				0.10	2.54[l]	1.61	1.61	2.54[l]	DME	—	298	336, 335
				u	2.49[l]	1.67	1.67	2.49[l]	DME	—	258	336
				u	2.37[l]	1.66	1.66	2.37[l]	DME	—	213	336
C317	No. 9		^{39}K(I)	u	2.58[l]	1.43	1.43	2.58[l]	$C_5H'_{11}OH$	Re K	273	50, 335
				u	2.18[l]	1.88	1.88	2.18[l]	DME	Re K	373	50
				u	2.21[l]	1.97	1.97	2.21[l]	DME	—[a]	298	336, 335
				u	2.32[l]	1.93	1.93	2.32[l]	DME	—	248	336
				u	2.24[l]	1.87	1.87	2.24[l]	DME	Re K	188	50, 336
				0.29	u	u	u	u	HMPA	Re K/KI[d]	RT	37
C318	No. 9		Rb(I)	u	2.56[l]	1.51	1.51	2.56[l]	$C_5H'_{11}OH$	—[a]	298	336, 335
				u	2.18[l]	2.00	2.00	2.18[l]	DME	—	298	336, 335
				u	2.14[l]	2.00	2.00	2.14[l]	DME	—	268	336

Table IV.1.7 (continued)
1,4-BENZOSEMIQUINONE-CATION COMPLEXES

No.	Complex	Central atom	Splitting constants/Gauss a_M	a_2	a_3	a_5	a_6	g-Factor	Solvent	Method	T [K]	Ref.
			u	2.00[1]	2.00	2.00	2.00[1]		DME	—	236	336
			u	2.18[1]	1.83	1.83	2.18[1]		DME	—	183	336
C319	No. 9	$^{133}Cs(I)$	0.17	2.56[1]	1.51	1.51	2.56[1]		$C_5H^{t}_{11}OH$	—[a]	298	336, 335
			0.24	2.14[1]	1.95	1.95	2.14[1]		DME	Re Cs	373	50, 336
			<0.1	2.17[1]	1.86	1.86	2.17[1]		DME	—	298	336
			0.19	2.20[1]	1.88	1.88	2.20[1]		DME	Re Cs	188	50, 336
C320	No. 10	IVA Si(IV)	u	5.5[1]	1.3	1.3	5.5[1]	2.0048	C_6H_6	Phot $Bu^t_2O_2$	RT	359

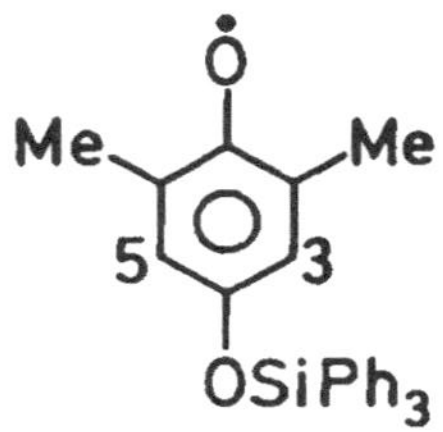

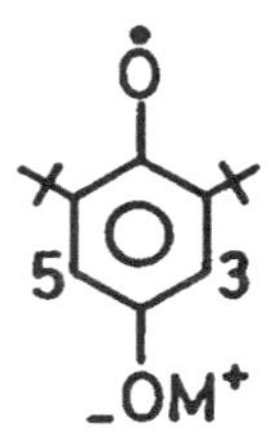

C321	No. 11	IA	Li(I)	u	u	1.02	1.02	u		$C_5H'_{11}OH$	—[a]	RT	335, 336
				u	u	1.20	1.20	u		C_6H_5/12Cw4[b]	Phot LiOMe	283	337
				u	u	1.23	1.23	u		C_6H_6/15Cw5	Phot LiOMe	283	337
				u	u	1.17	1.17	u		C_6H_6/18Cw6	Phot LiOMe	283	337
C322	No. 11		^{23}Na(I)	0.46	u	1.02	1.02	u		$C_5H'_{11}OH$	—[a]	RT	336
				u	u	1.20	1.20	u		C_6H_6/15Cw5	Phot NaOMe	283	337
				0.35	u	1.43	1.43	u		C_6H_6/18Cw6	Phot NaOMe	283	337
				0.30	u	1.50	1.50	u		THF/18Cw6	Re Na	298	338
				u	u	1.60	1.60	u		THF/18Cw6	Re Na	193	338
172	No. 11 (free anion)			—	u	2.10	2.10	u		THF/18Cw6	Re Na	193	338
C323	No. 11		K(I)	u	u	1.02	1.02	u		$C_5H'_{11}OH$	—[a]	RT	336
				u	u	1.28	1.28	u		C_6H_6/18Cw6	Phot KOMe	283	337
				u	u	1.71	1.71	u		DME	—	RT	335, 336
				u	u	1.81	1.81	u		DME	—	183	335, 336
				u	u	2.081	2.081	u		HMPA	Re K(0.361)[c]	296	354, 361
				u	u	2.194	2.194	u	2.004773	HMPA	Re K(0.101)	296	355-358
				u	u	2.223	2.223	u	2.004776	HMPA	Re K(0.067)	296	355
				u	u	2.243	2.243	u	2.004786	HMPA	Re K(0.050)	296	355
				u	u	2.267	2.267	u	2.004793	HMPA	Re K(0.034)	296	355
				u	u	2.298	2.298	u	2.004801	HMPA	Re K(0.018)	296	355
172	No. 11 (free anion)			—	0.037[2]	2.346	2.346	0.037[2]	2.004814	HMPA	Re K (0)	296	354, 37
C324	No. 11		Cs(I)	u	u	1.02	1.02	u		$C_5H'_{11}OH$	—[a]	RT	336
C325	No. 11		Rb(I)	u	u	1.02	1.02	u		$C_5H'_{11}OH$	—[a]	RT	336

Table IV.1.7 (continued)
1,4-BENZOSEMIQUINONE-CATION COMPLEXES

No.	Complex	Central atom	Splitting constants/Gauss a_M	a_2	a_3	a_5	a_6	g-Factor	Solvent	Method	T [K]	Ref.
	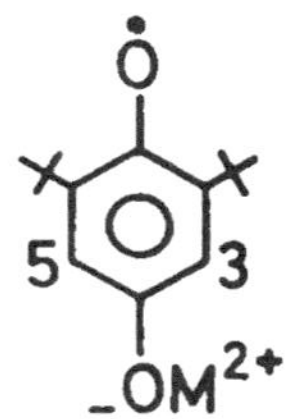											
C326	No. 12	IIA Mg(II)	u	u	0.86[j]	0.86	u	ΔH = 0.58[i]	HMPA	Re Na/Mg$(ClO_4)_2$[d]	RT	361
C327	No. 12	Ca(II)	u	u	1.28	1.28	u	ΔH = 0.33[i]	HMPA	Re Na/Ca$(ClO_4)_2$[d]	<318	361
			u	u	1.12[j]	1.12[k]	u		HMPA	—	~333	361
			u	u	1.46	1.46[k]	u		HMPA	—	~333	361
			u	0.03[z]	1.67	1.67	0.03[z]		HMPA	—	353	361
C328	No. 12	Ba(II)	u	u	1.42	1.42	u	ΔH = 0.41[i]	HMPA	Re Na/Ba$(ClO_4)_2$[d]	RT	361
	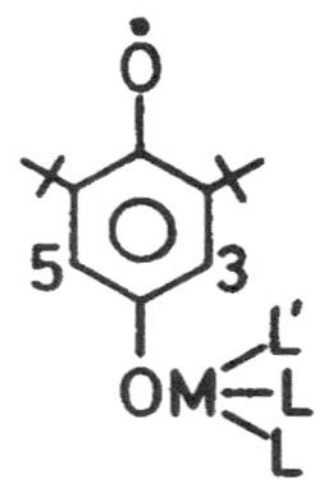											
C329	No. 13 (L = L′ = Me)	IVA C(IV)	—	u	0.98	0.98	u	2.00467	C_6H_6	Q/Butanoate[e]	293	360

C330	No. 13 (L = L′ = CH_3*)	^{29}Si(IV)	7.5^{Si} <0.02*	u	1.12	1.12	u	2.00464	C_6H_6	Phot Q/Me_3SiH^{m}	293	360
C331	No. 13 (L = L′ = Ph)	Si(IV)	7.46^{Si}	u	1.19	1.19	u	2.00469	C_6H_6	Phot Q/Ph_3SiH^{m}	293	360
			11.5^{Si}		1.3	1.3		2.0047	C_6H_6	Phot Q/Ph_3SiH^{m}	253	359
C332	No. 13 (L = Ph, L′ = H*)	Si(IV)	0.47*	u	1.19	1.19	u		C_6H_6	Phot Q/$Ph_2SiH_2^{m}$	293	360
			0.55*	u	1.4	1.4	u	2.0047	C_6H_6	Phot Q/$Ph_2SiH_2^{m}$	RT	359
C333	No. 13 (L = H*, L′ = Ph)	Si(IV)	0.40*	u	1.3	1.3	u	2.0047	C_6H_6	Phot Q/$PhSiH_3^{m}$	RT	359
C334	No. 13 (L = CH_3*, L′ = F)	Si(IV)	0.61^{F} 0.07*	u	1.31	1.31	u		C_6H_6	Phot Q/ $Me_2Si(F)H^{m}$	293	360

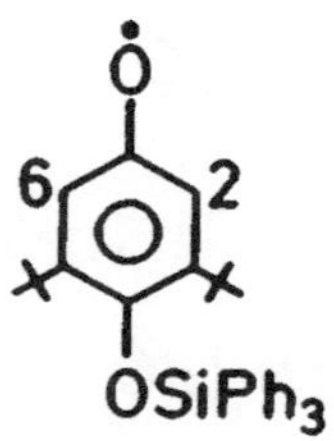

C335	No. 14	Si(IV)	u	4.1	u	u	4.1	2.0047	C_6H_6	Phot Q/Ph_3SiH^{m}	253	259

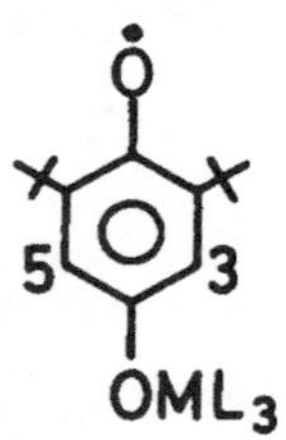

C336	No. 15 (L = Et)	IVA ^{73}Ge(IV)	11.0	u	0.76	0.76	u	2.00452	C_6H_6	Phot Q/Et_3GeH^{m}	293	360
C337	No. 15 (L = Me)	^{119}Sn(IV)	11.23	u	0.58	0.58	u	2.00437	C_6H_6	Phot Q/$Me_6Sn_2^{m}$	293	360, 362
C338	No. 15 (L = Ph)	Sn(IV)	13	u	0.65	0.65	u		C_6H_6	Phot Q/Ph_3SnH	283	362, 363

Table IV.1.7 (continued)
1,4-BENZOSEMIQUINONE-CATION COMPLEXES

No.	Complex	Central atom	Splitting constants/Gauss a_M	a_2	a_3	a_5	a_6	g-Factor	Solvent	Method	T [K]	Ref.
C339	No. 15 (L = Me)	^{207}Pb(IV)	16.96	u	<0.3	<0.3	u	2.00353	C_6H_6	Phot Q/Me_6Pb_2	293	360, 362
C340	No. 15 (L = Ph)	Pb(IV)	25	u	u	u	u		C_6H_6	Phot Q/Ph_6Pb_2	283	362, 363
	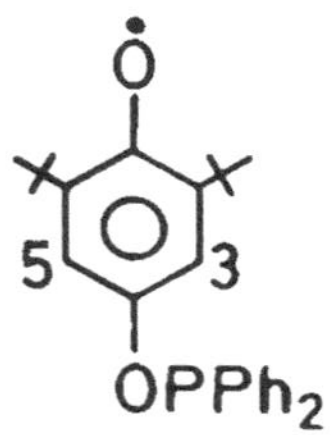											
C341	No. 16	VA ^{31}P(III)	2.99	u	1.18	1.18	u	2.00469	C_6H_6	Phot Q/Ph_4P_2	293	360, 362
	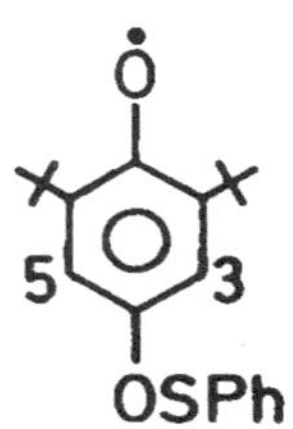											
C342	No. 17	VIA S(II)	—	u	1.03	1.03	u	2.00468	C_6H_6	Phot Q/Ph_2S_2^{m}	293	360

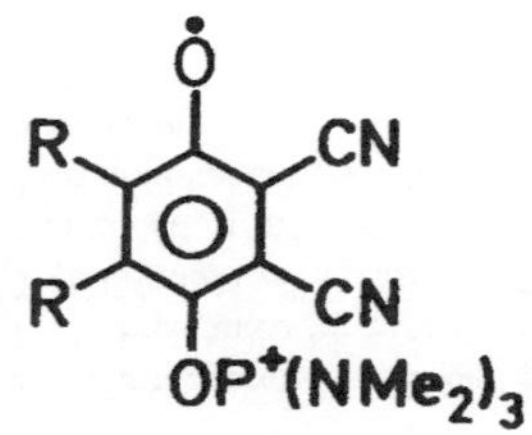

C343	No. 18 (R = H)	VA	[31]P(III)	10.61[P] 0.36[3N]	1.08[N]	1.08[N]	1.80	1.80	C_6H_6	Q/P$(NMe_2)_3$	RT	168
			P(III)	10.75[P] 0.42[3N]	1.19[N]	1.19[N]	1.95	1.95	C_6H_6/CH_2Cl_2	Q/P$(NMe_2)_3$	RT	168
C344	No. 18 (R = Cl)		P(III)	7.35[P] 0.39[3N]	0.49[N]	0.49[N]	—	—	C_6H_6	Q/P$(NMe_2)_3$	RT	168

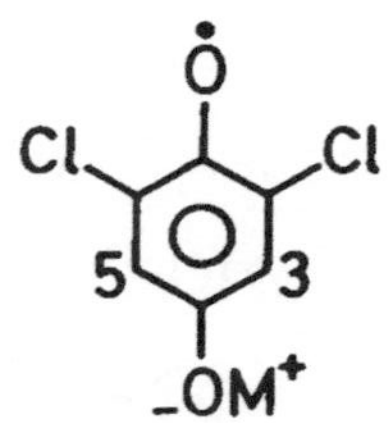

C345	No. 19	IA	Li(I)	u	—	1.40	1.40	—	Bu^{t}OH	Q/HQ LiOBu^t	RT	54
				u	—	1.54	1.54	—	$C_5H^t_{11}OH$	—[a]	RT	336, 335
				u	—	1.471	1.471	—	THF	Re Li	298	339

Table IV.1.7 (continued)
1,4-BENZOSEMIQUINONE-CATION COMPLEXES

No.	Complex	Central atom	Splitting constants/Gauss a_M	a_2	a_3	a_5	a_6	g-Factor	Solvent	Method	T [K]	Ref.
	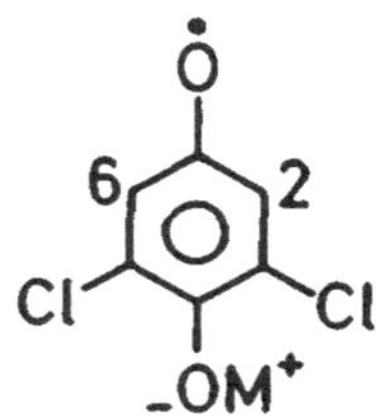											
C346	No. 20	Li(I)	u	2.85	—	—	2.85		Bu^tOH	Q/HQ $LiOBu^t$	RT	54
			u	3.23	—	—	3.23		$C_5H^t_{11}OH$	—[a]	RT	336, 335
C347	No. 19	$^{23}Na(I)$	0.35	—	1.54	1.54	—		$C_5H^t_{11}OH$	—[d]	RT	336
C348	No. 20	$^{23}Na(I)$	0.25	3.23	—	—	3.23		$C_5H^t_{11}OH$	—[d]	RT	336
			u	2.869	—	—	2.869		DME	Re Na	298	339
			u	2.139	—	—	2.139		DME	Re Na	198	339
C349	No. 20	K(I)	u	3.124	—	—	3.124		THF	Re K	198	339
			u	2.989	—	—	2.989		DME	Re K	298	339, 349
			u	2.943	—	—	2.943		DME	Re K	198	339

[a] Sample preparation included photolysis, electrolysis, and metal reduction. Method chosen in each case not stated in References 335 and 336.

[b] Lines broadened by exchange; only the sums $a_2 + a_3 = a_5 + a_6$ registered.

[c] 3.68 G might be a misprint: 2.68 seems appropriate (cf. the result for the potassium complex, no. C289)

[d] Salt added to make complex appear.

[e] Two simultaneously appearing spectra attributed a contact ion-pair (tight ion-pair) and a solvent shared ion-pair (loose ion-pair). See Reference 338 for a recent discussion.

[f] Upfield shift, when positive, in Gauss between center of particular spectrum and that of free ion.

[g] Sodium splitting obtained by simulating the observed lineshape.
[h] Crown ether; see list of abbreviations.
[i] Number in parenthesis amount of KI *M* added in order to observe the ion pair.
[j] Reference 361 attributes the triplet spectra (Mg 0.86 and Ca 1.12) to ion-pairs with the cation placed at the oxygen of the hindered position (C-1). The magnitudes of the splittings do not favor such an assignment. cf. the results of Reference 259 (Si-complex) included in the table.
[k] The two simultaneously observed triplets (Ca 1.12 and 1.46) may arise from a contact ion-pair and a solvent shared ion-pair. Cf. footnote e above.
[l] Adduct observed when Q and *t*-butyl-ββ-dimethylperbutanoate, prepared at − 190°C in the dark, was allowed to warm up in the cavity of the EPR spectrometer.
[m] Photolysis in the presence of $Bu^t_2O_2$.

Table IV.1.8
TETRASUBSTITUTED 1,4-BENZOSEMIQUINONE-CATION COMPLEXES

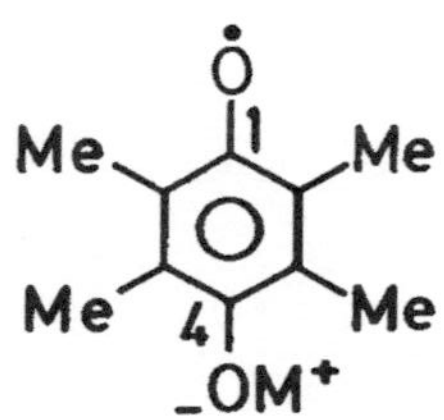

No.	Complex	Central Atom	a_M	Splitting Constants/Gauss $a_2^{Me} = a_6^{Me}$	$a_3^{Me} = a_5^{Me}$	g-Factor	Solvent	Method	T [K]	Ref
C350	No. 1	IA $^6Li(I)$	0.22	2.91	0.93		$C_5H^t_{11}OH$	Phot $LiOC_5H^t_{11}$	RT	366
C351	No. 1	$^7Li(I)$	0.59	2.91	0.93		$C_5H^t_{11}OH$	Phot $LiOC_5H^t_{11}$	298	336
			u	3.125	0.768		DME	Re Li	333	339
			u	2.981	0.863		DME	Re Li	298	339
			u	2.887	0.964		DME	Re Li	253	339
			u	2.825	1.017		DME	Re Li	223	339
			u	2.796	1.033		DME	Re Li	203	339, 371
			u	3.039	0.815		THF	Re Li	298	339, 365
			u	2.955	0.860		THF	Re Li	253	339
			u	2.952	0.892		THF	Re Li	233	339
			u	2.936	0.955		THF	Re Li	213	329

Table IV.1.8 (continued)
TETRASUBSTITUTED 1,4-BENZOSEMIQUINONE-CATION COMPLEXES

No.	Complex	Central Atom	Splitting Constants/Gauss a_M	$a_2^{Me} = a_6^{Me}$	$a_3^{Me} = a_5^{Me}$	g-Factor	Solvent	Method	T [K]	Ref.
C352	No. 1	$^{23}Na(I)$	1.03	2.88	1.01		C_6H_6/DBCw	Re Na	281—333	367
			0.71	2.75	1.12		C_6H_6/18Cw6	Re Na	282—328	367
			0.53	1.92[a]	1.92[a]		C_6H_6/15Cw5	Re Na	282—333	367
			0.47	2.47	1.37		PhMe: C_6H_6 (2:3)/15Cw5	Re Na	243	367
			0.65	1.93	1.93		DME	Re Na	458	50
			0.50	1.92	1.92		DME	Re Na	333	336
			0.37	1.92	1.92		DME	Re Na	298	336, 373
			0.29	2.55	1.28		DME	Re Na	243	336
			0.22	2.55	1.28		DME	Re Na	213	336, 373
			0.19	2.55	1.28		DME	Re Na	198	371
			0.346	2.739	1.056		THF	Re Na	298	339, 335
			0.257	2.653	1.198		THF	Re Na	253	339, 365
			0.208	2.623	1.208		THF	Re Na	233	339
C353	No. 1	$^{39}K(I)$	u	2.413	1.415		DME	Re K	223	339
			u	2.407	1.434		DME	Re K	203	339
			u	2.402	1.434		DME	Re K	193	339, 371
			0.05	u	u		DME	Re K	288	372, 378
			u	2.534	1.313		THF	Re K	253	339, 377
			u	2.485	1.312		THF	Re K	233	339
			u	2.458	1.350		THF	Re K	213	339
			u	2.505	1.381		THF	Re K	193	339
			u	2.45	1.47		THF	Re K	173	365
C354	No. 1	$^{85,87}Rb(I)$	u	2.84	2.04		DME	Re Rb	213—256	336
			0.11[b] 0.36	2.33	1.51		DME	Re Rb	198	371
			u	2.37	1.50		THF	Re Rb	158	365
C355	No. 1	$^{133}Cs(I)$	0.26	2.30	1.54		DME	Re Cs	198	371
			u	2.35	1.50		THF	Re Cs	153	365

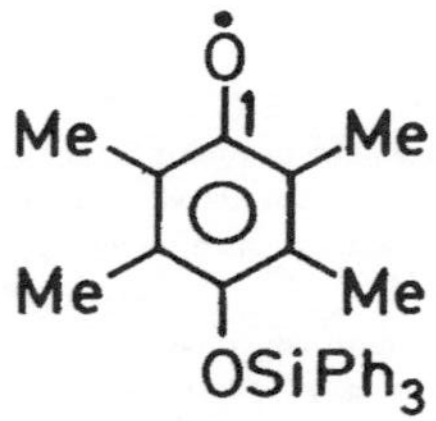

C356	No. 2		IVA	Si(IV)	—	5.9	1.5	2.0047	C_6H_6	Phot $HSiPh_3$	RT	359

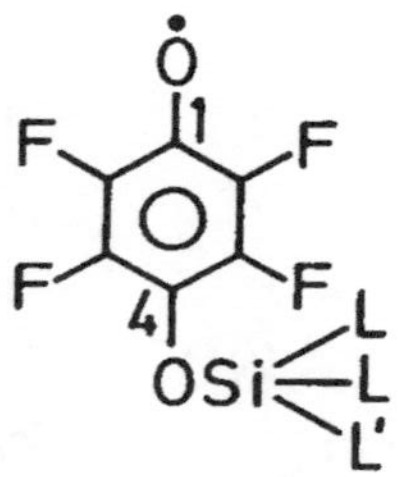

						$a_2^F = a_6^F$	$a_3^F = a_5^F$					
C357	No. 3	(L = L′ = Me)	IVA	Si(IV)	<0.03[c]	13.88	4.23		C_6H_6	Phot $HSiL_3$	293	360
C358	No. 3	(L = Me,L′ = F)		Si(IV)	0.65[F] <0.04[c]	14.42	4.57		C_6H_6	Phot $HSiL_3$	293	360
C359	No. 3	(L = L′ = Ph)		Si(IV)	u	14.21	4.42		C_6H_6	Phot $HSiL_3$	293	360
C360	No. 3	(L = Ph,L′ = H*)		Si(IV)	0.51*	14.15	4.34		C_6H_6	Phot $HSiL_3$	293	360

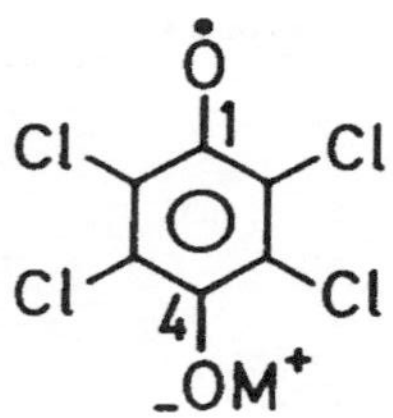

Table IV.1.8 (continued)
TETRASUBSTITUTED 1,4-BENZOSEMIQUINONE-CATION COMPLEXES

No.	Complex		Central Atom		a_M	Splitting Constants/Gauss $a_2^{Me} = a_6^{Me}$	$a_3^{Me} = a_5^{Me}$	g-Factor	Solvent	Method	T [K]	Ref.
C361	No. 4	$(M = P^+(NMe_2)_3)$	VA	P(III)	2.28[b]	—	—		C_6H_6	$Q/P(NMe_2)_3$	RT	168
C362	No. 4	$(M = Mn(CO)_5)$	VIIB	^{55}Mn(I)	8.33	0.53[c]	u	2.0043	PhMe	Phot $Mn_2(CO)_{10}$	293	374

Triple ions

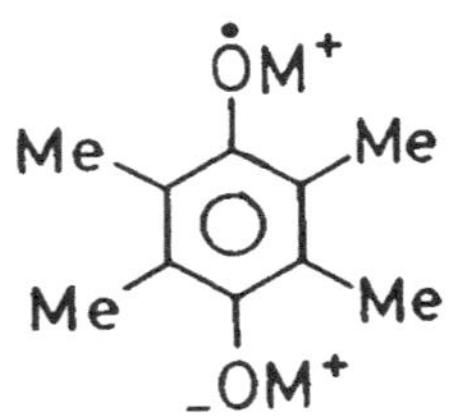

No.	Complex	Central Atom		a_M	a_i^{Me} (i = 2,3,5,6)	ΔH^d	Solvent	Method	T [K]	Ref.
C363	No. 5	IA	^{23}Na(I)	0.17[e]			THF	Re Na/BPh_4	253	411
C364	No. 5		K(I)	u	1.90	0.300	THF	Re K/BPh_4	293	377
				u	1.91	0.250	THF	Re K/BPh_4	253	377
				u	1.92	0.230	THF	Re K/BPh_4	233	377
				u	1.92	0.210	THF	Re K/VPh_4	203	377

[a] A bar indicates average value observed during fast exchange of cation.
[b] Splitting above line from ^{85}Rb, below from ^{87}Rb.
[c] a_H^{Si}-CH_3
[d] ΔH/Gauss, distance between center of triple ion spectrum (high field) and ion-pair spectrum (low field). Center of free ion spectrum occurs on the low-field side of the center of the ion-pair spectrum.
[e] Splitting from two equivalent Na nuclei.

Table IV.1.9
BENZOSEMIQUINONE-CATION COMPLEXES WITH THREE KEY OXYGENS [1,2,3] AND [1,2,4]

No.	Complex	Central atom	Splitting constants/Gauss $a_M^{1,2}$	$a_M^{2,3}$	a_3	a_4	a_5	a_6	g-Factor	Solvent	Ref.
	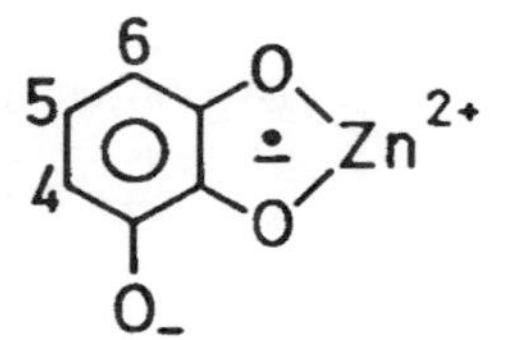										
C365	No. 1	IIB Zn(II)	—	—	—	0.5	5.2	0.5		w/pH = 6.5	299[a]
	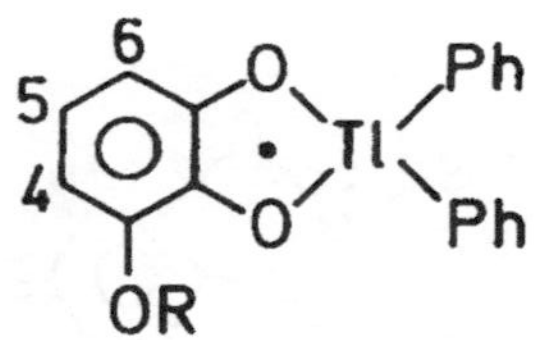										
C366	No. 2 (R = H*)	IIIA ^{203,205}Tl(III)	9.72	—	0.31*	1.31	4.29	0.39	2.00365	C_5H_5N	20[b]
			10.85	—	0.37*	1.55	4.22	0.37	2.00356	MeTHF	20
			12.35	—	0.37*	1.68	4.20	0.37	2.00353	Dioxane	20
C367	No. 2 (R = CH$_3^*$)	Tl(III)	10.70	—	0.64*	1.43	4.27	0.30	2.00375	C_5H_5N	20
			13.23	—	0.63*	1.57	4.24	0.28	2.00356	MeTHF	20
			14.40	—	0.64*	1.51	4.31	0.32	2.00356	Dioxane	20
C368	No. 2 (5–CONH$_2$, R = H)	Tl(III)	9.05	—	u	1.04	—	u	2.00395	C_5N_5N	20
			10.45	—	u	1.21	—	u	2.00379	MeTHF	20
			11.86	—	u	1.34	—	u	2.00387	Dioxane	20
C369	No. 2 (5–COOCH$_2^*$Me, R = H**)	Tl(III)	9.08	—	0.25**	0.88	0.25*	0.25	2.00409	C_5H_5N	20
			10.40	—	0.29**	1.13	0.25*	0.29	2.00397	MeTHF	20
			11.98	—	0.33**	1.24	0.21*	0.26	2.00395	Dioxane	20

Table IV.1.9 (continued)
BENZOSEMIQUINONE-CATION COMPLEXES WITH THREE KEY OXYGENS [1,2,3] AND [1,2,4]

No.	Complex	Central atom	Splitting constants/Gauss $a_M^{1,2}$	$a_M^{2,3}$	a_3	a_4	a_5	a_6	g-Factor	Solvent	Ref.
C370	No. 2 (5–COOTlPh$_2$, R = H)	Tl(III)	9.65	—	u	1.18	—	u	2.00403	C_5H_5N	20
			10.40	—	u	1.50	—	u	2.00384	MeTHF	20
			12.00	—	u	1.50	—	u	2.00397	Dioxane	20
	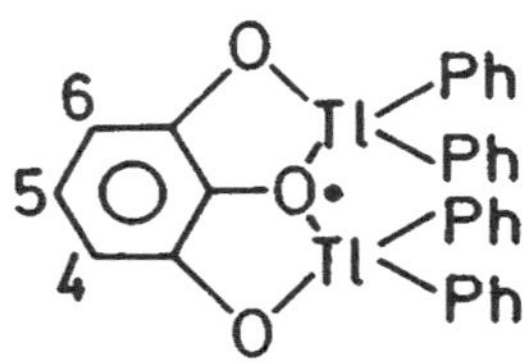										
C371	No. 3	^{203,205}Tl(III)	8.05	8.05	—	0.67	4.52	0.67	2.00295	C_5H_5N	20
			11.02	11.02	—	0.77	4.28	0.77	2.00287	Bu^tNH_2	20
			10.35	10.35	—	0.57	4.50	0.57	2.00262	Dioxane	20
C372	No. 3 (4–COMe)	Tl(III)	8.14	7.08	—	—	4.21	0.70	2.00334	C_5H_5N	20
			10.93	10.24	—	—	4.42	0.75	2.00319	Bu^tNH_2	20
			10.46	10.00	—	—	4.16	0.64		Dioxane	20
C373	No. 3 (4–COOMe)	Tl(III)	9.21	7.43	—	—	4.23	0.70	2.00326	C_5H_5N	20
			12.66	10.08	—	—	4.45	0.73	2.00319	Bu^tNH_2	20
C374	No. 3 (4,6–(di)–COOMe)	Tl(III)	8.96	8.96	—	—	3.96	—	2.00332	C_5H_5N	20
			11.98	11.98	—	—	4.15	—	2.00339	Bu^tNH_2	20
C375	No. 3 (4–COOEt)	Tl(III)	9.35	7.34	—	—	4.25	0.68	2.00318	C_5H_5N	20
			12.74	10.05	—	—	4.44	0.73	2.00317	Bu^tNH_2	20
C376	No. 3 (5–CONH$_2^*$)	Tl(III)	8.12	8.12	—	0.76	0.16^N	0.76	2.00326	C_5H_5N	20
						0.72	0.16*				
			7.65	7.65	—		0.21^N	0.72	2.00324	C_5H_5N/10% EtOD	20
			11.26	11.26	—	0.88	u	0.88	2.00315	Bu^tNH_2	20
C377	No. 3 (5–COOCH$_2^*$Me)	Tl(III)	8.35	8.35	—	0.84	0.28*	0.84	2.00344	C_5H_5N	20

				11.45	11.45	—	0.94	0.30*	0.94	2.00337	Bu^tNH_2	20
				10.78	10.78	—	0.75	u	0.75	2.00304	Dioxane	20
C378	No. 3 (5–$COOTlPh_2$)		Tl(III)	8.27	8.27	—	0.73	0.19[H]	0.73	2.00342	C_5H_5N	20
				7.93	7.93	—	0.71	0.32[H]	0.71	2.00344	C_5H_5N/10% EtOD	20
				10.98	10.98	—	0.82	0.27[H]	0.82	2.00327	Bu^tNH_2	20
C379	No. 4 (R = CH_3^*)	IA	Na(I)	—	—	—	0.4*	3.8	u		MeOH	131[c]
C380	No. 4 (R = CH_2^*Me)		Na(I)	—	—	u	0.35*	3.9	u		EtOH	131[d]
C381	No. 5	IIIA	^{203,205}Tl(III)	9.90	—	0.58	5.30[H]	5.30	1.24	2.00319	C_5H_5N	20
				12.20	—	0.67	8.22[H]	4.94	1.33		Bu^tNH_2	20
				14.05	—	0.73	10.60[H]	4.70	1.38	2.00382	Dioxane	20

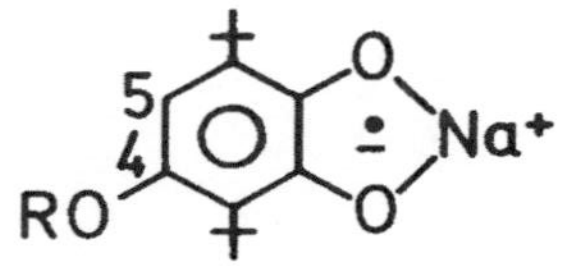

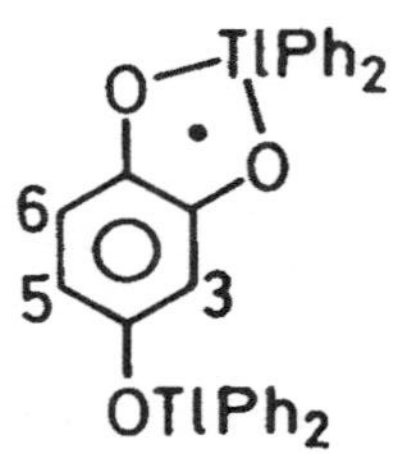

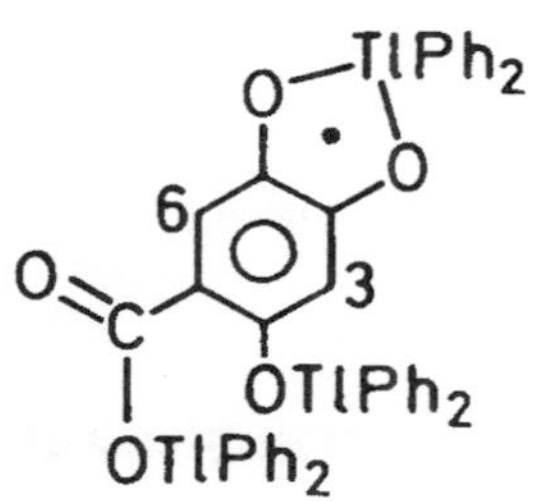

Table IV.1.9 (continued)
BENZOSEMIQUINONE-CATION COMPLEXES WITH THREE KEY OXYGENS [1,2,3] AND [1,2,4]

No.	Complex	Central atom	Splitting constants/Gauss $a_{M}^{1,2}$	$a_{M}^{2,3}$	a_3	a_4	a_5	a_6	g-Factor	Solvent	Ref.
C382	No. 6	$^{203,205}Tl(I)$	7.38	—	0.38	2.98^{H}	0.52^{H}	1.90	2.00334	C_5H_5N	20
			8.34	—	0.46	3.24^{H}	u	2.14	2.00398	Bu^tNH_2	20
			11.12	—	0.35	2.87^{H}	0.70^{H}	0.98	2.00375	MeTHF	20

[a] Complex generated by enzyme catalyzed oxidation (horseradish peroxidase, HRP) of phenol.
[b] Complex generated by reaction between diphenylthalliumhydroxide and pertinent phenol at 20°C.
[c] Complex generated by photochemical reaction of quinone in NaOMe.
[d] Exchange reaction, OMe ⇄ OEt (cf. footnote c).

Table IV.1.10
SOME 1,2-SEMIQUINONE TL-METAL COMPLEXES, SOLVENT EFFECTS

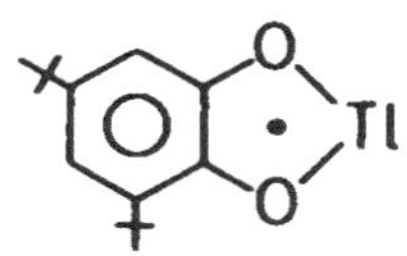

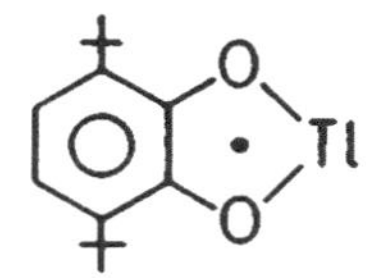

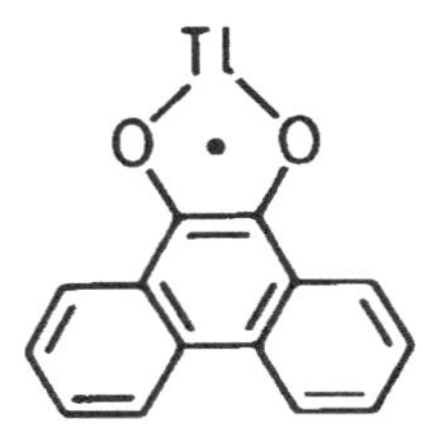

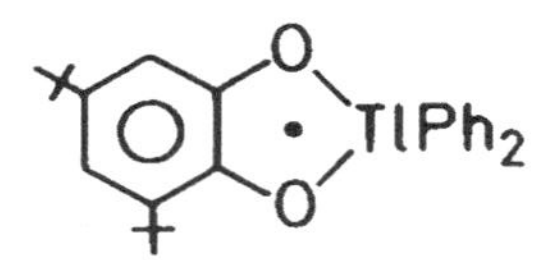

	No. C94	No. C383	No. C448	No. C44	
Solvent	$a_{Tl(I)}$	$a_{Tl(I)}$	$a_{Tl(I)}$	$a_{Tl(III)}$	g
Tetramethylethylenediamine	146.8^{a}	145.0^{a}	99.7^{a}	—	—

Piperidine	118.7	107.9	87.6	—	—
Tetramethylmethylenediamine	100.4	94.3	71.4	—	—
Triethylamine	99.1	96.4	74.0	—	—
Pyridine	91.1	86.7	75.5	15.25[b]	2.00349[b]
Ethanol	—	—	—	15.80	2.00331
Propylamin	—	—	—	15.90	2.00356
4-Methylmorpholine	90.4	81.5	64.7	—	—
Dimethyl sulfoxide	86.7	—	—	12.47	2.00343
Hexamethylphosphoric acid triamide	78.2	89.1	62.7	—	—
Dimethylformamide	64.1	66.7	51.5	12.98	2.00345
Dimethoxyethane	63.4	63.2	47.8	16.15	2.00335
Dipropylamine	—	—	—	17.13	2.00339
2-Propylamine	—	—	—	17.40	2.00352
Tetrahydrofuran	61.9	58.3	47.7	17.80	2.00339
2-Methyltetrahydrofuran	—	—	—	18.65	2.00334
Dioxane	61.7	55.8	46.6	18.70	2.00335
Di-iso-propylamine	—	—	—	19.60	2.00337
t-Butylamine	—	—	—	20.10	2.00351
Acetone	56.2	55.2	44.8	—	—
Ethoxyethane	55.8	54.0	43.3	—	—
Ethyl acetate	53.7	53.6	42.1	—	—
Dimethylaniline	53.7	51.6	42.3	—	—
Acetonitrile	50.9	51.9	43.0	16.40	2.00334
Dimethyl sulfate	44.2	—	—	—	—
Benzene	44.0	40.8	35.1	—	—
Dichloroethane	41.9	44.2	34.0	21.60	2.00317
1-Chlorobutane	37.0	40.6	30.1	—	—
Trichloromethane	35.8	37.3	25.5	21.36	2.00324
Toluene	—	—	—	23.30	2.00325
Tetrachloromethane	31.3	30.4	—	—	—
Tetrachloroethylene	31.1	42.5	24.2	—	—
Hexane	25.3	24.4	—	26.70	2.00307
Hexafluorobenzene	23.6	25.8	18.8	—	—

[a] Results from Reference 329 ($a_{Tl(I)}$). Complex generated by reduction with Tl metal at 20°C. Splitting constants in Gauss.

[b] Results from Reference 333 ($a_{Tl(III)}$, g). Complex generated by reaction of diphenylthalliumhydroxide with *o*-hydroquinone at 20°C.

Table IV.2.1
1,2-NAPHTHOSEMIQUINONE-CATION COMPLEXES

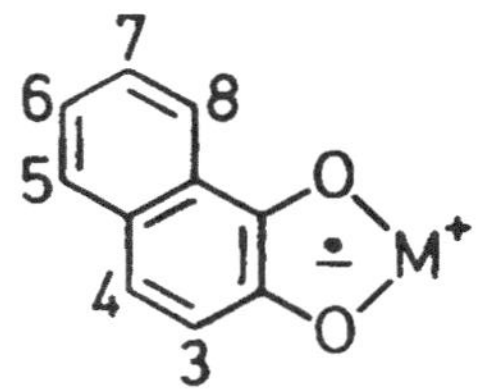

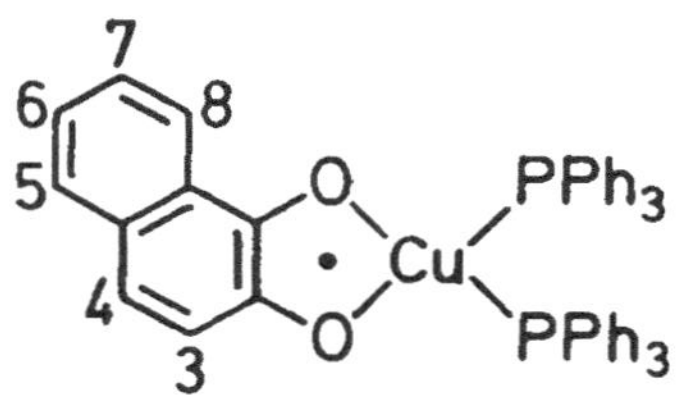

No.	Complex	Central atom		Splitting constants/Gauss a_M	a_3	a_4	a_5	a_6	a_7	a_8	g-Factor	Solvent	Method	T [K]	Ref.
C384	No. 1	IA	^{7}Li(I)	0.54[a]	0.27	4.29	0.21	1.57	u	1.31		DME	Re Li	243	124
C385	No. 1		^{23}Na(I)	0.49[a] 0.61[b]	0.29	4.11	0	1.51	u	1.20		DME	Re Na	243	124
C386	No. 1		K(I)	0	0.34	4.01	0	1.48	u	1.14		DME	Re K	243	124
C387	No. 1		^{85,87}Rb(I)	0.10[c] 0.29	0.36	4.03	0.07	1.53	u	1.17		DME	Re Rb	243	124
C388	No. 1		^{133}Cs(I)	0.56[a] 0.37[b]	0.35	4.01	0.09	1.48	u	1.13		DME	Re Cs	~243	124
C389	No. 2	IB	^{63}Cu(I)	9.5 13.3[d]	u	3.9	u	1.3	u	1.3	2.0041	THF	Q/Cu + PPh_3	RT	203, 384

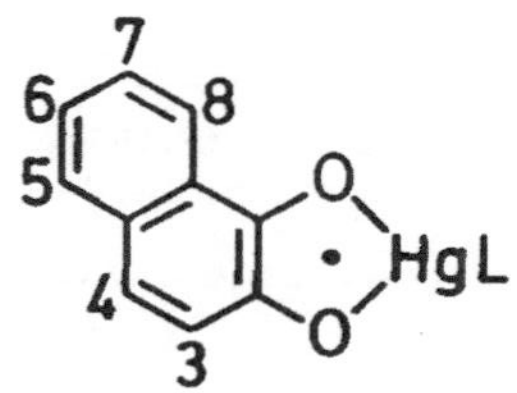

C390	No. 3 (L = Ph)	IIB	Hg(II)	—	0.54	5.60	0.54	1.66	u	1.66	1.9981	THF	Phot Q/L_2Hg	193	303
C391	No. 3 (L = Bz)		Hg(II)	—	0.49	5.52	0.49	1.70	0.18	1.70	1.9985	THF	Phot Q/L_2Hg	193	303
C392	No. 3 (L = Me)		Hg(II)	—	0.56	5.58	0.56	1.70	u	1.70	1.9982	THF	Phot Q/L_2Hg	193	303
C393	No. 3 (L = CH_2I)		Hg(II)	—	0.54	5.61	0.54	1.70	u	1.70	1.9983	THF	Phot Q/L_2Hg	193	303
C394	No. 3 (L = Bu)		^{199}Hg(II)	~6	u	5.26	1.24	1.24	u	1.24	1.9999	$Bu_2Hg(l)$	Thermolysis	298	303

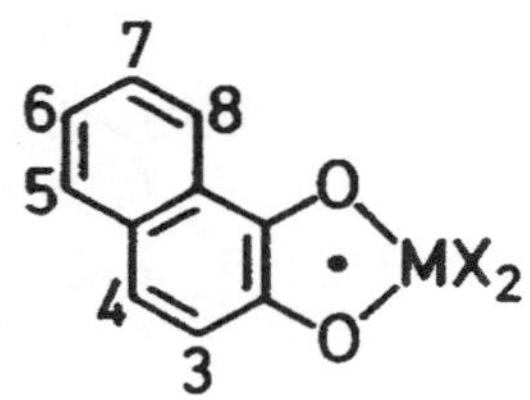

C395	No. 4 (X = Cl)	IIIA	^{27}Al(III)	2.4	u	2.4	u	2.4	u	2.4	2.0035	THF	Q $AlCl_3$	~268	381
C396	No. 4 (X = Br)		Al(III)	2.2	u	4.4	u	2.2	u	2.2	2.0030	THF	Q $AlBr_3$	~268	381
C397	No. 4 (X = I)		Al(III)	2.3	u	4.6	u	2.3	u	2.3	2.0030	THF	Q AlI_3	~268	381
C398	No. 4 (X = Br)		^{69,71}Ga(III)	$\frac{7.8^c}{10.5}$	u	4.7	u	2.3	u	2.3	2.0027	THF	Q $GaBr_3$	~268	381
C399	No. 4 (X = I)		Ga(III)	$\frac{8.1^c}{11.0}$	u	4.6	u	2.3	u	2.3	2.0019	THF	Q GaI_3	~268	381
C400	No. 4 (X = Cl)		^{113,115}In(III)	7.7	u	4.5	u	1.7	u	1.7	2.0025	THF	Q $InCl_3$	~268	381
C401	No. 4 (X = Br)			5.9	u	4.4	u	1.8	u	1.8	2.0023	THF	Q $InBr_3$	~268	381
			Tl	(see below)											
C402	No. 4 (X = Cl)	IIIB	^{45}Sc(III)	2.3	u	4.6	u	2.3	u	2.3	2.0034	THF	Q $ScCl_3$	~268	381
C403A	No. 4 (X = Cl)		Y(III)	—	u	4.4	u	2.2	u	2.2	2.0040	THF	Q YCl_3	~268	381
C403B	(X = Br)			—	u	4.4	u	2.2	u	2.2	2.0034	THF	Q YBr_3	~268	381
C404	No. 4 (X = Br)	LA	^{139}La(III)	3.0	u	1.5	u	1.5	u	1.5	2.0040	THF	Q $LaBr_3$	~268	381

Table IV.2.1 (continued)
1,2-NAPHTHOSEMIQUINONE-CATION COMPLEXES

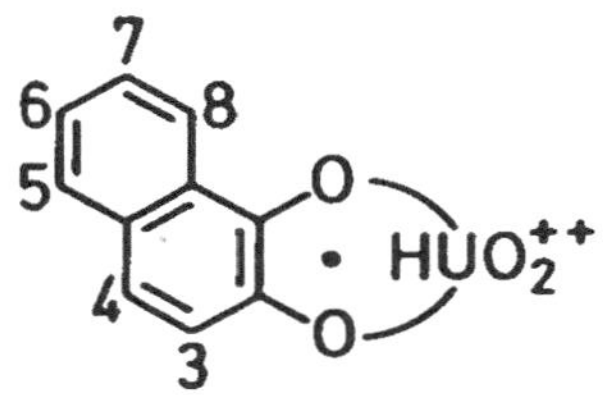

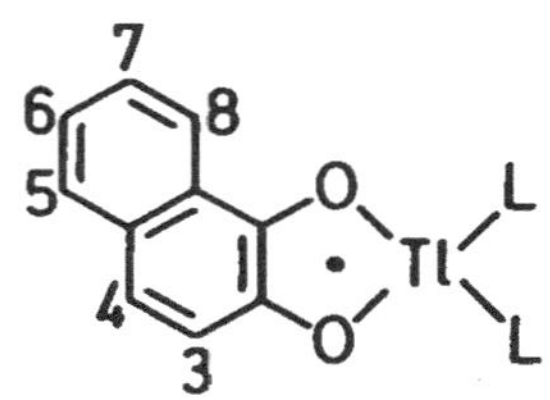

No.	Complex	Central atom	Splitting constants/Gauss a_M	a_3	a_4	a_5	a_6	a_7	a_8	g-Factor	Solvent	Method	T [K]	Ref.
C405	No. 5	AC U(VI)	0.73[i]	0.73	4.92	0.73	1.68	0.73	1.68	1.9950	THF		213	376, 303
C406	No. 1	IIIA Tl(I)	52	u	4.2	u	1.4	u	1.4	1.994	THF	Rc Tl	RT	328
C407	No. 6 (L = Et)	Tl(III)	16.9		u 4.3	u	1.5	u	1.5	2.0019	THF	NaSQ/ClTlEt$_2$	RT	22
C408	No. 6 (L = Ph)	Tl(III)	11.78	0.53	4.31	0.17	1.50	0.10	1.45	2.00333	Dioxane	Q/Ph$_2$TlOH:O$_2$	293	20
			10.68	0.48	4.27	0.14	1.49	u	1.40	2.00330	MeTHF		293	20
			8.73	0.47	4.30	0.20	1.48	0.10	1.38	2.00341	C_5H_5N		293	20
C409	No. 6 (4–CPh$_3$, L = Ph)	Tl(III)	10.78	0.53	—	u	1.36	u	1.36	2.00333	Dioxane	Q/Ph$_2$TlOH:O$_2$	293	20
			10.18	0.50	—	u	1.35	u	1.35	2.00331	MeTHF		293	20
			7.43	0.48	—	0.14	1.32	0.14	1.28	2.00347	C_5H_5N		293	20
C410	No. 6 (3–OH, L = Ph)	Tl(III)	9.35	(2.45	u	1.35	u	1.35	u)[g]	2.00294	Dioxane	Q/Ph$_2$TlOH:O$_2$	293	20
			7.03	(2.43	0.28	1.44	u	1.44	u)[g]	2.00325	C_5H_5N		293	20

C411	No. 6 (4 – OH*, L = Ph)		Tl(III)	9.16	0.39	u	0.20	1.95	0	1.52	2.00329	Bu^{t}NH$_2$	Q/Ph$_2$TlOH:O$_2$	293	20
				7.63	0.47	0.90[g]	0.21	1.65	0	1.33	2.00331	C$_5$H$_5$N		293	20
C412	No. 6 (4–O$^-$Tl$^+$Ph$_2$, L = Ph)		Tl(III)	9.10	0.38	8.82[h]	0.26	2.01	0	1.54	2.00307	Bu^{t}NH$_2$	Q/Ph$_2$TlOH:O$_2$	293	20
				7.19	0.31	5.10[h]	0.21	2.10	0	1.59	2.00297	C$_5$H$_5$N		293	20
C413	No. 6 (4–SO$_3^-$Na$^+$, L = Ph)		Tl(III)	11.10	u	—	u	1.32	u	1.32	2.00352	Dioxane	Q/Ph$_2$TlOH:O$_2$	293	20
				9.58	u	—	u	1.30	u	1.30	2.00352	MeTHF		293	20
				9.02	0.49	—	0.15	1.35	0.16	1.29	2.00352	C$_5$H$_5$N		293	20

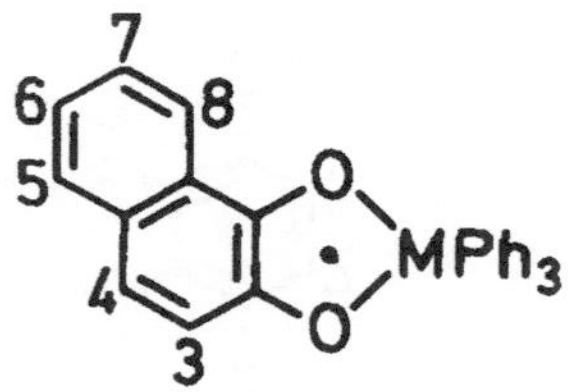

C414	No. 7	IVA	Sn(IV)	—	0.30	5.30	0.30	1.45	u	1.45	2.0049	PhMe	Phot Q/Ph$_3$SnH	193	303
			Pb(IV)	—	0.26	4.90	0.26	1.10	u	1.10	2.0039	PhMe	Phot Q/Ph$_3$PbH	193	303

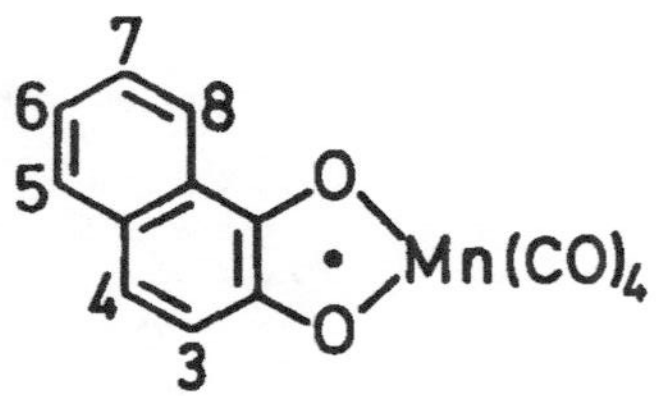

C415	No. 8	VIIB	^{55}Mn(I)	5.97	0.69	4.22	u	1.55	u	1.55	2.0040	PhMe	Phot Q/Mn$_2$(CO)$_{10}$	295	382, 303

[a] a_M measured at 272 K.
[b] a_M measured at 200 K.
[c] Splitting above line from ^{85}Rb, below from ^{87}Rb. The alkali metal splittings measured at 272 K.
[d] Splitting from two ^{31}P.
[e] Number above line from ^{69}Ga, below from ^{71}Ga.
[f] Extra proton in the complex assumed between the two carbonyl oxygens.
[g] Further unresolved splittings observed. No assignment given.
[h] Splitting from a second Tl-nucleus.

Table IV.2.2
1,4-NAPHTHOSEMIQUINONE-CATION COMPLEXES

No.	Complex	Central atom	a_M	a_3	a_4	a_5	a_6	a_7	a_8	g-Factor	Solvent	Method	T [K]	Ref.
			Splitting constants/Gauss											
	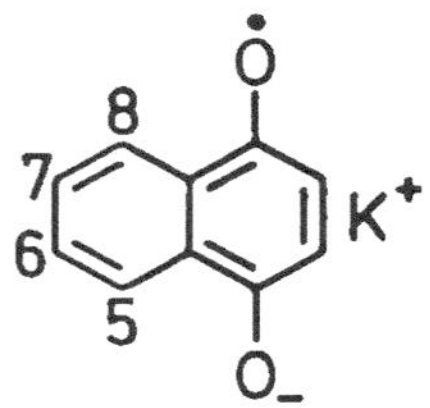									$\Delta g \times 10^5$ (K^+, M)[b]				
C416	No. 1	IA K(I)	u	3.28[a]	3.28	0.23	0.64	0.64	0.23	0 (0) 1.64(0.050) 2.57(0.075) 3.25(0.121) 6.30(0.357)	HMPA	Re K	293	37
	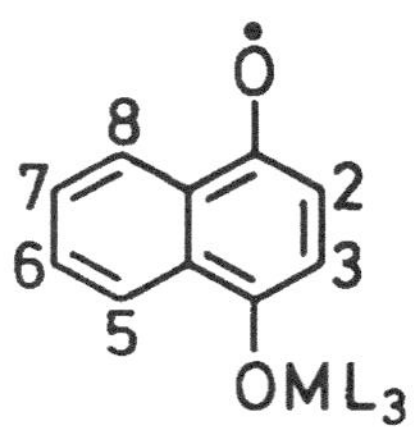													
C417	No. 2 (L = Ph)	IVA Si(IV)	—	8.45	1.52	2.14	0.64	2.14	0.64		PrⁱOH	Phot		346
C418	No. 2 (L = Ph)	Ge(IV)	—	8.05	1.16	1.98	0.54	1.86	0.51		PrⁱOH	Phot		346
	1122	(ML_3 = H*)	(1.59*	7.20	0.36	1.78	0.20	1.78	0.20)		PrⁱOH	Phot		346
C419	No. 2 (L = Bu)	Sn(IV)	—	6.59	0.25	1.58	0.25	1.30	0.25		THF:HMPA (10:1)	Phot		346

[a] The splitting constants are independent of the K^+ concentration.
[b] Due to changing K^+ concentrations (KI added) Δg ≡ change of g-factor relative to that of the free ion.

Table IV.3
9,10-ANTHRASEMIQUINONE-CATION COMPLEXES

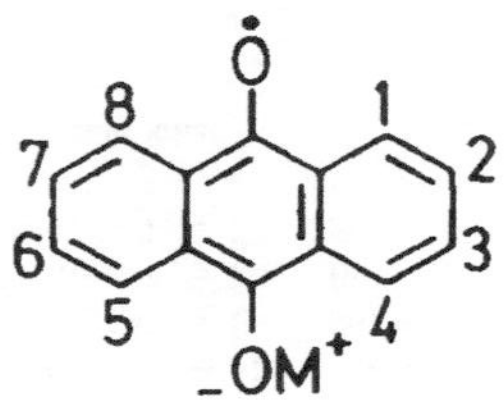

No.	Complex	Central atom	Splitting constants/Gauss a_M	a_L	$a_1 = a_8$	$a_2 = a_7$	$a_3 = a_6$	$a_4 = a_5$	g-Factor	Solvent	Method	T [K]	Ref.
C420	No. 1	IA ^{7}Li(I)	—	—	0.350	1.884	0.110	1.142		THF	Re Li	373	401, 402
			—	—	0.331	1.821	0.181	1.103		THF	Re Li	333	401
			0.14	—	0.309	1.804	0.170	1.093		THF	Re Li	293	162
			—	—	0.250	1.745	0.230	1.058		THF	Re Li	273	401
			—	—	0.250	1.733	0.250	1.038		THF	Re Li	233	401
			—	—	0.242	1.695	0.242	1.009		THF	Re Li	193	401
			—	—	0.220	1.694	0.278	0.996		THF	Re Li	173	162
C421	No. 1	^{23}Na(I)	0.45	—	0.197	1.600	0.390	0.857		THF	Re Na	293	162, 410
			—	—	0.12	1.51	0.38	0.78		THF	Re Na	263	401
			~0.15	—	0.075	1.460	0.530	0.725		THF	Re Na	208	410
			—	—	~0.0	1.38	0.43	0.68		THF	Re Na	183	401
			—	—	0.010	1.396	0.600	0.646		THF	Re Na	173	162
			0.41	—	0.0	1.31	0.66	0.66		THF	Re Na	293	351
C422	No. 1	K(I)	—	—	0.050	1.435	0.565	0.672		THF	Re K	293	162
			—	—	0.060	1.320	0.678	0.552		THF	Re K	173	162

Table IV.3 (continued)
9,10-ANTHRASEMIQUINONE-CATION COMPLEXES

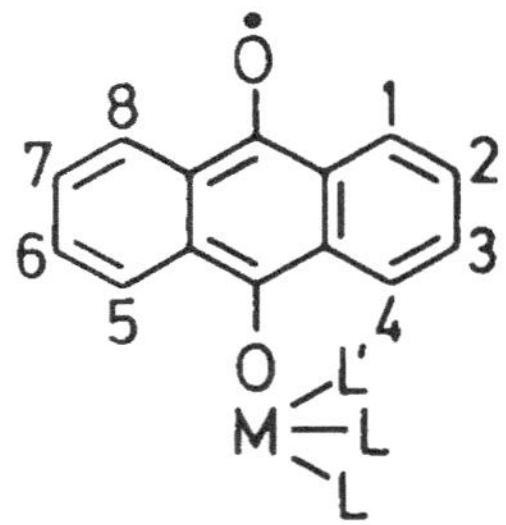

No.	Complex	Central atom	a_M	a_L	$a_1 = a_8$	$a_2 = a_7$	$a_3 = a_6$	$a_4 = a_5$	g-Factor	Solvent	Method	T [K]	Ref.
		IVA											
C423	No. 2 (L = L' = Me)	Si(IV)	—	—	1.10	3.14	1.05	2.75		C_6H_6	Phot Q/Me_3SiH + $Bu^t_2O_2$	293	360
C424	No. 2 (L = L' = Ph)	Si(IV)	—	—	0.87	3.19	0.87	2.83		C_6H_6	Phot Q/Ph_3SiH + $Bu^t_2O_2$	293	360, 346
			—	—	0.8	3.0	0.8	3.0	2.0046	C_6H_6	Phot Q Ph_3SiH + $Bu^t_2O_2$	—	359
C425	No. 2 (L = Ph, L' = H*)	Si(IV)	—	0.16*	0.86	3.19	0.86	2.82		C_6H_6	Phot Q Ph_2SiH_2 + $Bu^t_2O_2$	293	360
C246	No. 2 (L = Me, L' = F)	Si(IV)	—	0.81[f]	0.86	3.21	0.86	2.87		C_6H_6	Phot Q/Me_2SiF_2 + $Bu^t_2O_2$	293	360
C427	No. 2 (L = L' = Ph)	Ge(IV)	—	—	0.86	3.10	0.86	2.70		Melt	Q/Ph_3GeH	—	346
C428	No. 2 (L = L' = Bu)	Sn(IV)	—	—	0.82	3.00	0.82	2.50		Melt	Q Bu_6Sn_2	—	346
C429	No. 2 (L = L' = Ph)	Sn(IV)	—	—	0.74	2.68	0.74	2.34		Melt	Q/Ph_6Sn_2	—	346

Triple ions

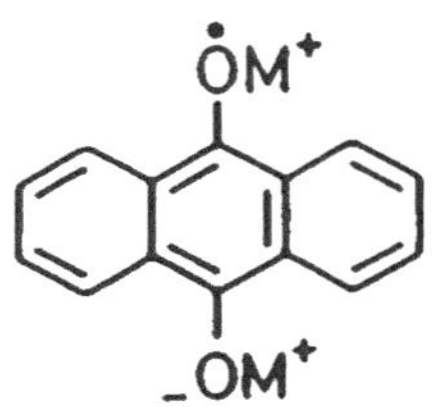

C430	No. 3	IA	Li(I)	—	—	0.479	0.977	0.977	0.479	THF	LiSQ/LiBPh$_4$	295	162
C431	No. 3		^{23}Na(I)	0.2	—	0.384	0.985	0.985	0.384	THF	NaSQ/NaBPh$_4$	226	162, 410
				—	—	0.41	0.98	0.98	0.41	MeTHF	NaSQ/NaBPh$_4$	—	410
C432	No. 3		K(I)	—	—	0.334	0.989	0.989	0.334	THF	KSQ/KBPh$_4$	293	162

Table IV.4
PHENANTHRENESEMIQUINONE-CATION COMPLEXES

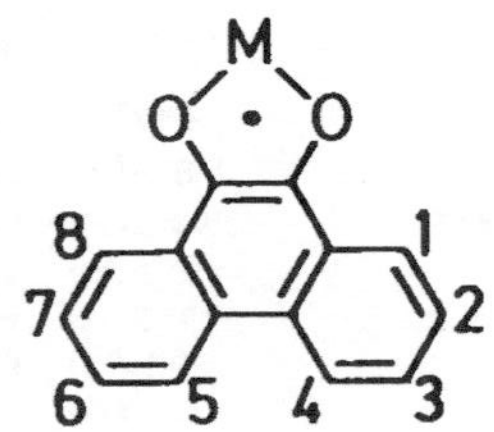

No.1, (SQ)M

No.	Complex		Central atom	Splitting constants/Gauss a_M	a_L	$a_{1,8}$	$a_{2,7}$	$a_{3,6}$	$a_{4,5}$	g-Factor	Solvent	Method	T [K]	Ref.
C433	(SQ)M	IA	^{23}Na(I)	0.48	—	1.43	0.28	1.72	0.43	2.00460	THF	L.Ex. KSQ/NaBPh$_4$	293	255, 119
			K(I)	u	—	1.43	0.26	1.68	0.44	2.00460	THF	Re K	293	255, 119
C434	(SQ)Cu(PPh$_3$)$_2$	IB	^{63}Cu(I)	8.8	11.9[2P]	1.5	u	1.5	u	2.0044	THF	Q Cu/2PPh$_3$	—	302
C435	(SQ)Cu(P(OEt)$_3$)$_2$		Cu(I)	9.2	17.0[2P]	u	u	u	u		THF	Q Cu/2P(OEt)$_3$	—	384
		IIA	Ba	(see Reference 409, showing spectrum of a Ba-complex)										
		IIB	Zn	(see Reference 409, showing spectrum of a Zn-complex)										
C436	(SQ)Hg–Me		Hg(II)	—	0.16[Me]	1.69	0.41	1.85	0.41	2.0052	THF	L.Ex. KSQ/MeHgCl	293	255
C437	(SQ)Hg–Ph		Hg(II)	—	—	—	—	—	—	2.004	THF	Phot Q/Ph$_2$Hg	213	303
C438	(SQ)Hg–Bz		Hg(II)	—	—	—	—	—	—	2.008	THF	Phot Q/Bz$_2$Hg	213	303
C439	(SQ)AlCl$_2$	IIIA	^{27}Al(III)	2.1	—	2.1	u	2.1	u	2.0033	THF	Q AlCl$_3$	RT	381
C440	(SQ)AlBr$_2$		Al(III)	2.3	—	2.3	u	2.3	u	2.0030	THF	Q AlBr$_3$	233—303	381
C441	(SQ)AlI$_2$		Al(III)	2.3	—	2.3	u	2.3	u	2.0029	THF	Q AlI$_3$	RT	381
C442	(SQ)GaCl$_2$		^{69,71}Ga(III)	6.5[a] / 8.2	—	2.1	u	2.1	u	2.0036	THF	Q GaCl$_3$	233—303	381
C443	(SQ)GaBr$_2$		Ga(III)	6.5[a] / 8.5	—	2.2	u	2.2	u	2.0034	THF	Q GaBr$_3$	233—303	381

Table IV.4 (continued)
PHENANTHRENESEMIQUINONE-CATION COMPLEXES

No.	Complex		Central atom	a_M	a_L	$a_{1,8}$	$a_{2,7}$	$a_{3,6}$	$a_{4,5}$	g-Factor	Solvent	Method	T [K]	Ref.
				Splitting constants/Gauss										
C444	(SQ)GaI_2		Ga(III)	7.1[a]/8.9	—	2.2	u	2.2	u	2.0031	THF	Q GaI_3	233—303	381
C445	(SQ)$InCl_2$		^{115}In(III)	6.7	—	2.2	u	2.2	u	2.0022	THF	Q $InCl_3$	RT	381
C446	(SQ)$InBr_2$		In(III)	5.7	—	1.9	u	1.9	u	2.0027	THF	Q $InBr_3$	RT	381
C447	(SQ)InI_2		In(III)	6.9	—	2.3	u	2.3	u	2.0034	THF	Q InI_3	RT	381
C448	(SQ)Tl		Tl(I)	47.7	—	1.6	u	1.6	u	1.9976	THF	Re Tl	RT	328
C449	(SQ)$TlPh_2$		Tl(III)	7.43	—	1.59	0.39	1.71	0.37	2.00338	C_5H_5N	HQ Ph_2TlOH	293	20
				9.08	—	1.60	0.39	1.70	0.38	2.00331	MeTHF	HQ Ph_2TlOH	293	20
				10.13	—	1.61	0.39	1.69	0.38	2.00334	Dioxane	HQ Ph_2TlOH	293	20
C450	(SQ)$TlEt_2$		Tl(III)	15.3	—	1.7	u	1.7	u	2.0025	THF	NaSQ $TlEt_2Cl$	RT	22
C451	(SQ)$ScCl_2$	IIIB	^{45}Sc(III)	2.3	—	2.3	u	2.3	u	2.0031	THF	Q $ScCl_3$	RT	381
C452	(SQ)YCl_2		Y(III)	—	—	2.1	u	2.1	u	2.0026	THF	Q YCl_3	233—303	381
C453	(SQ)YBt_2		Y(III)	—	—	2.0	u	2.0	u	2.0022	THF	Q YBr_3	RT	381
C454	(SQ)$LaCl_2$	LA	^{139}La(III)	3.2	—	1.6	u	1.6	u	2.0024	THF	Q $LaCl_3$	RT	381
C455	(SQ)$LaBr_2$		La(III)	3.0	—	1.5	u	1.5	u	2.0027	THF	Q $LaBr_3$	233—303	381
C456	(SQ)$SiPh_3$	IVA	Si(IV)	—	—	3.06	0.84	2.69	0.84		C_6H_6	Phot Q,$SiPh_4$/$Bu^t_2O_2$	273—353	346
C457	(SQ)$SiBz_3$		Si(IV)	—	—	3.07	0.90	2.69	0.90		C_6H_6	Phot Q,$SiBz_4$/$Bu^t_2O_2$	273—353	346
C458	(SQ)$GePh_3$		^{73}Ge(IV)	1.86	—	1.86	0.41	1.79	0.41		Hexane	Phot Q,$GePh_4$/$Bu^t_2O_2$	213—343	346
C459	(SQ)$SnMe_3$	IVA	^{117,119}Sn(IV)	8.40[b]/8.80	—	1.88	0.42	1.88	0.42	2.0030	C_6H_6	Phot Q/Me_6Sn_2	283	362, 363
				7.05[b]/7.43	—	1.84	0.43	1.84	0.43	2.00402	THF	L.Ex. KSQ/Me_6Sn_2	293	399 255
C460	(SQ)$SnBu_3$		Sn(IV)	8.53[b]/8.92	—	1.84	0.43	1.84	0.43	2.00401	PhMe	Phot Q/Bu_6Sn_2	293	374
				8.32[b]/8.68	—	1.68	0.34	1.76	0.34		Bu_6Sn_2	Heat Q/Bu_6Sn_2	313—373	346
C461	(SQ)$SnPh_3$		Sn(IV)	7.60[b]/7.95	—	1.92	0.44	1.92	0.44	2.00367	PhMe	Phot Q/Ph_6Sn_2	293	374
				6.02[b]/6.31	—	1.58	0.35	1.62	0.35		Ph_6Sn_2	Heat Q/Ph_6Sn_2	313—373	346
C462	(SQ)$PbMe_3$		^{207}Pb(IV)	2.81	—	1.75	0.40	1.75	0.40		C_6H_6	Phot Q/Me_6Ph_2		362, 363
				2.63	—	1.70	0.41	1.70	0.41	2.00433	THF	L.Ex. KSQ/Me_6Pb_2	293	255

C463	(SQ)$PbPh_3$		Pb(IV)	3.11	—	1.77	0.41	1.77	0.41	2.00352	MeTHF	L.Ex. KSQ/Ph_6Pb_2	293	255
		VA	^{31}P (see After uranium, below)											
C464	(SQ)SbF_2		^{121,123}Sb(III)	46.2[c] / 25.0	u	u	u	u	u		Diglyme	Heat Q/SbF_3	433	388
				41.6[c] / 22.5	8.0[2F]	2.1	u	2.1	u		$C_{10}H_8$	Heat Q/SbF_3	473	388
C465	(SQ)BiF_2		^{209}Bi(III)	61.5	u	u	u	u	u		Diglyme	Heat Q/BiF_3	433	388
C466	(SQ)$Mn(CO)_4$	VIIB	^{55}Mn(I)	5.44	—	1.70	0.40	1.70	0.40		PhMe	Phot Q/$Mn_2(CO)_{10}$	RT	385
				5.50	—	1.72	0.40	1.72	0.40	2.0038	C_6H_6	Phot Q/$Mn_2(CO)_{10}$	295	382, 399
C467	(SQ)$MnAsPh_3(CO)_3$		Mn(I)	7.15	32.5[1As]	1.62	<0.2	1.62	<0.2		PhMe	L.Ex. C466/Ph_3As	373	385
C468	(SQ)$MnSbPh_3(CO)_3$		Mn(I)	7.45	99.0[c] / 54.0	1.56	<0.2	1.56	<0.2		PhMe	L.Ex. C466/Ph_3Sb	393	385
C469	(SQ)$Re(CO)_4$		^{185}Re(I)	22.23	—	u	u	u	u	2.0031	C_6H_6	Phot Q/$Re_2(CO)_{10}$	293	382
C470	(SQ)$UO_2^{-\cdot}$	AC	^{235}U(VI)	—	—	1.86	0.43	1.86	0.43	1.9940	THF	Phot Q/$UO_2(NO_3)_2 \cdot 6H_2O_3$	213	400, 376
C471	(SQ)$H^*UO_2^{2+\cdot}$		U(VI)	—	1.43*[,d]	1.86	0.43	1.86	0.43	1.9945	THF	Phot Q/$UO_2(NO_3)_2 \cdot 6H_2O$	213	400, 376

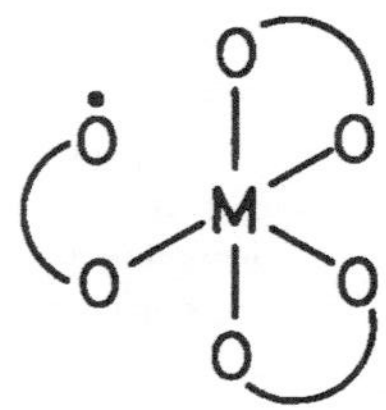

C472	No. 2[e]	VA	^{31}P(V)	4.35	—	u	u	u	u		PhMe	Phot Q/P_4		309
				4.0	—	1.75	u	1.75	u		PhMe or DEE	Phot Q/P_4	193	391
C473	No. 2		^{121,123}Sb(III)	54.5[c] / 29.4	—	u	u	u	u		PhMe	Phot Q/Sb_4		309

[a] Splitting above line from ^{69}Ga, below from ^{71}Ga.
[b] Splitting above line from ^{117}Sn, below from ^{119}Sn.
[c] Splitting above line from ^{121}Sb, below from ^{123}Sb.
[d] The extra proton is assumed to be placed between the two carbonyl oxygens.
[e] 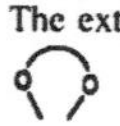 Represents the phenanthrasemiquinone moiety.

Table IV.5
ACENAPHTHENESEMIQUINONE-CATION COMPLEXES

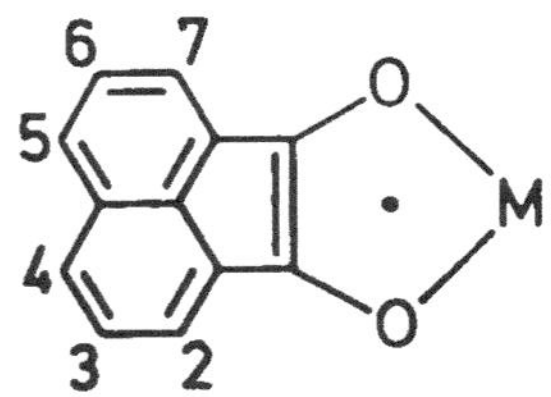

No.	Complex	Central atom	a_M	a_L	a_2	a_3	a_4	a_5	a_6	a_7	g-Factor	Solvent	Method	T [K]	Ref.
			Splitting constants/Gauss												
1322 [I]	Acenapthene-semiquinone (SQ, free anion)		—	—	1.15	0.28	1.30	1.30	0.28	1.15	2.00434	w/EtOH	Re alk	RT	8
C474	(SQ)M	IA	^{7}Li(I) 0.53	—	1.19	0.30	1.30	1.30	0.30	1.19		DME	Re Li	193—293	124
C475	(SQ)M		^{23}Na(I) 0.69	—	1.19	0.28	1.29	1.29	0.28	1.19		DME	Re Na	193—293	124
C476	(SQ)M		^{39}K(I) 0.09	—	1.18	0.26	1.27	1.27	0.26	1.18		DME	Re K	193—293	124
C477	(SQ)M		^{85,87}Rb(I) $\frac{0.27^{a}}{0.87}$	—	1.17	0.27	1.26	1.26	0.27	1.17		DME	Re Rb	193—293	124
C478	(SQ)M		^{133}Cs(I) 0.29	—	1.19	0.29	1.29	1.29	0.29	1.19		DME	Re Cs	193—293	124
C479	(SQ)Cu$(PPh_3)_2$	IB	^{63}Cu(I) 6.7	7.7^{31P}	1.1	u	1.1	1.1	u	1.1	2.0036	THF	NaSQ CuCl/Ph_3P	RT	302, 384
C480	(SQ)Cu$(P(OEt)_3)_2$		Cu(I) 6.9	9.8^{31P}	u	u	u	u	u	u		THF	NaSQ CuCl/$(OEt)_3P$		384
C481	(SQ)Ba(SQ)	IIA	Ba(II) —	—	1.50	0.22	1.63	1.03	0.37	0.86		DME	Re Ba	273^{b}	383
			—	—	1.18	0.30	1.30	1.30	0.30	1.18		DME	Re Ba	353^{c}	383
C482	(SQ)$TlPh_2$	IIIA	Tl(III) 7.70	—	1.14	0.28	1.30	1.30	0.28	1.14	2.00388	C_5H_5N	HQ Ph_2TlOH	293	20
			5.15	—	1.15	0.29	1.32	1.32	0.29	1.15	2.00394	Dioxane	HQ Ph_2TlOH	293	20

C483	(SQ)$SiPh_3$	IVA	Si(IV)	—	—	4.00	1.02	4.00	0.22	0.00	0.22	$HSiPh_3$	Melt	473[b]—523	346
C484	(SQ)$GeBu_3$		Ge(IV)	—	—	3.32	0.80	3.20	0.18	0.20	0.18	$HGeBu_3$	Melt	423[b]	346
				—	—	1.84	0.50	1.77	1.77	0.50	1.84	$HGeBu_3$	Melt	529[c]	346
C485	(SQ)$GePh_3$		Ge(IV)	—	—	3.53	0.86	3.34	0.18	0.19	0.18	$HGePh_3$	Melt	333[b]	346
				—	—	1.92	0.55	1.82	1.82	0.55	1.92	$HGePh_3$	Melt	533[c]	346
C486	(SQ)$SnMe_3$	IVA	^{117,119}Sn(IV)	3.12[d]	—	1.47	0.42	1.55	1.55	0.42	1.47	$HSnMe_3$	Melt	356	346
C487	(SQ)$SnBu_3$		Sn(IV)	—	—	2.57	0.60	2.49	0.37	0.19	0.50	$HSnBu_3$	Melt	218[b]	346
				—	—	1.47	0.42	1.55	1.55	0.42	1.47	$HSnBu_3$	Melt	356[c]	346
C488	(SQ)$SnBu_3$		Sn(IV)	—	—	2.18	0.52	2.02	0.35	0.10	0.45	THF:HMPTA	(1:1)	(?)[b]	346
				—	—	1.36	0.36	1.44	1.44	0.36	1.36	THF:HMPTA	(1:1)	(?)[c]	346
C489	(SQ)$SnPh_3$		Sn(IV)	2.7[d]	—	1.34	0.36	1.52	1.52	0.36	1.34	Ph_3SnH	Melt	404	380, 346
C490	(SQ)$SnPh_2$(Cl)		Sn(IV)	6.5[d]	—	1.25	0.37	1.55	1.55	0.37	1.25	—	Melt	419	380
C491	(SQ)Sn(Ph)Cl_2		Sn(IV)	5.3[d]	0.32[e] / 0.27	1.15	0.32	1.47	1.47	0.32	1.15	Diglyme	—	393	380
C492	(SQ)$SnCl_3$		Sn(IV)	6.9[f] / 7.2	0.34[e] / 0.28	1.18	0.34	1.52	1.52	0.34	1.18	Diglyme	—	366	380

[a] Splitting above line from ^{85}Rb, below from ^{87}Rb.
[b] Spectrum exhibiting slow exchange of cation between carbonyl oxygens at the lowered temperature.
[c] Spectrum exhibiting fast exchange of cation between carbonyl oxygens at the elevated temperature.
[d] Average splittings from ^{117}Sn and ^{119}Sn.
[e] Splitting from a single Cl nucleus, value above line from ^{35}Cl, value below from ^{37}Cl.
[f] Splitting above line from ^{117}Sn, below from ^{119}Sn.

Table IV.6.1
DITHIENO-1,2-BENZOSEMIQUINONE-CATION COMPLEXES

No.	Complex	Central atom	Splitting constants/Gauss a_M	a_L	a_2	a_3	a_4	a_5	g-Factor	Solvent	Method	T [K]	Ref.
	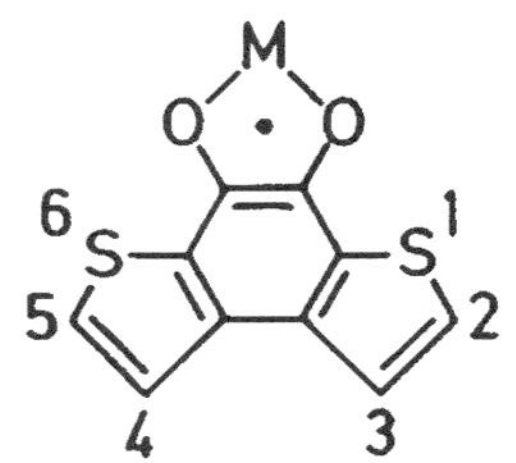												
C493	No. 1	IA $^{7}Li(I)$	0.53	—	1.72	0.22	0.22	1.72		THF	Re Li	RT	408
C494	No. 1	$^{23}Na(I)$	0.39	—	1.66	0.25	0.25	1.66		THF	Re Na	RT	408
C495	No. 1	K(I)	—	—	1.66	0.25	0.25	1.66		DMSO	Re KOBu^t	RT	408
C496	No. 1	$^{133}Cs(I)$	0.51	—	1.60	0.25	0.25	1.60		THF	Re Cs	RT	408
	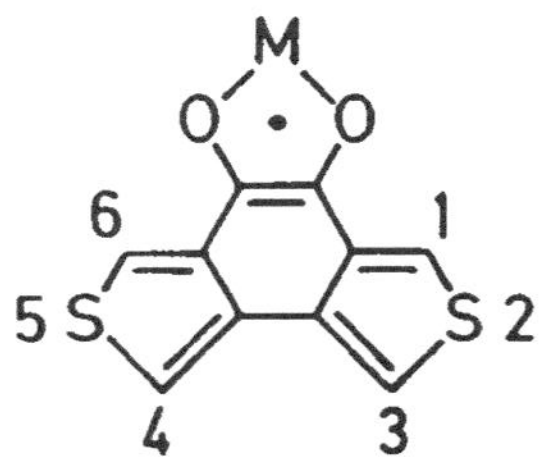												
					a_1	a_3	a_4	a_6					
C497	No. 2	$^{23}Na(I)$	0.47	—	2.87	0.28	0.28	2.87		THF	Re Na	RT	408
C498	No. 2	K(I)	—	—	2.88	0.30	0.30	2.88		DMSO	Re KOBu^t	RT	408
C499	No. 2	$^{133}Cs(I)$	0.40	—	2.78	0.30	0.30	2.78		THF	Re Cs	RT	408

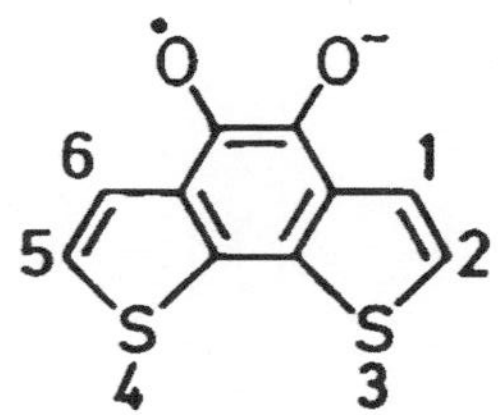

					a_1	a_2	a_5	a_6				
1322	No. 3 (free anion)				<0.06	0.45	0.45	<0.06	DMSO	Re LiOBu^t	RT	408

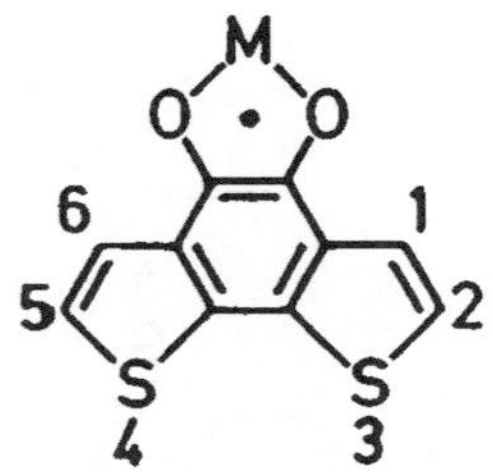

C500	No. 4	^{7}Li(I)	0.48	—	0.094	0.61	0.61	0.094	DMSO	Re LiOBu^t	RT	408
			0.58		0.12	0.62	0.62	0.12	THF	Re Li	RT	408
C501	No. 4	^{23}Na(I)	u	—	0.13	0.64	0.64	0.13	MeOH	Phot NaOMe	RT	408
			0.48	—	<0.08	0.52	0.52	<0.08	THF	Re Na	RT	408
C502	No. 4	K(I)	—	—	<0.06	0.45	0.45	<0.06	DMSO	Re KOBu^t	RT	408
			—	—	<0.08	0.50	0.50	<0.06	THF	Re K	RT	408
C503	No. 4	^{133}Cs(I)	0.56	—	<0.08	0.50	0.50	<0.08	THF	Re Cs	RT	408

Table IV.6.1 (continued)
DITHIENO-1,2-BENZOSEMIQUINONE-CATION COMPLEXES

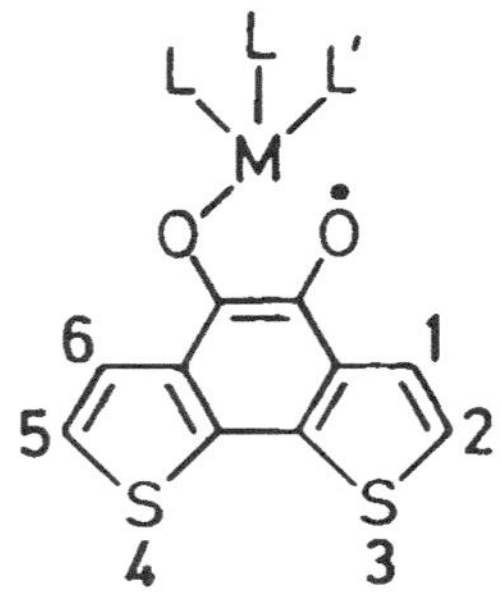

No.	Complex	Central atom		Splitting constants/Gauss a_M	a_L	a_2	a_3	a_4	a_5	g-Factor	Solvent	Method	T [K]	Ref.
C504	No. 5 (L = L' = Ph)	IVA	Si(IV)	—	—	0.53[a]	1.25[a]	0.25[a]	0.71[a]		$PhBu^t$:$Bu^t_2O_2$(2:1)	Phot Q $SiPh_3H$	283	380
C505	No. 5 (L = L' = Ph)		Ge(IV)	—	—	0.52[a]	1.17[a]	0.36[a]	0.52[a]		$PhBu^t$:$Bu^t_2O_2$(2:1)	Phot Q $GePh_3H$	221	380
				—	—	~0.0	0.76	0.76	~0.0		—	Phot Q $GePh_3H$	366	380
C506	N. 5 (L = L' = Ph)		^{117,119}Sn(IV)	7.6[b]/8.0	—	0.25	0.91	0.91	0.25		Ph_3SnH	Phot Q $SnPh_3H$	298	380
C507	No. 5 (L = Ph, L' = Cl)		Sn(IV)	11.7[b]/12.25	—[c]	0.30	1.00	1.00	0.30		Diglyme	Phot Q $SnPh_3Cl$	341	380
C508	No. 5 (L = Cl, L' = Ph)		Sn(IV)	3.7[d]	0.40[c]/0.33	0.23	0.80	0.80	0.23		Diglyme	Phot Q $SnPh_2Cl_2$	373	380
C509	No. 5 (L = L' = Cl)		Sn(IV)	8.1[b]/8.5	0.42[c]/0.33	0.25	0.85	0.85	0.25		C_6H_6	Phot Q $SnCl_4$	298	380

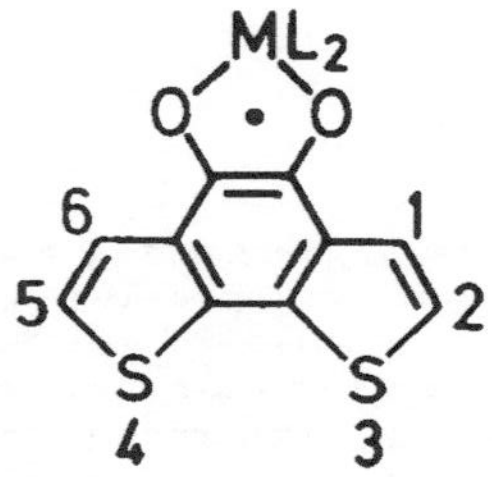

C510	No. 6 (L = F)	VA	^{121,123}Sb(III)	39.2'/21.0	6.8[21]	u	u	u	u		Diglyme	Heat Q SbF$_3$	433	388
C511	No. 6 (L = CL)		Sb(III)	46.9'/25.3	—	u	u	u	u		Diglyme	Heat Q/SbCl$_3$	433	388
C512	No. 6 (L = Ph)		^{209}Bi(III)	63.5	—	u	u	u	u		Diglyme	Heat Q/BiPh$_3$	433	388

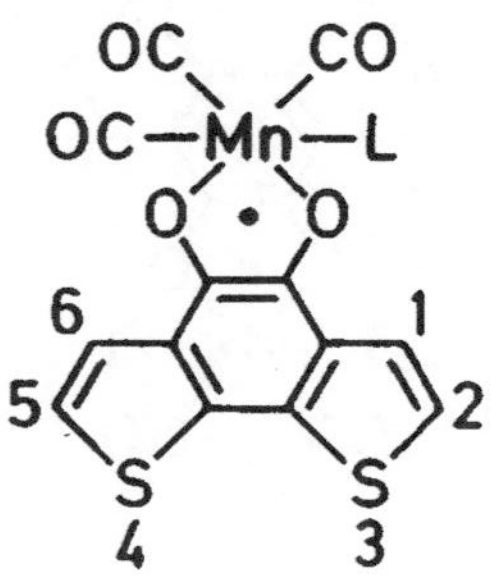

C513	No. 7 (L = CO)	VIIB	^{55}Mn(I)	5.74	—	0.23	0.79	0.79	0.23	2.00397	PhMe	Phot Q/Mn$_2$(CO)$_{10}$	RT	385
C514	No. 7 (L = C$_5$H$_5$N)		Mn(I)	4.80	1.60[1N]	<0.3	0.76	0.76	<0.3		PhMe	L.Ex.C513	RT	385
C515	No. 7 (L = NPh$_3$)		Mn(I)	4.81	1.62[1N]	<0.3	0.76	0.76	<0.3		PhMe	L.Ex.C513	373	385
C516	No. 7 (L = isopropylidencytidine)		Mn(I)	5.45	2.35[1N]	<0.3	0.71	0.71	<0.3		PhMe	L.Ex.C513	RT	385
C517	No. 7 (L = Ph$_3$P)		Mn(I)	8.2	26.3[1P]	<0.2	0.71	0.71	<0.2	2.00357	PhMe	L.Ex.C513	213—333	385
C518	No. 7 (L = ((OEt)$_3$P)		Mn(I)	8.10	32.6[1P]	<0.2	0.50	0.50	<0.2		PhMe	L.Ex.C513	243—333	385

Table IV.6.1 (continued)
DITHIENO-1,2-BENZOSEMIQUINONE-CATION COMPLEXES

No.	Complex	Central atom	Splitting constants/Gauss a_M	a_L	a_2	a_3	a_4	a_5	g-Factor	Solvent	Method	T [K]	Ref.
C519	No. 7 (L = HMPA)	Mn(I)	3.10	0.72^{1P}	<0.3	0.82	0.82	<0.3		PhMe	L.Ex.C513	RT	385
C520	No. 7 (L = Ph_3As)	Mn(I)	7.47	33.5^{1As}	<0.3	0.74	0.74	<0.3		PhMe	L.Ex.C513	373	385
C521	No. 7 (L = Ph_3Sb)	Mn(I)	7.95	102.0^{f} 56.0	<0.3	<0.3	<0.3	<0.3		PhMe	L.Ex.C513	393	385
C522	No. 8 (L = Ph_3P)	^{55}Mn(I)	15.0	32.0^{2P}	<0.3	<0.3	<0.3	<0.3	2.00243	PhMe	L.Ex.C513	333	385
C523	No. 8 (L = $(OEt)_3P$)	Mn(I)	13.5	41.7^{2P}	<0.3	<0.3	<0.3	<0.3		PhMe	L.Ex.C513	333	385

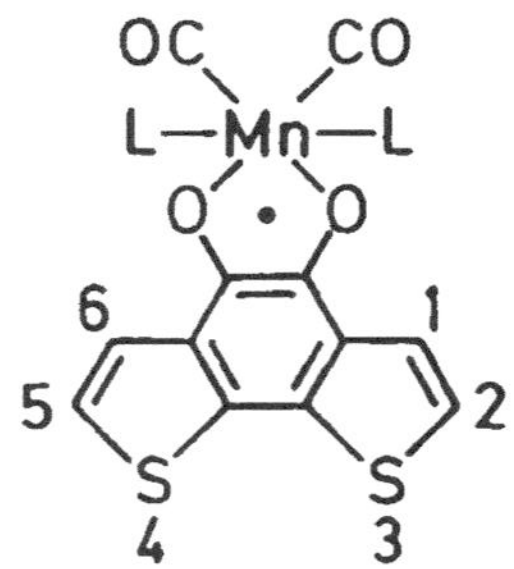

[a] Spectrum indicates that adduct is asymmetric.
[b] Splitting above line from ^{117}Sn, below from ^{119}Sn.
[c] The absence of a chlorine splitting suggests that the chlorine atom lies in the nodal plane.
[d] Average value reported only of 117,119 isotope splitting.
[e] Splitting from one chlorine nucleus (^{35}Cl above the line, ^{37}Cl below) suggests one chlorine atom lies outside the modal plane.
[f] Splitting above line from ^{121}Sb, below from ^{123}Sb.

Table IV.6.2
DITHIENO-1,4-BENZOSEMIQUINONE-CATION COMPLEXES

No.	Complex	Central atom	Splitting constants/Gauss a_M	a_2	a_3	a_6	a_7	Solvent	Method	T [K]	Ref.
	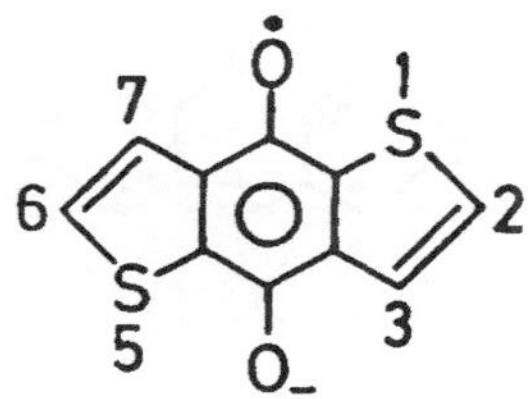										
1324	No. 1 (free anion)[a]		—	0.66	0.17	0.66	0.17	DMSO	Re Bu^{t}OK	RT	408
			—	0.65	0.32	0.65	0.32	MeOH	Phot MeONa	RT	408
	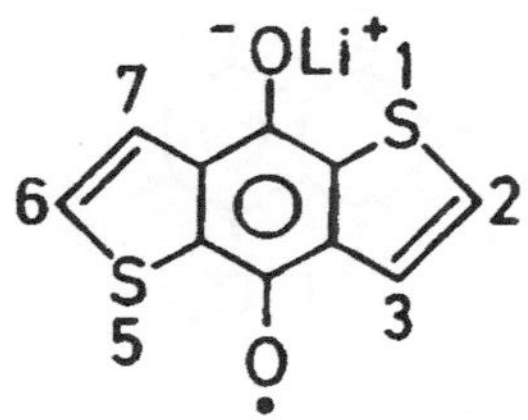										
C524	No. 2	IA ^{7}Li(I)	0.08[b]	1.23	~0	0.43	0.19	THF	Re Li	RT	408
	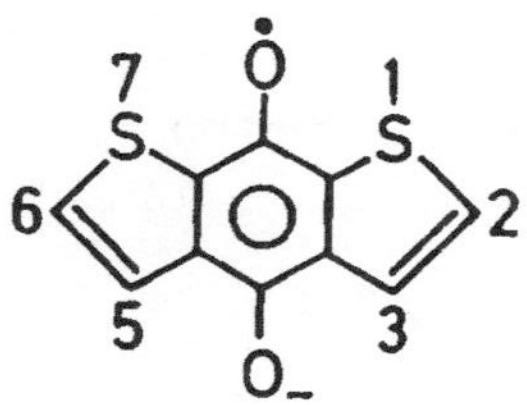										

Table IV.6.2 (continued)
DITHIENO-1,4-BENZOSEMIQUINONE-CATION COMPLEXES

No.	Complex	Central atom	Splitting constants/Gauss a_M	a_2	a_3	a_6	a_7	Solvent	Method	T [K]	Ref.
				a_2	a_3	a_5	a_6				
1325	No. 3 (free anion)[a]		—	0.86	0.07	0.07	0.86	DMSO	Re Bu^{t}OK	RT	408
			—	0.82	0.22	0.22	0.82	MeOH	Phot MeONa	RT	408
C525	No. 4	IA ^{7}Li(I)	0.11[b]	1.44	0.18	0.18	1.44	THF	Re Li	RT	408
1326	No. 5		(Spectrum not observed)[c]					EtOH	Phot[c]	RT	408

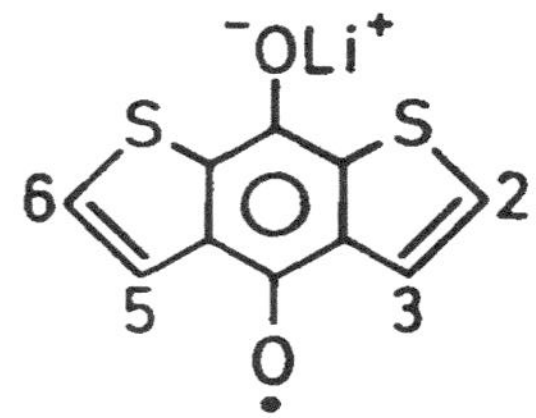

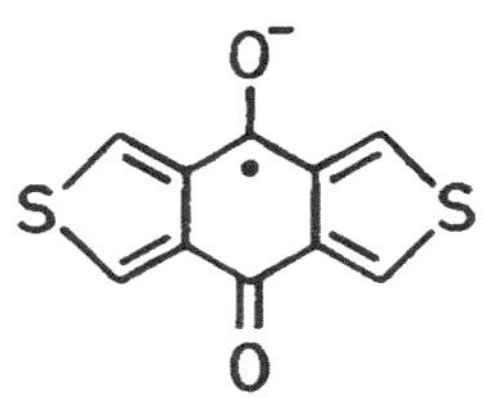

[a] Spectrum suggests the presence of the free anion radical.
[b] a_{Li} decreases with decreasing temperature.
[c] A 1:4:6:4:1 quintet with plitting 2.37 G, observed after prolonged irradiation in presence of ethoxide, is attributed to *p*-benzosemiquinone.

V. ISOTOPES: ^{13}C AND ^{17}O

Table V.1
1,2-BENZOSEMIQUINONES

^{13}C and ^{17}O Splittings

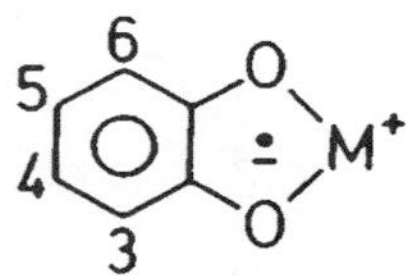

No.	Radical	M	$a_3^H = a_6^H$	$a_4^H = a_5^H$	$a_1^C = a_2^C$	$a_3^C = a_6^C$	$a_4^C = a_5^C$	$a_1^O = a_2^O$	a_M	g-Factor	Solvent	Method	Ref.
			Splitting constants/Gauss										
C3	No. 1	K(I)	−0.94	−3.42	−1.22	−2.94	2.16	—	—		Bu^{t}OH/MeTHF	Ox K_2Q	6
			−0.96	−3.50	−1.20	−2.94	2.16	—	—		Bu^{t}OH	Ox K_2Q	6
C2	No. 1	^{23}Na(I)	−0.96	−3.53	u	−2.93	2.20	—	−0.27		Bu^{t}OH	Ox Na_2Q	6
C1	No. 1	^{7}Li(I)	−0.93	−3.62	u	−3.14	2.38	—	−0.52		Bu^{t}OH/MeTHF	Ox Li_2Q	6
			−0.87	−3.64	u	−3.08	2.32	—	−0.48		Bu^{t}OH	Ox Li_2Q	6
1	No. 1 (free anion)	—	−0.75	−3.67	u	−3.35	2.40	—	—	2.00455	w	Ox phot	30

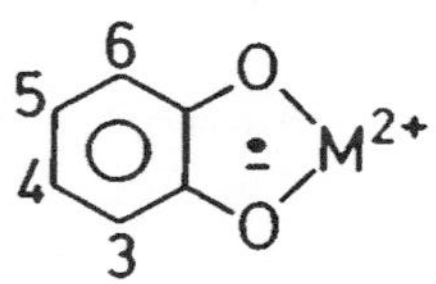

No.	Radical	M	$a_3^H = a_6^H$	$a_4^H = a_5^H$	$a_1^C = a_2^C$	$a_3^C = a_6^C$	$a_4^C = a_5^C$	$a_1^O = a_2^O$	a_M	g-Factor	Solvent	Method	Ref.
C8	No. 2	Ba(II)	−0.71	−3.71	u	−3.50	2.47	—	—	2.0040	w/alk	Ox phot	30
C7	No. 2	Sr(II)	−0.68	−3.75	u	−3.65	2.55	—	—	2.0042	w/alk	Ox phot	30
C5	No. 2	Mg(II)	−0.51	−3.93	u	−3.95	2.90	—	—	2.0042	w/alk	Ox phot	30

V. ISOTOPES: ^{13}C AND ^{17}O

Table V.1 (continued)
1,2-BENZOSEMIQUINONES

^{13}C and ^{17}O Splittings

No.	Radical	M	Splitting constants/Gauss $a_3^H = a_6^H$	$a_4^H = a_5^H$	$a_1^C = a_2^C$	$a_3^C = a_6^C$	$a_4^C = a_5^C$	$a_7^O = a_8^O$	a_M	g-Factor	Solvent	Method	Ref.
	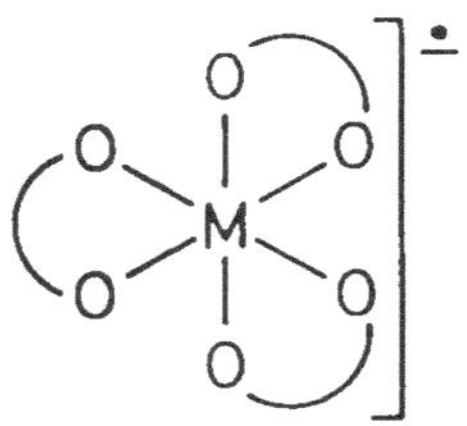												
C209	No. 3[c]	Si(IV)	u	−3.75	−1.4	−4.35	2.9	—	—		THF	Re K	324
	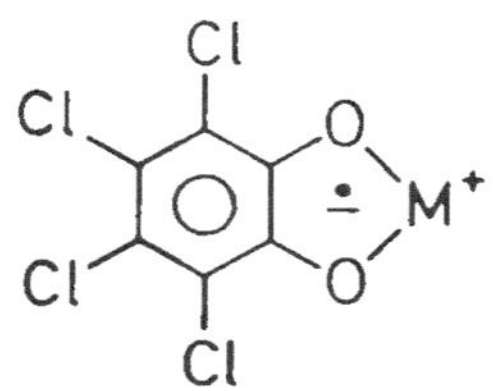												
C249	No. 4	Cs(I)	—	—	−2.23	−3.21	2.70	—	—[d]	2.00453	DME	Re Cs	221[c]
C248	No. 4	Rb(I)	—	—	u	−3.26	2.66	—	—[d]	2.00509	DME	Re Rb	221[c]
C247	No. 4	K(I)	—	—	−2.04	−3.31	2.61	—	—[d]	2.00525	DME	Re K	221[c], 459
C246	No. 4	Na(I)	—	—	−1.65	−3.53	2.65	—	—[d]	2.00523	DME	Re Na	221[c]

C245	No. 4	Li(I)	—	—	u	−4.02	2.85	—	—[d]	2.00507	DME	Re Li	221
89	No. 4 (free anion)	—	—	—	−1.42	−2.59	2.59	—	—	2.00527	DMF/THF	Re el	221
		—	—	—	u	u	u	−8.14	—		DMF	Re el	123

[a] Radical complex obtained by O_2 (or I_2) oxidation of alkali metal quinolate.

[b] Radical generated by continuous photolysis of a solution containing catechol and a chloride or a sulfate salt of the metal ion in question, as the solution was passed through the cavity at 301 K (1 to 2 mℓ/min).

[c]  Represents 3,6-di-*t*-butylcatechol, cf. the result in Table IV.1.6.

[d] See Table IV.1.6 for alkali metal splittings.

Table V.2
1,3-BENZOSEMIQUINONES[a]

^{13}C Splittings

No.	Radical	Splitting constants/Gauss									
		a^H_2	a^H_4	a^H_5	a^H_6	a^C_1	a^C_2	a^C_3	a^C_4	a^C_5	a^C_6
	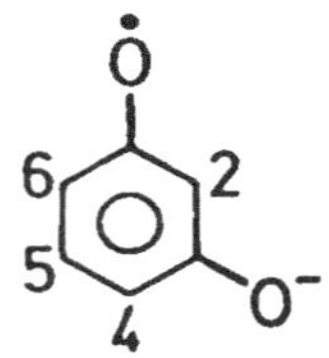										
94	1,3-Benzosemiquinone	0.68	11.44[b]	2.43	11.44	(5.9	1.3	5.9	15.0	2.3	15.0)[c]
102	-,2-carboxylato-	—	11.34	2.41	11.34	6.18	0.99	6.18	15.08	12.32	15.08
103	-,4-carboxylato-	0.68	—	2.31	10.99	5.69	1.39	6.20	15.86 5.38[d]	12.08	14.36
104	-,5-carboxylato-	0.68	11.41	—	11.41	5.86	1.28[e]	5.85	14.96	12.33 1.20[e,f]	14.96

[a] All results from Reference 132. Spectra observed during continuous electron irradiation of N_2O saturated 10^{-3} *M* solutions (w,pH~11) of pertinent resorcinol. Splittings of ^{13}C observed from natural abundance. Sign of the splittings not reported.

[b] Splittings >10 G corrected for second-order shifts.

[c] Splittings estimated from observations on the carboxylated derivatives as described in Reference 132.

[d] Splitting from C-atom of carboxylato group [a^C_4(COO)]. Assignment of the splittings 5.69 and 5.38 may be the opposite of the one shown.

[e] The 0.08-G field difference between the constants (1.20, 1.28) determined by simulation.

[f] Splitting from C-atom of the 5-COO group.

Table V.3.1
1,4-BENZOSEMIQUINONE

^{13}C and ^{17}O Splittings

No.	Radical	Splitting constants/Gauss $-a_2^H$	$-a_1^C$	$-a_2^C$	$-a_7^O$	Solvent	x_w^a	Method	Ref.
	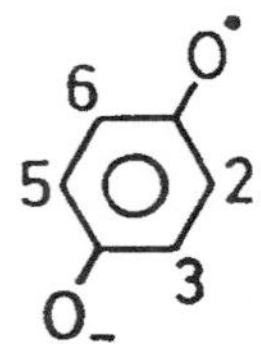								
107	1,4-Benzosemiquinone[i]	2.357	−0.24	0.713	8.70[b]	w	1.000	Re el	223, 224
		u	−0.20	u	u	w/DMSO	0.990	—	224
		u	±0.06	u	u	—	0.909	—	224
		u	0.36	u	u	—	0.832	—	224
		u	0.632	u	u	—	0.703	—	224
		u	0.837	u	u	—	0.613	—	224
		u	0.989	u	u	—	0.543	—	224
		u	1.19	u	u	—	0.442	—	224
		u	1.348	u	u	—	0.357	—	224
		u	1.57	u	u	—	0.240	—	224
		u	1.759	u	u	—	0.137	—	224
		u	1.841	u	u	—	0.100	—	224
		u	1.938	u	u	—	0.060	—	224
		u	2.040	u	u	—	0.031	—	224
		u	2.120	u	u	—	0.006	—	224
		2.42	2.130	u	9.46	DMSO	0.000	—	224, 223
		2.385	0.647	0.497	u	w/DMSO	0.81	Re el	226, 39
		2.393	1.12	(0.380)[c]	u	—	0.59	—	226, 39
		2.41	1.71	(0.230)	u	—	0.17	—	226, 39

Table V.3.1 (continued)
1,4-BENZOSEMIQUINONE

^{13}C and ^{17}O Splittings

No.	Radical	Splitting constants/Gauss				Solvent	x_w^a	Method	Ref.
		$-a_2^H$	$-a_1^C$	$-a_2^C$	$-a_7^O$				
		2.419	2.13	(0.123)	9.46	DMSO	0.00	—	226, 39
		2.40	1.06	0.45	u	29% D_2O/DMSO		Re el	226
		2.41	1.77	u	u	4% D_2O/DMSO		—	226
							`EtOH		
		2.40	1.19	0.44	u	EtOH/DMSO	0.80	Re el	226
		2.41	1.47	(0.283)[d]	u	—	0.55	—	226
		2.41	2.00	(0.162)	u	—	0.09	—	226
						`w			
		2.357	−0.24	0.713	8.70	w	1.000	Re el	223, 224
		2.361	−0.162	u	8.80	w/MeCN	0.943	—	223
		2.364	−0.10	u	8.83	—	0.897	—	223
		2.373	0.109	u	8.93	—	0.744	—	223
		2.385	0.378	u	8.93	—	0.553	—	223
		2.386	0.39	u	u	—	0.538	—	224
		2.393	0.476	u	u	—	0.431	—	224
		2.397	0.53	u	u	—	0.368	—	224
		2.408	0.843	u	u	—	0.189	—	224
		2.413	1.09	u	u	—	0.104	—	224
		2.414	1.157	u	9.34	—	0.089	—	223
		2.415	1.193	u	u	—	0.081	—	224
		2.42	1.236	u	9.32	—	0.071	—	223
		2.42	1.282	u	u	—	0.060	—	224
		2.42	1.318	u	9.31	—	0.055	—	223
		2.42	1.451	u	9.32	—	0.037	—	223

2.42	1.473	u	u	w/MeCN	0.034	Re el	224
2.42	1.625	u	9.41	—	0.019	—	223
2.42	1.686	u	u	—	0.013	—	224
2.42	1.760	u	9.43	—	0.010	—	223
2.42	1.907	u	u	w/MeCN	0.004	—	224
2.42	2.057	u	9.48	MeCN	0.000	—	223
					x_{EtOH}		
u	u	0.458	9.12	EtOH abs	1.000	Re el	223, 39
u	u	u	9.12	EtOH/MeCN	0.545	—	223
u	u	u	9.19	—	0.473	—	223
u	u	u	9.18	—	0.376	—	223
2.42	2.057	u	9.48	MeCN	0.000	—	223
					x_{MeOH}		
u	u	u	9.22	MeOH/MeCN	0.464	Re el	223
u	u	u	9.20	—	0.409	—	223
u	u	u	9.18	—	0.341	—	223
u	u	u	9.21	—	0.306	—	223
u	u	u	9.18	—	0.257	—	223
u	u	u	9.25	—	0.149	—	223
u	u	u	9.31	—	0.071	—	223
u	u	u	9.30	—	0.025	—	223
u	u	u	9.41	—	0.013	—	223
u	u	u	9.44	—	0.004	—	223
2.42	2.057	u	9.48	MeCN	0.000	—	223
2.368	0.40	0.594	u	w and/or EtOH		Re el	36, 39, 222
u	u	0.54	9.02	10% w/EtOH		Ox alk	225, 228
2.357	u	0.630	u	32% w/EtOH		Ox alk	8
u	u	u	9.53	DMF		Re el	228, 123
2.395	2.21	u	9.54	DME		Re el	39, 223
2.395	2.04	u	u	DME (298 K)		Re K	340, 378
2.410	2.45	u	u	DME (198 K)		Re K	340, 378

[a] x_w, Mole fraction.
[b] All ^{17}O splittings observed from enriched samples only.

Table V.3.2
1,4-BENZOSEMIQUINONES, CATIONS AND NEUTRAL SPECIES

^{13}C and ^{17}O Splittings

No.	Radical	Splitting constants/Gauss							Solvent	Method	Ref.
		a^H_{OH}	a^H_2	a^H_3	a^C_1	a^C_2	a^C_4	a^O_7			
C289	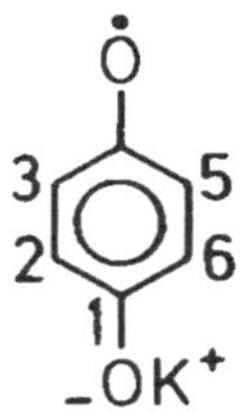	—	−1.935	−2.885	−0.79	u	−3.50	—	DME(198 K)	Re K	340, 378
1012	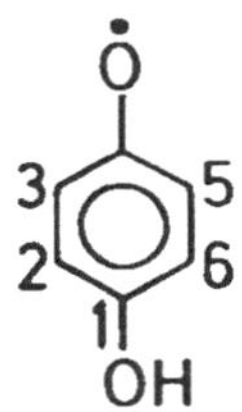	−1.563	−5.638	0.731	4.4	u	−4.8	—	Bu_3PO_4(330 K)	Phot UV	460
		−1.875	−5.163	0.288	4.1	u	−3.5	—	Pr^iOH (300 K)	Phot UV	460

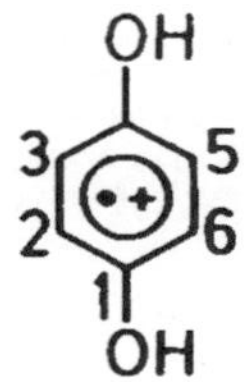

1220	−3.294[a]	−2.253	−2.253	4.23	−1.65	4.23	−7.83	$MeNO_2/AlCl_3$	Ox	278, 203

[a] From Reference 203.

Table V.3.3
SUBSTITUTED 1,4-BENZOSEMIQUINONES

^{13}C and ^{17}O Splittings

No.	Radical	a_3^H	a_5^H	a_6^H	$-a_1^C$	a_2^C	$-a_3^C$	$-a_4^C$	$-a_5^C$	$-a_6^C$	$-a_7^O$	$-a_8^O$	Solvent	Method	Ref.
	Splitting constants/Gauss														
	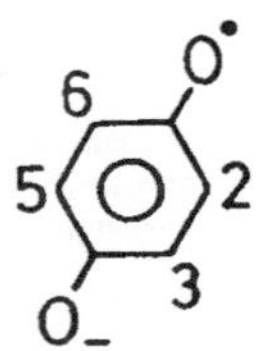 1,4-Benzosemiquinone														
115	-,2-*t*-butyl-	1.65	−2.10	2.85	u	0.74[a]	u	u	u	u	u	u	MeCN	Ox alk	61
163	-,2,5-diphenyl-	2.47	0.12[b]	2.47	u	u	u	u	u	u	9.07	9.07	DMF	Re el	228
176	-,tetramethyl-	1.90[l]	1.90[l]	1.90[l]	1.067	0.721	0.721	1.067	0.721	0.721	u	u	EtOH		76, 222
189	-,2,3-dicyano-	0.61^N	−1.05	1.05	3.89	4.83	−4.83	3.89	u	u	u	u	DME	Ox I_2	60
		0.66^N	−1.09	−1.09	u	u	u	u	u	u	u	u	DME	Re el	168
190	-,5,6-dichloro-2,3-dicyano-	0.57^N	u	u	4.11	4.58	4.58	−4.11	−2.80	−2.80	u	u	DME Me_2CO	Re Me_4NI	60

Table V.3.3 (continued)
SUBSTITUTED 1,4-BENZOSEMIQUINONES

^{13}C and ^{17}O Splittings

No.	Radical	a_3^H	a_5^H	a_6^H	$-a_1^C$	a_2^C	$-a_3^C$	$-a_4^C$	$-a_5^C$	$-a_6^C$	$-a_7^O$	$-a_8^O$	Solvent	x_{EtOH}	Method	Ref.
		Splitting constants/Gauss														
		0.59[N]	u	u	2.86	−2.42	−2.42	2.86	u	u	u	u	C_6H_6		Q TDAP	168
265	-,2,5-difluoro-	1.41	5.35[F]	−1.41	u	u	u	u	u	u	9.41	9.41	MeCN		Re el	65
266	-,2,5-dichloro-	u	—	u	u	u	u	u	u	u	9.28	9.28	DMF		Re el	228, 123
270	-,tetrafluoro-	u	u	u	u	u	u	u	u	u	9.53	9.53	DMF		Re al	123
		3.95[F]	3.95[F]	3.95[F]	2.19	0.80	0.80	2.19	0.80	0.80	9.46	9.46	DME	0.000	Re el	65
		u	u	u	u	u	u	u	u	u	9.45	9.45	DME/EtOH	0.021	—	65
		u	u	u	u	u	u	u	u	u	9.37	9.37	—	0.048	—	65
		u	u	u	u	u	u	u	u	u	9.34	9.34	—	0.079	—	65
		3.91[F]	3.91[F]	3.91[F]	1.90	0.91	0.91	1.90	0.91	0.91	u	u	—	0.085	—	65
		u	u	u	1.80	0.92	0.92	1.80	0.92	0.92	u	u	—	0.116	—	65
		u	u	u	u	u	u	u	u	u	9.33	9.33	—	0.120	—	65
		u	u	u	u	u	u	u	u	u	9.33	9.33	—	0.162	—	65
		3.90[F]	3.90[F]	3.90[F]	1.74	0.96	0.96	1.74	0.96	0.96	u	u	—	0.210	—	65
		u	u	u	1.56	0.97	0.97	1.56	0.97	0.97	u	u	—	0.227	—	65
		u	u	u	u	u	u	u	u	u	9.33	9.33	—	0.231	—	65
		u	u	u	u	u	u	u	u	u	9.33	9.33	—	0.290	—	65
		u	u	u	u	u	u	u	u	u	9.31	9.31	—	0.355	—	65
271	-,tetrachloro-	—	—	—	2.57	u	u	2.57	u	u	8.80	8.80	DME		Re el	65
		—	—	—	u	u	u	u	u	u	8.89	8.89	DMF		Re el	228, 123
		—	—	—	u	u	u	u	u	u	8.73	8.73	DMSO		Re el	228
		—	—	—	u	u	u	u	u	u	8.69	8.69	MeCN		Re el	228
		—	—	—	u	u	u	u	u	u	8.60	8.60	w EtOH		Re el	228
		—	—	—	(4.8	2.8	2.8	4.8	2.8	2.8)[c]	8.5	8.5	w alk(0.1 *M*)		Re DTN	229
		—	—	—	u	u	u	u	u	u	8.16	8.16	w		Re el	228
272	-,tetrabromo-	—	—	—	u	u	u	u	u	u	8.67	8.67	DMF		Re el	123
200	-2,5-bis(trimethylsilyl)-	2.79	1.51[Si]	2.79	$(a_{C(Si)} = 2.06, 2.54)$[d]						u	u	DMF		Re el	72[e]

[a] ^{13}C splitting from t-butyl-β-C, 48% enriched.
[b] A fine structure of at least 15 lines separated by 0.12 to 0.13 G observed upon each line of the main triplet and assigned to the phenyl protons.
[c] Numerical values reported in Reference 229 only.
[d] ^{13}C splittings for which assignment is unknown.
[e] Sign of splitting constants not quoted in Reference 73.

Table V.3.4
2,5-DIOXIDO-1,4-BENZOSEMIQUINONES

^{13}C and ^{17}O Splittings

No.	Radical	a_3^H	a_1^C	a_3^C	a_1^O	Solvent	Method	Ref.
		Splitting constants/Gauss						
	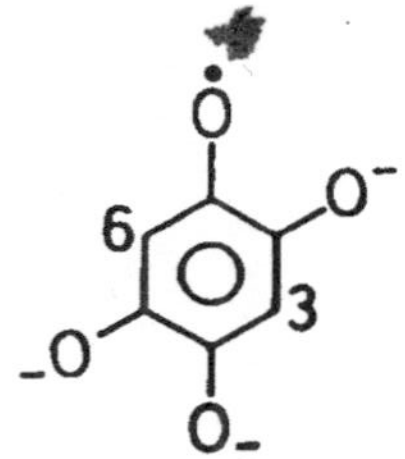							
501	2,5-Dioxido-1,4-benzosemiquinone	0.79	2.82	−6.62	u	unknown	unknown	227
		0.79	2.63	−6.66	u	w/KOH	Re DTN	39, 230
		u	u	u	−4.57	w/alk	Re DTN	228, 230
504	-,3,6-difluoro-	3.85^{F}	2.55	−8.02	−4.78	w/0.5 *M* NaOH	Re DTN	65
		3.7^{F}	2.5	−7.7	−4.6	w/5 *N* NaOH	Re DTN	229[a]
505	-,3,6-dichloro-	u	2.70	−8.17	−4.79	w/alk	Re DTN	228, 230
		0.18^{Cl}	2.71	−8.17	u	w/alk	Re DTN	134, 230
		u	2.7	−7.9	−4.6	w/5 *N* NaOH	Re DTN	229
		u	2.5	−7.9	−4.7	90% w/Me_4NOH	Re DTN	229
520	-,3,6-dioxido-	—	−0.643	−0.643	—	w(pH ~ 13)	Ox alk	8

[a] A doublet splitting of ~0.3 G of unknown origin is reported for no. 504 in Reference 229.

Table V.4
MISCELLANEOUS RADICALS

^{13}C and ^{17}O Splittings

No.	Radical	EPR data	Solvent	Method	Ref.
288	5-Carboxylatopyrogallolsemiquinone	(see Figure 11)			8
559	1,4-Naphthosemiquinone	$a^O = -8.58$	DMF	Re el	123, 228
579	-,2,3-dichloro-	$a^O = -8.02$	DMF	Re el	123
683	9,10-Anthrasemiquinone	$a^O = -7.53$	DMF	Re el	123, 228
		$a^O = -7.53$, $a_9^C = -0.47^a$, $(a_1^C$ [or $a_{11}^C] = -1.21)^b$	DME	Re el	39, 230
		$a_9^C = 0.70^a$, $(a_{11}^C = -1.37)^b$	w/EtOH	Re el	39
756	9,10-Phenanthrasemiquinone	$a^O = -8.19$	DMF	Re el	123
1322	Acenaphthenesemiquinone	$a^O = -8.84$	DMF	Re el	123

[a] Labeled position ($^{13}C - 9$).
[b] Assignment and sign uncertain.

REFERENCES

1. **Pedersen, J. A.,** *J. Chem. Soc. Perkin Trans. 2*, 424, 1973.
2. **Pilař, J.,** *J. Phys. Chem.*, 74, 4029, 1970.
3. **Adams, M., Blois, M. S., Jr., and Sands, R. H.,** *J. Chem. Phys.*, 28, 774, 1958.
4. **Holton, D. M. and Murphy, D.,** *J. Chem. Soc. Faraday Trans. 1*, 78, 1223, 1982.
5. **Hewgill, F. R., Stone, T. J., and Waters, W. A.,** *J. Chem. Soc.*, 408, 1964.
6. **Brustolon, M., Pasimeni, L., and Corvaja, C.,** *J. Chem. Soc. Faraday Trans.*, 2, 71, 193, 1975.
7. **Lee-Ruff, E., Lever, A. B. P., and Rigaudy, J.,** *Can. J. Chem.*, 54, 1837, 1976.
8. **Pedersen, J. A.,** unpublished laboratory results, 1983.
9. **Stone, T. J. and Waters, W. A.,** *J. Chem. Soc.*, 1488, 1965.
10. **Pilař, J., Buben, I., and Pospíšil, J.,** *Collect. Czech. Chem. Commun.*, 35, 489, 1970.
11. **Trapp, C., Tyson, C. A., and Giacometti, G.,** *J. Am. Chem. Soc.*, 90, 1394, 1968.
12. **Ryba, O., Pilař, J., and Petranek, J.,** *Collect. Czech. Chem. Commun.*, 33, 26, 1968.
13. **Müller, E., Günther, F., Scheffler, K., Ziemek, P., and Rieker, A.,** *Justus Liebigs Ann. Chem.*, 688, 134, 1965.
14. **Felix, C. C. and Sealy, R. C.,** *Photochem. Photobiol.*, 34, 423, 1981.
15. **Pedersen, J. A. and Øllgaard, B.,** *Biochem. Syst. Ecol.*, 10, 3, 1982.
16. **Pedersen, J. A.,** *Phytochemistry*, 17, 775, 1978.
17. **Ashworth, P.,** *J. Org. Chem.*, 41, 2920, 1976.
18. **Neta, P. and Fessenden, R. W.,** *J. Phys. Chem.*, 78, 523, 1974.
19. **Dixon, W. T., Moghimi, M., and Murphy, D.,** *J. Chem. Soc. Faraday Trans. 2*, 70, 1713, 1974.
20. **Stegman, H. B., Ulmschneider, K. B., and Scheffler, K.,** *J. Organomet. Chem. (Amsterdam)*, 101, 145, 1975.
21. **Stegman, H. B., Ulmschneider, K. B., and Scheffler, K.,** *J. Organomet. Chem. (Amsterdam)*, 72, 41, 1974.
22. **Klimov, E. S., Abakumov, G. A., Gladyshev, E. N., Baysuhkin, P. Ya., Muraev, V. A., and Razuvaev, G. A.,** *Dokl. Akad. Nauk SSR*, 218, 844, 1974.
23. **Leblond, J., Papon, P., and Uebersfeld, J.,** *Magn. Res. Relat. Phenom. Proc. Congr. Ampère*, 16, 961, 1971.
24. **Stegman, H. B., Bergler, H. U., and Scheffler, K.,** *Angew. Chem.*, 93, 398, 1981.
25. **Stegman, H. B., Stolze, K., Bergler, H. U., and Scheffler, K.,** *Tetrahedron Lett.*, 22, 4057, 1981.
26. **Dixon, W. T., Moghimi, M., and Murphy, D.,** *J. Chem. Soc. Perkin Trans. 2*, 1189, 1975.
27. **Stone, T. J. and Waters, W. A.,** *J. Chem. Soc.*, 4302, 1964.
28. **Eaton, D. R.,** *Inorg. Chem.*, 3, 1268, 1964.
29. **Kuwata, K. and Shimizu, Y.,** *Bull. Chem. Soc. Jpn.*, 42, 864, 1969.
30. **Felix, C. C. and Sealy, R. C.,** *J. Am. Chem. Soc.*, 104, 1555, 1982.
31. **Oakes, J. and Symons, M. C. R.,** *Trans. Faraday Soc.*, 64, 2579, 1968.
32. **Pedersen, J. A.,** *Mol. Phys.*, 28, 1031, 1974.
33. **Ashworth, P. and Dixon, W. T.,** *J. Chem. Soc. Perkin Trans. 2*, 1130, 1972.
34. **Yao, H. C. and Heller, H. C.,** *Anal. Chem.*, 41, 1540, 1969.
35. **Ashworth, P.,** *Mol. Phys.*, 30, 313, 1975.
36. **Stone, E. W. and Maki, A. H.,** *J. Chem. Phys.*, 36, 1944, 1962.
37. **Stevenson, G. R., Alegria, A. E., and Block, A. McB.,** *J. Am. Chem. Soc.*, 97, 4859, 1975.
38. **Pedersen, J. B., Hansen, C. E. M., Parbo, H., and Muus, L. T.,** *J. Chem. Phys.*, 63, 2398, 1975.
39. **Das, M. R. and Fraenkel, G. K.,** *J. Chem. Phys.*, 42, 1350, 1965.
40. **Segal, B. G., Kaplan, M., and Fraenkel, G. K.,** *J. Chem. Phys.*, 43, 4191, 1965.
41. **Zandstra, P. J.,** *J. Chem. Phys.*, 41, 3655, 1964.
42. **Venkataraman, B., Segal, B. G., and Fraenkel, G. K.,** *J. Chem. Phys.*, 30, 1006, 1959.
43. **Pilař, J., Buben, I., and Pospíšil, J.,** *Collect. Czech. Chem. Commun.*, 37, 3599, 1972.
44. **Kosman, D. and Stock, L. M.,** *J. Am. Chem. Soc.*, 91, 2011, 1969.
45. **Johnson, C. S. and Tully, J. C.,** *J. Chem. Phys.*, 40, 1744, 1964.
46. **Ryba, O., Pilař, J., and Petránek, J.,** *Collect. Czech. Chem. Commun.*, 34, 2581, 1969.
47. **Huntington, J. L. and Davis, D. G.,** *J. Magn. Res.*, 3, 456, 1970.
48. **Ashworth, P. and Dixon, W. T.,** *J. Chem. Soc. Perkin Trans. 2*, 2264, 1972.
49. **Ashworth, P. and Dixon, W. T.,** *J. Chem. Soc. Perkin Trans. 2*, 1533, 1973.
50. **Gill, P. S. and Gough, T. E.,** *Trans. Faraday Soc.*, 64, 1997, 1968.
51. **Ashworth, P. and Dixon, W. T.,** *Chem. Commun.*, 1150, 1971.
52. **Ackermann, P., Babarin, F., Germain, J. P., Fabre, C., and Tchoubar, B.,** *Tetrahedron*, 30, 1019, 1974.
53. **Sullivan, P. D.,** *J. Am. Chem. Soc.*, 97, 3992, 1975.

54. **Lucken, E. A. C.,** *J. Chem. Soc.*, 4234, 1964.
55. **Poupko, R. and Rosenthal, I.,** *J. Phys. Chem.*, 77, 1722, 1973.
56. **Brunton, G., Gilbert, B. C., and Mawby, R. J.,** *J. Chem. Soc. Perkin Trans. 2*, 1267, 1976.
57. **Colegate, S. M., Hewgill, F. R., and Howie, G. B.,** *Aust. J. Chem.*, 28, 343, 1975.
58. **Kohl, D. H., Wood, P. M., Weissman, M., and Wright, J. R.,** in *Magnetic Resonance in Biological Research*, Franconi, C., Ed., Gordon & Breach, New York, 1971, 189.
59. **Sullivan, A. B., and Reynolds, G. F.,** *J. Phys. Chem.*, 80, 2671, 1976.
60. **Corvaja, C., Pasimeni, L., and Brustolon, M.,** *Chem. Phys.*, 14, 177, 1976.
61. **Stock, L. M. and Suzuki, J.,** *J. Am. Chem. Soc.*, 87, 3909, 1965.
62. **Dixon, W. T., Hoyle, P. M., and Murphy, D.,** *J. Chem. Soc. Faraday Trans. 2*, 2027, 1978.
63. **Matsunaga, Y. and McDowell, C. A.,** *Can. J. Chem.*, 38, 1167, 1960.
64. **Yonezawa, T., Kawamura, T., Ushio, M., and Nakao, Y.,** *Bull. Chem. Soc. Jpn.*, 43, 1022, 1970.
65. **Geiger, W. E. and Gulick, W. M., Jr.,** *J. Am. Chem. Soc.*, 91, 4657, 1969.
66. **Fischer, P. H. H. and Simmermann, H.,** *Z. Naturforsch.*, 23a, 1399, 1968.
67. **Anderson, D. H., Frank, P. J., and Gutowsky, H. S.,** *J. Chem. Phys.*, 32, 196, 1960.
68. **Hudson, A. and Lewis, J. W.,** *J. Chem. Soc. B*, 531, 1969.
69. **Matsuura, T., Kon, H., and Cahnmann, H. J.,** *J. Org. Chem.*, 29, 3058, 1964.
70. **Lucken, E. A. C.,** *J. Chem. Soc.*, 5123, 1963.
71. **Lucken, E. A. C.,** *Z. Naturforsch.*, 18b, 166, 1963.
72. **Müller, E., Eggensberger, H., Teissier, B., and Scheffler, K.,** *Z. Naturforsch.*, 18b, 984, 1963.
73. **Gerson, F., Krynitz, U., and Bock, H.,** *Helv. Chim. Acta*, 52, 2512, 1969.
74. **Lucken, E. A. C.,** *J. Chem. Soc.*, 4240, 1964.
75. **Stevenson, G. R. and Alegria, A. E.,** *J. Phys. Chem.*, 77, 3100, 1973.
76. **Das, M. R., Connor, H. D., Leniart, D. S., and Freed, J. H.,** *J. Am. Chem. Soc.*, 92, 2258, 1970.
77. **Fritsch, J. M., Tatwawadi, S. V., and Adams, R. N.,** *J. Phys. Chem.*, 71, 338, 1967.
78. **Pedersen, J. A. and Thomson, R. H.,** *J. Magn. Res.*, 43, 373, 1981.
79. **Sharma, S. C. and Torssell, K.,** *Acta Chem. Scand. Ser. B*, 32, 347, 1978.
80. **Clay, J. D. R. and Murphy, D.,** *J. Chem. Soc. Faraday Trans. 2*, 77, 1589, 1981.
81. **Oshima, T. and Nagai, T.,** *Tetrahedron Lett.*, 3715, 1977.
82. **Klingler, R. J., Mochida, K., and Kochi, J. K.,** *J. Am. Chem. Soc.*, 101, 6626, 1979.
83. **Dixon, W. T. and Murphy, D.,** *J. Chem. Soc. Perkin Trans. 2*, 850, 1975.
84. **Dixon, W. T. and Murphy, D.,** *J. Chem. Soc. Faraday Trans. 2*, 135, 1976.
85. **Dixon, W. T. and Murphy, D.,** *J. Chem. Soc. Faraday Trans. 2*, 1221, 1976.
86. **Carrington, A. and Smith, I. C. P.,** *Mol. Phys.*, 8, 101, 1964.
87. **Matsunaga, Y. and McDowell, C. A.,** *Can. J. Chem.*, 38, 1158, 1960.
88. **Lott, K. A., Short, E. L., and Waters, D. N.,** *J. Chem. Soc. B*, 1232, 1969.
89. **Dixon, W. T. and Murphy, D.,** *J. Chem. Soc. Perkin Trans. 2*, 1430, 1974.
90. **Hewgill, F. R. and Lee, S. L.,** *J. Chem. Soc. C*, 2080, 1969.
91. **Ashworth, P.,** *Tetrahedron Lett.*, 1435, 1975.
92. **Adams, R. N., Murrill, E., McCreery, R., Blank, L., and Karolczak, M.,** *Eur. J. Pharmacol.*, 17, 287, 1972.
93. **Hewgill, F. R. and Mullings, L. R.,** *J. Chem. Soc. B*, 1155, 1969.
94. **Matsuura, T., Nagamachi, T., Nishinaga, A., Kon, H., and Cahnmann, H. J.,** *J. Org. Chem.*, 34, 2554, 1969.
95. **Rabold, G. P., Ogata, R. T., Okamura, M., Piette, L. H., Moore, R. E., and Scheuer, P. J.,** *J. Chem. Phys.*, 46, 1161, 1967.
96. **Fitzpatrick, J. D. and Steelink, C.,** *J. Org. Chem.*, 37, 762, 1972.
97. **Pilař, J., Buben, I., and Pospíšil, J.,** *Tetrahedron Lett.*, 4203, 1968.
98. **Atherton, N. M. and Henshaw, P. A.,** *J. Chem. Soc. Perkin Trans. 2*, 258, 1975.
99. **McKeown, E. and Waters, W. A.,** *J. Chem. Soc. B*, 679, 1966.
100. **Leaver, I. H.,** *Aust. J. Chem.*, 24, 753, 1971.
101. **Fitzpatrick, J. D. and Steelink, C.,** *Tetrahedron Lett.*, 5041, 1969.
102. **Holton, D. M. and Murphy, D.,** *J. Chem. Soc. Perkin Trans. 2*, 1757, 1980.
103. **Clare, S. I. and Steelink, C.,** *Tappi*, 56, 119, 1973.
104. **Atherton, N. and Blackhurst, A. J.,** *J. Chem. Soc. Faraday Trans. 2*, 68, 470, 1972.
105. **Hewgill, F. R. and Mullings, L. R.,** *Aust. J. Chem.*, 28, 355, 1975.
106. **Kosman, D. and Stock, L. M.,** *Chem. Commun.*, 551, 1968.
107. **Kosman, D. and Stock, L. M.,** *J. Org. Chem.*, 40, 2255, 1975.
108. **Nelsen, S. F. and Trost, B. M.,** *Tetrahedron Lett.*, 5737, 1966.
109. **Kosman, D. and Stock, L. M.,** *Tetrahedron Lett.*, 1511, 1967.
110. **Kosman, D. and Stock, L. M.,** *J. Am. Chem. Soc.*, 88, 843, 1966.

111. **Russell, G. A., Holland, G. W., and Chang, K.-Y.,** *J. Am. Chem. Soc.*, 89, 6629, 1967.
112. **Anderson, K. E., Kosman, D., Mayers, C. J., Ruekberg, B. P., and Stock, L. M.,** *J. Am. Chem. Soc.*, 90, 7168, 1968.
113. **Kosman, D. and Stock, L. M.,** *J. Am. Chem. Soc.*, 92, 409, 1970.
114. **Iwasaki, M.,** *J. Am. Chem. Soc.*, 92, 6348, 1970.
115. **Steenken, S. and O'Neill, P.,** *J. Phys. Chem.*, 81, 505, 1977.
116. **Kozima, S., Fujita, H., Hitomi, T., Kobayashi, K., Kobayashi, K., and Kawanisi, M.,** *Bull. Chem. Soc. Jpn.*, 53, 2953, 1980.
117. **Dehl, R. and Fraenkel, G. K.,** *J. Chem. Phys.*, 39, 1793, 1963.
118. **Brandon, R. W. and Lucken, E. A. C.,** *J. Chem. Soc.*, 4273, 1961.
119. **Maruyama, K.,** *Bull. Chem. Soc. Jpn.*, 37, 553, 1964.
120. **Maruyama, K. and Otsuki, T.,** *Bull. Chem. Soc. Jpn.*, 41, 444, 1968.
121. **Maruyama, K., Suzue, S., and Osugi, J.,** *Bull. Chem. Soc. Jpn.*, 44, 1161, 1971.
122. **Trost, B. M. and Nelsen, S. F.,** *J. Am. Chem. Soc.*, 88, 2876, 1966.
123. **Broze, M. and Luz, Z.,** *J. Chem. Phys.*, 51, 738, 1969.
124. **Warhurst, E. and Wilde, M.,** *Trans. Faraday Soc.*, 65, 1413, 1969.
125. **Stone, E. W. and Maki, A. H.,** *J. Chem. Phys.*, 41, 284, 1964.
126. **Russell, G. A. and Lokensgard, J.,** *J. Am. Chem. Soc.*, 89, 5059, 1967.
127. **Ikegami, Y. and Seto, S.,** *Bull. Chem. Soc. Jpn.*, 48, 140, 1975.
128. **Yoshioka, M., Kirino, Y., Tamura, Z., and Kwan, T.,** *Chem. Pharm. Bull.*, 25, 75, 1977.
129. **Loth, K., Graf, F., and Günthard, Hs. H.,** *Chem. Phys. Lett.*, 45, 191, 1977.
130. **Pedersen, J. A.,** Ph.D. Thesis, University of Aarhus, Denmark, 1973.
131. **Prokof'ev, A. I., Vol'eva, V. B., Novikova, I. A., Belostotskaya, I. S., Ershov, V. V., and Kabachnik, M. I.,** *Izv. Akad. Nauk SSR Ser. Khim.*, 2707, 1980.
132. **Madden, K. P., McManus, H. J., and Schuler, R. H.,** *J. Phys. Chem.*, 86, 2926, 1982.
133. **Kotake, Y. and Kuwata, K.,** *Bull. Chem. Soc. Jpn.*, 45, 2663, 1972.
134. **Broze, M. and Luz, Z.,** *J. Phys. Chem.*, 71, 3690, 1967.
135. **Vincow, G. and Fraenkel, G. K.,** *J. Chem. Phys.*, 34, 1333, 1961.
136. **Ashworth, P. and Dixon, W. T.,** *J. Chem. Soc. Perkin Trans. 2*, 739, 1974.
137. **Yoshida, H., Hayashi, K., and Warashina, T.,** *Bull. Chem. Soc. Jpn.*, 45, 3515, 1972.
138. **Yoshida, H., Sohma, J., and Warashina, T.,** *Bull. Chem. Soc. Jpn.*, 47, 1396, 1974.
139. **Piette, L. H., Okamura, M., Rabold, G. P., Ogata, R. T., Moore, R. E., and Scheuer, P. J.,** *J. Phys. Chem.*, 71, 29, 1967.
140. **Dixon, W. T., Kok, P. M., and Murphy, D.,** *Tetrahedron Lett.*, 623, 1976.
141. **Rieke, R. D. and Rich, W. E.,** *J. Am. Chem. Soc.*, 92, 7349, 1970.
142. **Petránek, J., Pilař, J., and Ryba, O.,** *Collect. Czech. Commun.*, 35, 2571, 1970.
143. **Bullock, A. T. and Howard, C. B.,** *Trans. Faraday Soc.*, 66, 1861, 1970.
144. **Hollocher, T. C., Tooney, N. M., and Adman, R.,** *Nature (London)*, 197, 74, 1963.
145. **Hollocher, T. C. and Weber, M. M.,** *Nature (London)*, 195, 247, 1962.
146. **Binks, J. H., Bullock, A. T., and Hopkin, D. J.,** *J. Chem. Soc. Faraday Trans. 2*, 68, 814, 1972.
147. **Gendell, J., Miller, W. R., and Fraenkel, G. K.,** *J. Am. Chem. Soc.*, 91, 4369, 1969.
148. **Freed, J. H. and Fraenkel, G. K.,** *J. Chem. Phys.*, 38, 2040, 1963.
149. **Edwards, T. G., Grinter, R., and Parlett, G. R.,** *Chem. Ind. (London)*, 200, 1970.
150. **Brockmann, H., Greve, H., and Hoyermann, K.,** *Tetrahedron Lett.*, 1493, 1971.
151. **Smeyers, Y. G., Sieiro, C., and Sanchez-Palacios, A.,** *J. Mol. Struct.*, 78, 113, 1982.
152. **Sieiro, C., Sanchez, A., and Echavarren, A.,** *J. Mol. Struct.*, 77, 305, 1981.
153. **Brodskii, A. I., Gordienko, L. L., Chukhlantseva, A. G., Balandin, A. A., Alieva, R. Yu., Klabunovskii, E. I., and Antik, L. V.,** *Zh. Strukt. Khim.*, 11, 604, 1970.
154. **Brumby, S.,** *J. Magn. Res.*, 34, 317, 1979.
155. **Elofson, R. M., Schulz, K. F., Galbraith, B. E., and Newton, R.,** *Can. J. Chem.*, 43, 1553, 1965.
156. **Schlossel, R. H., Geske, D. H., and Gulick, W. M., Jr.,** *J. Phys. Chem.*, 73, 71, 1969.
157. **Brodskli, A. I. and Gordienko, L. L.,** *Teor. Eksp. Khim. Adad. Nauk. Ukr. SSR*, 1, 451, 1965.
158. **Yasukoushi, K., Yamaguchi, H., and Nagayama, S.,** *Nippon Kagaku Zasshi*, 91, 213, 1970.
159. **Umemoto, K.,** *Bull. Chem. Soc. Jpn.*, 40, 1058, 1967.
160. **Rinkel, H. and Windsch, W.,** *Z. Phys. Chem. (Leipzig)*, 227, 281, 1964.
161. **Haustein, H., Möbius, K., and Dinse, K. P.,** *Z. Naturforsch.*, 24a, 1768, 1969.
162. **Chen, K. S. and Hirota, N.,** *J. Am. Chem. Soc.*, 94, 5550, 1972.
163. **Lutskii, A. E. and Romodanov, I. S.,** *Teor. Eksp. Khim.*, 2, 830, 1966.
164. **Russell, G. A. and Neugebauer, F. A.,** *Org. Magn. Res.*, 1, 125, 1969.
165. **Elschner, B. and Neubert, R.,** *Z. Chem.*, 1, 361, 1961.
166. **Baugh, P. J., Phillips, G. O., and Arthur, J. C., Jr.,** *J. Phys. Chem.*, 70, 3061, 1966.

167. **Kano, K., and Matsuo, T.,** *Chem. Lett.*, 11, 1974.
168. **Boekestein, G. and Buck, H. M.,** *Phosphorus and Sulphur*, 5, 61, 1978.
169. **Loth, K., Graf, F., and Günthard, Hs. H.,** *Chem. Phys.*, 13, 95, 1976.
170. **Loth, K., Andrist, M., Graf, F., and Günthard, Hs. H.,** *Chem. Phys. Lett.*, 29, 163, 1974.
171. **Loth, K. and Graf, F.,** *Helv. Chim. Acta*, 64, 1910, 1981.
172. **Smith, I. C. P. and Carrington, A.,** *Mol. Phys.*, 12, 439, 1967.
173. **Stone, T. J. and Waters, W. A.,** *J. Chem. Soc.*, 213, 1964.
174. **Prokof'ev, A. I., Bubnov, N. N., Solodovnikov, S. P., and Kabachnik, M. I.,** *Tetrahedron Lett.*, 2479, 1973.
175. **Gough, T. E.,** *Trans. Faraday Soc.*, 62, 2321, 1966.
176. **Weiner, S. A.,** *J. Am. Chem. Soc.*, 94, 581, 1972.
177. **Huysmans, W. G. B. and Waters, W. A.,** *J. Chem. Soc. B*, 1047, 1966.
178. **Jensen, O. N. and Pedersen, J. A.,** unpublished laboratory results, 1982.
179. **Warashina, T., Edlund, O., and Yoshida, H.,** *Bull. Chem. Soc. Jpn.*, 48, 636, 1975.
180. **Gough, T. E.,** *Can. J. Chem.*, 47, 331, 1969.
181. **Perez-Reyes, E. and Mason, R. P.,** *Mol. Pharmacol.*, 18, 594, 1980.
182. **Wilson, R.,** *J. Chem. Soc. B*, 1581, 1968.
183. **Claxton, T. A., Oakes, J., and Symons, M. C. R.,** *Trans. Faraday Soc.*, 63, 2125, 1967.
184. **Foster, T., Elliot, A. J., Adeleke, B. B., and Wan, J. K. S.,** *Can. J. Chem.*, 56, 869, 1978.
185. **Claxton, T. A., Gough, T. E., and Symons, M. C. R.,** *Trans. Faraday Soc.*, 62, 279, 1966.
186. **Noda, S., Sanehira, K., Doba, T., and Yoshida, H.,** *Bull. Chem. Soc. Jpn.*, 55, 1705, 1982.
187. **Bowers, P. R., McLauchlan, K. A., and Sealy, R. C.,** *J. Chem. Soc. Perkin Trans.*, 915, 1976.
188. **Pannell, J.,** *Chem. Ind. (London)*, 1797, 1962.
189. **Fitzpatrick, J. D., Steelink, C., and Hansen, R. E.,** *J. Org. Chem.*, 32, 625, 1967.
190. **Adderley, C. J. R. and Hewgill, F. R.,** *J. Chem. Soc. C*, 1443, 1968.
191. **van den Hoek, W. J., de Winter, J. F. Th., and Smidt, J.,** *J. Magn. Res.*, 6, 15, 1972.
192. **van den Hoek, W. J., Huysmans, W. G. B., and van Gemert, M. J. C.,** *J. Magn. Res.*, 3, 137, 1970.
193. **Petránek, J. and Pilař, J.,** *Collect. Czech. Chem. Commun.*, 35, 830, 1970.
194. **Gilbert, L. and Kreilick, R.,** *J. Chem. Phys.*, 48, 3377, 1968.
195. **Atherton, N. M., Blackhurst, A. J., and Cook, I. P.,** *Trans. Faraday Soc.*, 67, 2510, 1971.
196. **Dixon, W. T., Foster, W. E. J., and Murphy, D.,** *J. Chem. Soc. Perkin Trans. 2*, 2124, 1973.
197. **Wong, S. K., Sytnyk, W., and Wan, J. K. S.,** *Can. J. Chem.*, 50, 3052, 1972.
198. **Ayscough, P. B. and Sealy, R. C.,** *J. Chem. Soc. Perkin Trans. 2*, 543, 1973.
199. **Kambara, Y. and Yoshida, H.,** *Bull. Chem. Soc. Jpn.*, 50, 1367, 1977.
200. **Bolton, J. R. and Carrington, A.,** *Proc. Chem. Soc. (London)*, 174 and 385, 1961.
201. **Bolton, J. R., Carrington, A., and dos Santos-Veiga, J.,** *Mol. Phys.*, 5, 465, 1962.
202. **Sullivan, P. D.,** *J. Phys. Chem.*, 75, 2195, 1971.
203. **Barabas, A. B., Forbes, W. B., and Sullivan, P. D.,** *Can. J. Chem.*, 45, 267, 1967.
204. **Sullivan, P. D. and Bolton, J. R.,** *J. Am. Chem. Soc.*, 90, 5366, 1968.
205. **Sullivan, P. D.,** *J. Phys. Chem.*, 76, 3943, 1972.
206. **Sullivan, P. D., and Bolton, J. R.,** *J. Phys. Chem.*, 73, 4387, 1969.
207. **Sullivan, P. D. and Brette, N. A.,** *J. Phys. Chem.*, 79, 474, 1975.
208. **Sullivan, P. D.,** *J. Magn. Res.*, 5, 438, 1971.
209. **Devolder, P. and Goudmand, P.,** *C.R. Acad. Sci. (Paris)*, 279, 1001, 1974.
210. **Singer, L. S., Lewis, I. C., Richerzhagen, T., and Vincow, G.,** *J. Phys. Chem.*, 75, 290, 1971.
211. **Agranat, I., Rabinovitz, M., Falle, H. R., Luckhurst, G. R., and Ockwell, J. N.,** *J. Chem. Soc. B*, 294, 1970.
212. **Caldwell, E. S. and Steelink, C.,** *Biochem. Biophys. Acta*, 184, 420, 1969.
213. **Campbell, E. M. and Hewgill, F. R.,** *Aust. J. Chem.*, 27, 2279, 1974.
214. **Ackermann, P., Berthet, G., and Robert, H.,** *C. R. Acad. Sci. (Paris)*, 260, 2174, 1965.
215. **Huntington, J. L. and Davies, D. G.,** *J. Electrochem. Soc.*, 118, 57, 1971.
216. **Kohl, D. H., Wright, J. R., and Weissman, M.,** *Biochem. Biophys. Acta*, 180, 536, 1969.
217. **Atherton, N. M., Horsewill, A. J., and Kennedy, P. A.,** *J. Phys. Chem.*, 82, 1097, 1978.
218. **Nair, M. K. V., Santhanam, K. S. V., and Venkataraman, B.,** *Mol. Phys.*, 19, 585, 1970.
219. **Nair, M. K. V., Santhanam, K. S. V., and Venkataraman, B.,** *J. Magn. Res.*, 9, 229, 1973.
220. **Pedulli, G. F., Spisni, A., Vivarelli, P., Dembech, P., and Seconi, G.,** *J. Magn. Res.*, 12, 331, 1973.
221. **Pasimeni, L., Brustolon, M., and Corvaja, C.,** *J. Magn. Res.*, 21, 259, 1976.
222. **Strauss, H. L. and Fraenkel, G. K.,** *J. Chem. Phys.*, 35, 1738, 1961.
223. **Gulick, W. M., Jr. and Geske, D. H.,** *J. Am. Chem. Soc.*, 88, 4119, 1966.
224. **Stone, E. W. and Maki, A. H.,** *J. Am. Chem. Soc.*, 87, 454, 1965.
225. **Silver, B. L., Luz, Z., and Eden, C.,** *J. Chem. Phys.*, 44, 4258, 1966.

226. **Gendell, J., Freed, J. H., and Fraenkel, G. K.,** *J. Chem. Phys.*, 37, 2832, 1962.
227. **Reitz, D. C., Dravnieks, F., and Wertz, J. E.,** *J. Chem. Phys.*, 33, 1880, 1960.
228. **Broze, M., Luz, Z., and Silver, B. L.,** *J. Chem. Phys.*, 46, 4891, 1967.
229. **Melamud, E. and Silver, B. L.,** *J. Phys. Chem.*, 78, 2140, 1974.
230. **Broze, M. and Luz, Z.,** *J. Chem. Phys.*, 51, 749, 1969.
231. **Komenda, J., Kubáček, P., and Balusek, J.,** *Scr. Fac. Sci. Nat. Ujep Brun. Chem.* 2, 4, 81, 1974.
232. **Stone, T. J. and Waters, W. A.,** *Proc. Chem. Soc. (London)*, 253, 1962.
233. **Hansen, R. L.,** *J. Org. Chem.*, 33, 3968, 1968.
234. **Ackermann, P., Berthet, G., and Robert, H.,** *C. R. Acad. Sci.*, 261, 4721, 1965.
235. **Russell, G. A., Konaka, R., Strom, E. T., Danen, W. C., Chang, K. Y., and Kaupp, G.,** *J. Am. Chem. Soc.*, 90, 4646, 1968.
236. **Büchner, W. and Lucken, E. A. C.,** *Helv. Chim. Acta*, 47, 2113, 1964.
237. **Weiner, S. A. and Mahoney, L. R.,** *J. Am. Chem. Soc.*, 94, 5029, 1972.
238. **Imbaud, J.-P. and Berthet, G.,** *C. R. Acad. Sci.*, 261, 953, 1965.
239. **Bazhin, N. M., Schastnev, P. V., Chuvylkin, N. D., Berdnikov, V. M., Fedorov, V. K., Merkulov, A. P., and Shteingarts, V. D.,** *Zh. Strukt. Khim.*, 11, 998, 1970.
240. **Lloyd, R. V. and Wood, D. E.,** *J. Am. Chem. Soc.*, 96, 659, 1974.
241. **Holton, D. M. and Murphy, D.,** *J. Chem. Soc. Faraday Trans. 2*, 1637, 1979.
242. **Shiozawa, M., Yamamoto, H., and Fujita, Y.,** *Bull. Chem. Soc. Jpn.*, 50, 2177, 1977.
243. **Pokhodenko, V. D. and Khizhnyi, V. A.,** *Teor. Eksp. Khim.*, 2, 700, 1966.
244. **Loth, K., Graf, F., and Günthard, Hs. H.,** *Chem. Phys. Lett.*, 45, 191, 1977.
245. **Wong, S. K., Fabes, L., Green, W. J., and Wan, J. K. S.,** *J. Chem. Soc. Faraday Trans. 1*, 68, 2211, 1972.
246. **Yoshida, H., Kambara, Y., and Rånby, B.,** *Bull. Chem. Soc. Jpn.*, 47, 2599, 1974.
247. **Hales, B. J. and Bolton, J. R.,** *Photochem. Photobiol.*, 12, 239, 1970.
248. **Pokhodenko, V. D. and Bidzilya, V. A.,** *Teor. Eksp. Khim.*, 1, 801, 1965.
249. **Holton, D. M. and Murphy, D.,** *J. Chem. Soc. Faraday Trans. 2*, 77, 383, 1981.
250. **Kano, K. and Matsuo, T.,** *Tetrahedron Lett.*, 4323, 1974.
251. **Brauer, O., Köll, P., and Voss, J.,** *Tetrahedron Lett.*, 2311, 1974.
252. **Dixon, W. T. and Murphy, D.,** *J. Chem. Soc. Perkin Trans. 2*, 1630, 1974.
253. **Huysmans, W. G. B. and Waters, W. A.,** *J. Chem. Soc. B*, 1163, 1967.
254. **Mukai, K., Nishiguchi, H., Ishizu, K., Deguchi, Y., and Takaki, H.,** *Bull. Chem. Soc. Jpn.*, 40, 2731, 1967.
255. **Chen, K. S., Smith, R. T., and Wan, J. K. S.,** *Can. J. Chem.*, 56, 2503, 1978.
256. **Dixon, W. T. and Murphy, D.,** *J. Chem. Soc. Faraday Trans. 2*, 432, 1978.
257. **Dixon, W. T. and Murphy, D.,** *J. Chem. Soc. Faraday Trans. 2*, 1475, 1977.
258. **Adeleke, B. B. and Wan, J. K. S.,** *J. Chem. Soc. Faraday Trans. 1*, 1799, 1976.
259. **Adeleke, B. B., Choo, K. Y., and Wan, J. K. S.,** *J. Chem. Phys.*, 62, 3822, 1975.
260. **Rieker, A.,** *Z. Naturforsch.*, 21b, 647, 1966.
261. **Rieker, A., Zeller, N., Schurr, K., and Müller, E.,** *Justus Liebigs Ann. Chem.*, 697, 1, 1966.
262. **Dixon, W. T., Holton, D. M., and Murphy, D.,** *J. Chem. Soc. Faraday Trans. 2*, 521, 1978.
263. **Nemoto, F., Mukai, K., Tsuzuki, N., and Ishizu, K.,** *Chem. Lett.*, 307, 1981.
264. **van den Hoek, W. J.,** Thesis, Delft University of Technology, The Netherlands, 1972.
265. **Nemoto, F., Tsuzuki, N., Mukai, K., and Ishizu, K.,** *J. Phys. Chem.*, 85, 2450, 1981.
266. **O'Neill, P., Steenken, S., and Schulte-Frohlinde, D.,** *J. Phys. Chem.*, 79, 2773, 1975.
267. **Shimoda, F., Hanafusa, T., Nemoto, F., Mukai, K., and Ishizu, K.,** *Bull. Chem. Soc. Jpn.*, 52, 3743, 1979.
268. **Mukai, K. and Inagaki, N.,** *Bull. Chem. Soc. Jpn.*, 53, 2695, 1980.
269. **Dixon, W. T. and Murphy, D.,** *J. Chem. Soc. Perkin Trans. 2*, 182, 1976.
270. **Bolton, J. R. and Carrington, A.,** *Mol. Phys.*, 5, 161, 1962.
271. **Dixon, W. T., Kok, P. M., and Murphy, D.,** *J. Chem. Soc. Faraday Trans. 2*, 709, 1977.
272. **Holton, D. M. and Murphy, D.,** *J. Chem. Soc. Faraday Trans. 2*, 1185, 1979.
273. **Bullock, A. T. and Howard, C. B.,** *Mol. Phys.*, 27, 949, 1974.
274. **Fairhurst, S. A., Smith, I. M., Sutcliffe, L. H., and Taylor, S. M.,** *Org. Magn. Res.*, 18, 231, 1981.
275. **Sullivan, P. D.,** *J. Phys. Chem.*, 77, 1853, 1973.
276. **Ondercin, D. G. and Sullivan, P. D.,** *J. Phys. Chem.*, 78, 130, 1974.
277. **Ondercin, D. G. and Sullivan, P. D.,** *J. Magn. Res.*, 23, 39, 1976.
278. **Sullivan, P. D., Bolton, J. R., and Geiger, W. E., Jr.,** *J. Am. Chem. Soc.*, 92, 4176, 1970.
279. **Sullivan, P. D.,** *J. Am. Chem. Soc.*, 89, 4294, 1967.
280. **Zweig, A., Hodgson, W. G., and Jura, W. H.,** *J. Am. Chem. Soc.*, 86, 4124, 1964.
281. **Tozer, T. N. and Tuck, L. D.,** *J. Chem. Phys.*, 38, 3035, 1963.

282. **Yang, G. C. and Pohland, A. E.,** *J. Phys. Chem.*, 76, 1504, 1972.
283. **Nishinaga, A., Ziemek, P., and Matsuura, T.,** *Tetrahedron Lett.*, 4905, 1969.
284. **Nishinaga, A., Ziemek, P., and Matsuura, T.,** *J. Chem. Soc. C*, 2613, 1970.
285. **Baciu, I., Hillebrand, M., and Sahini, V. E.,** *J. Chem. Soc. Perkin Trans. 2*, 986, 1974.
286. **Zarubin, M. Y., Kutnevich, A. M., and Rudenko, A. P.,** *Z. Org. Khim.*, 11, 1284, 1975.
287. **Forbes, W. F. and Sullivan, P. D.,** *J. Chem. Phys.*, 48, 1411, 1968.
288. **Forbes, W. F. and Sullivan, P. D.,** *J. Am. Chem. Soc.*, 88, 2862, 1966.
289. **Harrah, L. A. and Becker, R.,** *J. Phys. Chem.*, 69, 2487, 1965.
290. **Sieiro, C., Sanchez, A., Crouigneau, P., and Lamy, C.,** *J. Chem. Soc. Perkin Trans. 2*, 1069, 1982.
291. **Hewgill, F. R. and Legge, F.,** *J. Chem. Soc. Perkin Trans. 1*, 2863, 1982.
292. **Byrne, L. T., Hewgill, F. R., Legge, F., Skelton, B. W., and White, A. H.,** *J. Chem. Soc. Perkin Trans. 1*, 2855, 1982.
293. **Sullivan, P. D.,** *J. Phys. Chem.*, 74, 2563, 1970.
294. **Volanschi, E., Vlădescu, C., and Volanschi, C.,** *Rev. Rom. Chim.*, 19, 755, 1974.
295. **Rieker, A., Scheffler, K., Mayer, R., Narr, B., and Müller, E.,** *Justus Liebigs Ann. Chem.*, 693, 10, 1966.
296. **Stegmann, H. B., Uber, W., and Scheffler, K.,** *Tetrahedron Lett.*, 2697, 1977.
297. **Avela, E., Törmälä, P., and Melander, B.,** *Finn. Chem. Lett.*, 123, 1974.
298. **Ulmschneider, K. B., Stegmann, H. B., Scheffler, K., and Viertel, G.,** *Z. Naturforsch.*, 33b, 237, 1978.
299. **Kalyanaraman, B. and Sealy, R. C.,** *Biochem. Biophys. Res. Commun.*, 106, 1119, 1982.
300. **Abakumov, G. A. and Klimov, E. S.,** *Izv. Akad. Nauk SSR Ser. Khim.*, 1199, 1972.
301. **Kukes, S., Prokof'ev, A. I., Masalimov, A. S., Bubnov, N. N., Solodovnikov, S. P. and Kabachnik, M. I.,** *Izv. Akad. Nauk SSR Ser. Khim.*, 1519, 1978.
302. **Muraev, V. A., Cherkasov, V. K., Abakumov, G. A., and Razuvaev, G. A.,** *Dokl. Akad. Nauk SSR*, 236, 620, 1977.
303. **Wier, D., Hutchinson, D. A., Russell, J., Wan, J. K. S.,** *Can. J. Chem.*, 60, 703, 1982.
304. **Abakumov, G. A., Gladyshev, E. N., Vyazankin, N. S., Razuvaev, G. A., Bayushkin, P. Ya., and Muraev, V. A.,** *J. Organomet. Chem. (Amsterdam)*, 64, 327, 1974.
305. **Abakumov, G. A. and Klimov, E. S.,** *Dokl. Akad. Nauk SSR*, 202, 827, 1972.
306. **Razuvaev, G. A., Abakumov, G. A., and Klimov, E. S.,** *Dokl. Akad. Nauk SSR*, 201, 624, 1971.
307. **Abakumov, G. A., Klimov, E. S., Ershov, V. V., and Belostotskaya, I. S.,** *Izv. Akad. Nauk SSR Ser. Khim.*, 927, 1975.
308. **Abakumov, G. A., Nevodchikov, V. I., Cherkasov, V. K., and Razuvaev, G. A.,** *Dokl. Akad. Nauk SSR*, 242, 609, 1978.
309. **Prokof'ev, A. I., Malysheva, N. A., Tumanskii, B. L., Bubnov, N. N., Solodovnikov, S. P., and Kabachnik, M. I.,** *Izv. Akad. Nauk SSR Ser. Khim.*, 1976, 1978.
310. **Prokof'ev, A. I., Malysheva, N. A., Bubnov, N. N., Solodovnikov, S. P., and Kabachnik, M. I.,** *Izv. Akad. Nauk SSR Ser. Khim.*, 1969, 1978.
311. **Creber, K. A. M. and Wan, J. K. S.,** *Chem. Phys. Lett.*, 81, 453, 1981.
312. **Creber, K. A. M. and Wan, J. K. S.,** *J. Am. Chem. Soc.*, 103, 2101, 1981.
313. **Prokof'ev, A. I., Solodovnikov, S. P., Belostotskaya, I. S., and Ershov, V. V.,** *Izv. Akad. Nauk SSR Ser. Khim.*, 199, 1974.
314. **Malysheva, N. A., Prokof'ev, A. I., Solodovnikov, S. P., Bubnov, N. N., and Kabachnik, M. I.,** *Izv. Akad. Nauk SSR Ser. Khim.*, 997, 1979.
315. **Prokof'ev, A. I., Bubnov, N. N., Solodovnikov, S. P., and Kabachnik, M. I.,** *Dokl. Akad. Nauk SSR*, 245, 1123, 1979.
316. **Chekalov, A. K., Prokof'ev, A. I., Bubnov, N. N., Solodovnikov, S. P., Zhdanov, A. A., and Kabachnik, M. I.,** *Dokl. Akad. Nauk SSR*, 262, 622, 1982.
317. **Prokof'eva, T. I., Prokof'ev, A. I., Lavrukhin, B. D., Belostotskaya, I. S., Ershov, V. V., and Kabachnik, M. I.,** *Dokl. Akad. Nauk SSR*, 257, 1412, 1981.
318. **Davies, A. G., Hawari, J. A. A., Gaffney, C., and Harrison, P. G.,** *J. Chem. Soc. Perkin Trans. 2*, 631, 1982.
319. **Prokof'ev, A. I., Prokof'eva, T. I., Bubnov, N. N., Solodovnikov, S. P., Belostotskaya, I. S., Ershov, V. V., and Kabachnik, M. I.,** *Dokl. Akad. Nauk SSR*, 245, 1393, 1979.
320. **Prokof'ev, A. I., Masalimov, S. A., Bubnov, N. N., Solodovnikov, S. P., and Kabachnik, M. I.,** *Izv. Akad. Nauk SSR Ser. Khim.*, 193, 1976.
321. **Prokof'ev, A. I., Bubnov, N. N., Solodovnikov, S. P., and Kabachnik, M. I.,** *Izv. Akad. Nauk SSR Ser. Khim.*, 2213, 1974.
322. **Prokof'ev, A. I., Malysheva, N. A., Bubnov, N. N., Solodovnikov, S. P., Belostotskaya, I. S., Komissarova, N. L., Ershov, V. V., and Kabachnik, M. I.,** *Dokl. Akad. Nauk SSR*, 229, 1396, 1976.

323. **Abakumova, L. G., Lobanov, A. V., and Abakumov, G. A.**, *Izv. Akad. Nauk SSR Ser. Khim.*, 204, 1978.
324. **Prokof'ev, A. I., Prokof'eva, T. I., Bubnov, N. N., Solodovnikov, S. P., Belostotskaya, I. S., Ershov, V. V., and Kabachnik, M. I.**, *Izv. Akad. Nauk SSR*. Ser. Khim., 1512, 1978.
325. **Ho, T.-I., Creber, K. A. M., and Wan, J. K. S.**, *J. Am. Chem. Soc.*, 103, 6524, 1981.
326. **Wicklund, P. A., Beckmann, L. S., and Brown, D. G.**, *Inorg. Chem.*, 15, 1996, 1976.
327. **Prokof'ev, A. I., Prokof'eva, T. I., Bubnov, N. N., Solodovnikov, S. P., Belostotskaya, I. S., Ershov, V. V., and Kabachnik, M. I.**, *Izv. Akad. Nauk SSR Ser. Khim.*, 1961, 1978.
328. **Muraev, V. A., Abakumov, G. A., and Razuvaev, G. A.**, *Dokl. Akad. Nauk SSR*, 217, 1083, 1974.
329. **Abakumov, G. A. and Muraev, V. A.**, *Dokl. Akad. Nauk SSR*, 217, 1313, 1974.
330. **Prokof'ev, A. I., Prokof'eva, T. I., Belostotskaya, I. S., Bubnov, N. N., Solodovnikov, S. P., Ershov, V. V., and Kabachnik, M. I.**, *Tetrahedron*, 35, 2471, 1979.
331. **Prokof'ev, A. I., Prokof'eva, T. I., Bubnov, N. N., Solodovnikov, S. P., Belostotskaya, I. S., Ershov, V. V., and Kabachnik, M. I.**, *Dokl. Adad. Nauk SSR*, 239, 1367, 1978.
332. **Prokof'ev, A. I., Prokof'eva, T. I., Bubnov, N. N., Solodovnikov, S. P., Belostotskaya, I. S., Ershov, V. V., and Kabachnik, M. I.**, *Dokl. Akad. Nauk SSR*, 236, 897, 1977.
333. **Stegmann, H. B., Ulmschneider, K. B., Heike, K., and Scheffler, K.**, *J. Organomet. Chem. (Amsterdam)*, 118, 259, 1976.
334. **Balch, A. L.**, *J. Am. Chem. Soc.*, 95, 2723, 1973.
335. **Claxton, T. A., Oakes, J., and Symons, M. C. R.**, *Trans. Faraday Soc.*, 64, 596, 1968.
336. **Oakes, J. and Symons, M. C. R.**, *Trans. Faraday Soc.*, 66, 10, 1970.
337. **Chen, K. S. and Wan, J. K. S.**, *Spectrosc. Lett.*, 12, 647, 1979.
338. **Konishi, S., Niizuma, S., and Kokubun, H.**, *Chem. Phys. Lett.*, 71, 164, 1980.
339. **Khakhar, M. P., Prabhananda, B. S., and Das, M. R.**, *J. Am. Chem. Soc.*, 89, 3100, 1967.
340. **Prabhananda, B. S., Khakhar, M. P., and Das, M. R.**, *J. Am. Chem. Soc.*, 90, 5980, 1968.
341. **Chen, D. H., Warhurst, E., and Wilde, A. M.**, *Trans. Faraday Soc.*, 63, 2561, 1967.
342. **Chippendale, J. C. and Warhurst, E.**, *Trans. Faraday Soc.*, 64, 2332, 1968.
343. **Oakes, J., Symons, M. C. R., and Claxton, T. A.**, *J. Am. Chem. Soc.*, 93, 3769, 1971.
344. **Felix, C. C. and Sealy, R. C.**, *J. Am. Chem. Soc.*, 103, 2831, 1981.
345. **Gill, P. S. and Gough, T. E.**, *Can. J. Chem.*, 46, 656, 1968.
346. **Alberti, A. and Hudson, A.**, *J. Chem. Soc. Perkin Trans. 2*, 1098, 1978.
347. **Chen, K. S. and Hirota, N.**, *Chem. Phys. Lett.*, 13, 190, 1972.
348. **Rutter, A. W. and Warhurst, E.**, *Trans. Faraday Soc.*, 66, 1866, 1970.
349. **Khakhar, M. P., Prabhananda, B. S., and Das, M. R.**, *J. Chem. Phys.*, 45, 2327, 1966.
350. **Alegria, A. E., Fontanez, F., and Stevenson, G. R.**, *J. Phys. Chem.*, 80, 1113, 1976.
351. **Rutter, A. W. and Warhurst, E.**, *Trans. Faraday Soc.*, 64, 2338, 1968.
352. **Claxton, T. A. and Oakes, J.**, *Trans. Faraday Soc.*, 64, 607, 1968.
353. **Claxton, T. A. and Oakes, J.**, *Chem. Phys. Lett.*, 1, 189, 1967.
354. **Alegria, A. E., Concepcion, R., and Stevenson, G. R.**, *J. Phys. Chem.*, 79, 361, 1975.
355. **Stevenson, G. R. and Alegria, A. E.**, *J. Phys. Chem.*, 80, 69, 1976.
356. **Stevenson, G. R. and Alegria, A. E.**, *J. Phys. Chem.*, 78, 1771, 1974.
357. **Stevenson, G. R. and Alegria, A. E.**, *J. Phys. Chem.*, 79, 1042, 1975.
358. **Stevenson, G. R. and Alegria, A. E.**, *J. Am. Chem. Soc.*, 97, 3869, 1975.
359. **Adeleke, B. B. and Wan, J. K. S.**, *J. Chem. Soc. Perkin Trans. 2*, 225, 1980.
360. **Chen, K. S., Foster, T., and Wan, J. K. S.**, *J. Chem. Soc. Perkin Trans. 2*, 1288, 1979.
361. **Echegoyen, L., Nieves, I., and Stevenson, G. R.**, *J. Phys. Chem.*, 86, 1611, 1982.
362. **Kochi, J. K., Chen, K. S., and Wan, J. K. S.**, *Chem. Phys. Lett.*, 73, 557, 1980.
363. **Chen, K. S., Wan, J. K. S., and Kochi, J. K.**, *J. Phys. Chem.*, 85, 1726, 1981.
364. **Chen, K. S. and Hirota, N.**, *J. Phys. Chem.*, 82, 1133, 1978.
365. **Pasimeni, L., Brustolon, M., and Corvaja, C.**, *J. Chem. Soc. Faraday Trans. 2*, 68, 223, 1972.
366. **Gough, T. E. and Symons, M. C. R.**, *Trans. Faraday Soc.*, 62, 269, 1966.
367. **Eastman, M. P., Bruno, G. V., McGuyer, C. A., Gutierrez, A. R., and Shannon, J. M.**, *J. Phys. Chem.*, 83, 2523, 1979.
368. **Warhurst, E. and Wilde, M.**, *Trans. Faraday Soc.*, 67, 605, 1971.
369. **Claxton, T. A. and Oakes, J.**, *Chem. Phys. Lett.*, 1, 189, 1967.
370. **Adams, R. F., Staples, T. L., and Szwarc, M.**, *Chem. Phys. Lett.*, 5, 474, 1970.
371. **Allendoerfer, R. D. and Papez, R. J.**, *J. Phys. Chem.*, 76, 1012, 1972.
372. **Gill, P. S. and Gough, T. E.**, *Can. J. Chem.*, 45, 2112, 1967.
373. **Claxton, T. A., Oakes, J., and Symons, M. C. R.**, *Nature (London)*, 216, 914, 1967.
374. **Bowmaker, G. A. and Campbell, G. K.**, *Aust. J. Chem.*, 32, 1897, 1979.
375. **Abakumov, G. A., Teplova, I. A., Cherkasov, V. K., and Shalnova, K. G.**, *Inorg. Chim. Acta*, 32, L57, 1979.

376. **Hutchinson, D. A., Chen, K. S., Russell, J., and Wan, J. K. S.,** *J. Chem. Phys.*, 73, 1862, 1980.
377. **Gough, T. E. and Hindle, P. R.,** *Can. J. Chem.*, 49, 1529, 1971.
378. **Prabhananda, B. S., Khakhar, M. P., and Das, M. R.,** *J. Indian Chem. Soc.*, 46, 659, 1969.
379. **Prokof'ev, A. I., Bubnov, N. N., Solodovnikov, S. P., and Belostotskaya, I. S.,** *Dokl. Akad. Nauk SSR*, 210, 361, 1973.
380. **Alberti, A., Hudson, A., and Pedulli, G. F.,** *J. Chem. Soc. Faraday Trans. 2*, 76, 948, 1980.
381. **Klimov, E. S., Lobanov, A. V., and Abakumov, G. A.,** *Izv. Akad. Nauk SSR Ser. Khim.*, 2028, 1981.
282. **Foster, T., Chen, K. S., and Wan, J. K. S.,** *J. Organomet. Chem. (Amsterdam)*, 184, 113, 1980.
383. **Warhurst, E. and Wilde, M.,** *Trans. Faraday Soc.*, 66, 2124, 1970.
384. **Razuvaev, G. A., Cherkasov, V. K., and Abakumov, G. A.,** *J. Organomet. Chem. (Amsterdam)*, 160, 361, 1978.
385. **Alberti, A. and Camaggi, C. M.,** *J. Organomet. Chem. (Amsterdam)*, 181, 355, 1979.
386. **Nevodchikov, V. I., Abakumov, G. A., Cherkasov, V. K., and Razuvaev, G. A.,** *J. Organomet. Chem. (Amsterdam)*, 214, 119, 1981.
387. **Girgis, A. Y., Sohn, Y. S., and Balch, A. L.,** *Inorg. Chem.*, 14, 2327, 1975.
388. **Alberti, A. and Hudson, A.,** *J. Organomet. Chem. (Amsterdam)*, 182, C49, 1979.
389. **Abakumov, G. A., Muraev, V. A., and Razuvaev, G. A.,** *Dokl. Akad. Nauk SSR*, 215, 1113, 1974.
390. **Prokof'ev, A. I., Khodak, A. A., Malysheva, N. A., Petrovskii, P. V., Bubnov, N. N., Solodovnikov, S. P., and Kabachnik, M. I.,** *Dokl. Akad. Nauk SSR*, 240, 92, 1978.
391. **Prokof'ev, A. I., Khodak, A. A., Malysheva, N. A., Bubnov, N. N., Solodovnikov, S. P., Bolostotskaya, I. S., Ershov, V. V., and Kabachnik, M. I.,** *Dokl. Akad. Nauk SSR*, 240, 358, 1978.
392. **Abakumov, G. A. and Klimov, E. S.,** *Zh. Strukt. Khim.*, 13, 922, 1972.
393. **Teplova, I. A., Shal'nova, K. G., Razuvaev, G. A., and Abakumov, G. A.,** *Dokl. Akad. Nauk SSR*, 235, 1323, 1977.
394. **Lobanov, A. V., Abakumov, G. A., and Razuvaev, G. A.,** *Dokl. Akad. Nauk SSR*, 235, 824, 1977.
395. **Masalimov, A. S., Prokof'ev, A. I., Bubnov, N. N., Solodovnikov, S. P, and Kabachnik, M. I.,** *Dokl. Akad. Nauk SSR*, 236, 116, 1977.
396. **Prokof'ev, A. I., Prokof'eva, T. I., Bubnov, N. N., Solodovnikov, S. P., Belostotskaya, I. S., Ershov, V. V., and Kabachnik, M. I.,** *Dokl. Akad. Nauk SSR*, 234, 845, 1977.
397. **Prokof'ev, A. I., Prokof'eva, T. I., Bubnov, N. N., Solodovnikov, S. P., Belostotskaya, I. S., Ershov, V. V., and Kabachnik, M. I.,** *Dokl. Akad. Nauk SSR*, 234, 603, 1977.
398. **Kukes, S. G., Prokof'ev, A. I., Bubnov, N. N., Solodovnikov, S. P., Korniets, E. D., Kravtsov, D. N., and Kabachnik, M. I.,** *Dokl. Akad. Nauk SSR*, 229, 877, 1976.
399. **Mochida, K., Kochi, J. K., Chen, K. S., and Wan, J. K. S.,** *J. Am. Chem. Soc.*, 100, 2927, 1978.
400. **deBoer, J. W. M., Chen, K. S., Chung, Y. C. C., and Wan, J. K. S.,** *J. Am. Chem. Soc.*, 101, 5425, 1979.
401. **Takeshita, T. and Hirota, N.,** *J. Am. Chem. Soc.*, 93, 6421, 1971.
402. **Chen, K. S., Takeshita, T., Nakamura, K., and Hirota, N.,** *J. Phys. Chem.*, 77, 708, 1973.
403. **Lukashevich, V. O. and Dokunikhina, E. N.,** *Zh. Org. Khim.*, 4, 1985, 1968.
404. **Razuvaev, G. A., Abakumov, G. A., Klimov, E. S., Gladyshev, E. N., and Bayushkin, P. Ya.,** *Izv. Akad. Nauk SSR Ser. Khim.*, 1128, 1977.
405. **Haas, O. and von Zelewsky, A.,** *J. Chem. Res. (S)*, 78, 1980.
406. **Buchanan, R. M., Downs, H. H., Shorthill, W. B., Pierpont, C. G., Kessel, S. L., and Hendrickson, D. N.,** *J. Am. Chem. Soc.*, 100, 4318, 1978.
407. **Pierpont, C. G. and Downs, H. H.,** *J. Am. Chem. Soc.*, 98, 4834, 1976.
408. **Pedulli, G. F., Alberti, A., Testaferri, L., and Tiecco, M.,** *J. Chem. Soc. Perkin Trans. 2*, 1701, 1974.
409. **Krinitsyna, N. A., Kachan, A. A., and Shrubovich, V. A.,** *Ukr. Khim. Zh.*, 42, 381, 1976.
410. **Gough, T. E. and Hindle, P. R.,** *Trans. Faraday Soc.*, 66, 2420, 1970.
411. **Gough, T. E. and Hindle, P. R.,** *Can. J. Chem.*, 47, 1698, 1969.
412. **Gough, T. E. and Hindle, P. R.,** *Can. J. Chem.*, 47, 3393, 1969.
413. **Forbes, W. F., Sullivan, P. D., and Wang, H. M.,** *J. Am. Chem. Soc.*, 89, 2705, 1967.
414. **Forbes, W. F. and Sullivan, P. D.,** *Can. J. Chem.*, 44, 1501, 1966.
415. **Bolton, J. R., Carrington, A., and Todd, P. F.,** *Mol. Phys.*, 6, 169, 1963.
416. **Sullivan, P. D. and Bolton, J. R.,** in *Advances in Magnetic Resonance*, Academic Press, New York, 1970, 39.
417. **Russell, G. A., Dodd, J. R., Ku, T., Tanger, C., and Chung, C. S. C.,** *J. Am. Chem. Soc.*, 96, 7255, 1974.
418. **Russell, G. A., Tanger, C. M., and Kosugi, Y.,** *J. Org. Chem.*, 43, 3278, 1978.
419. **Russell, G. A. and Suleman, N. K.,** *J. Am. Chem. Soc.*, 103, 1560, 1981.
420. **Sakurai, H., Okada, A., Shoji, Y., and Itô, S.,** *Chem. Lett.*, 539, 1975.
421. **Ikegami, Y., Watanabe, H., and Seto, S.,** *Bull. Chem. Soc. Jpn.*, 45, 1976, 1972.

422. **Ikegami, Y., Watanabe, H., and Seto, S.,** *Chem. Lett.*, 877, 1972.
423. **Russell, G. A. and Stevenson, G. R.,** *J. Am. Chem. Soc.*, 93, 2432, 1971.
424. **Kirste, B. and Kurreck, H.,** *J. Am. Chem. Soc.*, 102, 6180, 1980.
425. **El'tsov, A. V., Studzinskii, D. P., Sendyurev, M. V., and Levental, Yu. K.,** *Zh. Obsch. Khim.*, 48, 2799, 1978.
426. **Günther, K., Filby, W. G., and Eiben, K.,** *Tetrahedron Lett.*, 251, 1971.
427. **Lunney, D., Bailes, J., and Memory, J. D.,** *J. Chem. Phys.*, 53, 3387, 1970.
428. **Hocking, M. B. and Mattar, S. M.,** *J. Magn. Res.*, 47, 187, 1982.
429. **Arshava, N. V.,** *Ukr. Khim. Zh.*, 46, 863, 1980.
430. **Lutskii, A. E., Romodanov, I. S., and Kruglyak, Ya. A.,** *Teor. Eksp. Khim. Akad. Nauk Ukr. SSR*, 2, 616, 1966.
431. **Lutskii, A. E., Romodanov, I. S., and Kruglyak, Ya. A.,** *Teor. Eksp. Khim. Akad. Nauk Ukr. SSR*, 2, 624, 1966.
432. **Arshava, N. V., Romodanov, I. S., and Lutskii, A. E.,** *Vestn. Khar'k. Politekh. Inst.*, 83, 44, 1973.
433. **Bulatov, A. V., Nikidaev, A. T., and Khidekel, M. L.,** *Izv. Akad. Nauk SSR. Ser. Khim.*, 437, 1982.
434. **Kruglyak, Yu. A.,** *Arm. Khim. Zh.*, 22, 559, 1969.
435. **Lutskii, A. E., Romodanov, I. S., and Ganyuk, L. N.,** *Teor. Eksp. Khim. Akad. Nauk Ukr. SSR*, 1, 540, 1965.
436. **Lutskii, A. E. and Romodanov, I. S.,** *Teor. Eksp. Khim. Acad. Nauk Ukr. SSR*, 2, 832, 1966.
437. **Hocking, M. B., Bolker, H. I., and Fleming, B. I.,** *Can. J. Chem.*, 58, 1983, 1980.
438. **Shapovalov, V. A., Gavar, R. A., Korunova, A. F., Stradyn, Ya. P., and Bezuglyi, V. D.,** *Zh. Obsch. Khim.*, 47, 900, 1977.
439. **Romodanov, I. S., Arshava, N. V., and Lutskii, A. E.,** *Teor. Eksp. Khim. Akad. Nauk. Ukr. SSR*, 9, 68, 1973.
440. **Ackermann, P.,** *C.R. Acad. Sci. (Paris)*, 265, 635, 1967.
441. **Misra, B. N., Sharma, S. F., and Gupta, S. K.,** *Acta Phys. Pol. A*, 49, 41, 1976.
442. **Misra, B. N., Gupta, S. K., and Sharma, S. D.,** *Z. Naturforsch.*, 27b, 1360, 1972.
443. **Romodanov, I. S., Arshava, N. V., and Semyanovskii, V. N.,** *Teor. Eksp. Khim. Acad. Nauk Ukr. SSR*, 10, 236, 1974.
444. **Brunton, G., Gilbert, B. C., and Mawby, R. J.,** *J. Chem. Soc., Perkin Trans. 2*, 1267, 1976.
445. **Matsuo, M. and Matsumoto, S.,** *Lipids*, 18, 81, 1983.
446. **Tsuchiya, J., Niki, E., and Kamiya, Y.,** *Bull. Chem. Soc. Jpn.*, 56, 229, 1983.
447. **Martinmaa, J., Törmälä, P., and Avela, E.,** *Finn. Chem. Lett.*, 30, 1979.
448. **Mukai, K., Tsuzuki, N., Ishizu, K., Ouchi, S., and Fukuzawa, K.,** *Chem. Phys. Lipids*, 29, 129, 1981.
449. **Mukai, K., Tsuzuki, N., Ouchi, S., and Fukazawa, K.,** *Chem. Phys. Lipids*, 30, 337, 1982.
450. **Schröder, B.,** M. S. thesis, University of Aarhus, Denmark, 1972.
451. **Hewgill, F. R. and Legge, F.,** *Tetrahedron Lett.*, 1075, 1977.
452. **Hewgill, F. R. and Kennedy, B. R.,** *J. Chem. Soc. C*, 362, 1966.
453. **Waiss, A. C., Jr., Kuhnle, J. A., Windle, J. J., and Wiersema, A. K.,** *Tetrahedron Lett.*, 6251, 1966.
454. **Kuhnle, J. A., Windle, J. J., and Waiss, A. C., Jr.,** *J. Chem. Soc. B*, 613, 1969.
455. **Jensen, O. N.,** M.S. thesis, University of Aarhus, Denmark, 1982.
456. **Jensen, O. N. and Pedersen, J. A.,** *Tetrahedron*, 39, 1609, 1983.
457. **Dixon, W. T., Moghimi, M., and Murphy, D.,** *J. Chem. Soc. Perkin Trans. 2*, 101, 1975.
458. **Stegmann, H. B., Stolze, K., and Scheffler, K.,** *Z. Naturforsch.*, 38b, 243, 1983.
459. **Pasimeni, L., Brustolon, M., and Corvaja, C.,** *Chem. Phys. Lett.*, 30, 249, 1975.
460. **Gough, T. E. and Taylor, G. A.,** *Can. J. Chem.*, 47, 3717, 1969.
461. **Cass, M. E., Greene, D. L., Buchnanan, R. M., and Pierpont, C. G.,** *J. Am. Chem. Soc.*, 105, 2680, 1983.
462. **Pedersen, J. A.,** *J. Magn. Res.*, in press.

EPR MONOGRAPHS

Atherton, N. M., *Electron Spin Resonance, Theory and Applications*, Halsted, New York, 1973.
Ayscough, P. B., *Electron Spin Resonance in Chemistry*, Methuen, London, 1967.
Carrington, A. and McLachlan, A. D., *Introduction to Magnetic Resonance*, Harper & Row, New York, 1967.
Gerson, F., *Hochauflösende ESR-Spectroscopie*, Verlag Chemie, Weinheim, 1967.
Scheffler, K. and Stegmann, H. B., *Electronenspinresonanz*, Springer-Verlag, New York, 1970.
Swarts, H. M., Bolton, J. R., and Borg, D. C., *Biological Applications of Electron Spin Resonance*, John Wiley & Sons, New York, 1972.
Wertz, J. E. and Bolton, J. R., *Electron Spin Resonance*, McGraw-Hill, New York, 1972.
Dorio, M. M. and Freed, J. H., Eds., *Multiple Electron Resonance Spectroscopy*, Plenum Press, New York, 1979.

ABBREVIATIONS

Magnetic Resonance Spectroscopy

CIDEP	Chemically induced dynamic electron polarization
CIDNP	Chemically induced dynamic nuclear polarization
ELDOR	Electron electron double resonance
ENDOR	Electron nuclear double resonance
EPR	Electron paramagnetic resonance
ESR	Electron spin resonance
NMR	Nuclear magnetic resonance

General

abs	Absolute
a_i	Scalar (isotropic) hyperfine splitting constants (G)
alk	Alkaline, NaOH, if not otherwise stated
el	Electrolytic/electrolysis
g	g-Factor, isotropic
G	Gauss ($1G = 10^{-4}$ tesla)
hfs	Hyperfine splittings
K	Kelvin/also potassium
L	Ligand (Table IV)
L.Ex.	Ligand exchange (Table IV)
Lw	Linewidth (peak-to-peak, Figure 7)
neut	Neutral
Ox	Oxidation
phot	Photolytic/photolysis
R	Alkyl/aryl substituent
Re	Reduction/also rhenium
RT	Room temperature (~300 K)
TW	Total width, spectral field separation between two outermost lines (G)

Chemical

Ac	Acetyl
AC	Actinide
Acac	Acetylacetonate
AcOH	Acetic acid
Bu	Butyl
$BuNH_2$	Butylamine
BuOH	Butanol
Bz	Benzyl
BzOH	Benzylalcohol
Cat	Fully reductive form of pertinent quinone
C_3H_5	Allyl
C_6H_6	Benzene
$C_{10}H_7$	Naphthyl
$C_{10}H_8$	Naphthalene
C_5H_5N	Pyridine
$(CH_2OH)_2$	Ethylene glycol
$C_5H_{11}OH$	Pentanol
Cod (1.5 –)	1.5-Cyclooctadiene
Cw	Crown ether (12Cw4 = 12-crown-4, etc.)
D or d	Deuterium
DBCw	Dibenzo-18-crown-6
DCCw	Dicyclohexyl-18-crown-6
DEA	Diethylamine
DEE	Diethylether
DG	Diethylene glycol (diglyme)

DME	1,2-Dimethoxyethane
DMF	Dimethylformamide
DMSO	Dimethylsulfoxide
DOPA	3,4-Dihydroxyphenylalanine
DPPH	2,2-Diphenyl-1-picrylhydrazyl
DTN	Sodium dithionite
Et	Ethyl
Et_3N	Triethylamine
$EtNO_2$	Nitroethane
EtOH	Ethanol
HMPA HMPT	Hexamethylphosphoric acid triamide
HRP	Horseradish peroxidase
K	Potassium/also Kelvin
KCO	"Potassium carbonyl", see no. 517
KO_2	Potassium superoxid
LA	Lanthanide
M	Central atom/metal ion (Table IV)
Me	Methyl
MeCN	Acetonitrile
Me_2CO	Acetone
$MeNO_2$	Nitromethane
MeOH	Methanol
Mes	Mesityl (2,4,6-$Me_3C_6H_2$)
MeTHF	2-Methyltetrahydrofuran
ND_3	Deuterated NH_3
NaN_3	Sodium azide
NaOD	Deuterated NaOH
Ph	Phenyl
PhMe	Toluene
$PhNH_2$	Aniline
PhOH	Phenol
Pr	Propyl
PrOH	Propanol
Q	Quinone (also Li_2Q, Na_2Q, K_2Q, for the alkali metal salts)
QH	Hydroquinone (also Q)
Re	Rhenium/also reduction
SpP	Spirophosphorane
SpSi	Spirosilane
SQ or Q	Semiquinone (also LiSQ, KSQ, NaSQ for alkali metal semiquinolates)
TDAP	Tris(dimethylamino)phosphine
THBCw	Tetrahydrobenzo-18-crown-6
THF	Tetrahydrofuran
Trien	Triethylenetetramine
w	Water

COMPOUND INDEX

The index gives names of parent compounds of the basic radical structures in the tables. Further given are several synonyms of naturally occurring compounds for which radicals have been observed. The numbers refer to the consecutive numbers of the radicals in the tables. For general location of individual radicals (substituted derivatives), see **Explanation to Tables.**

Note: Anions: 1—878, 882—885, 1106, 1322—1326
Neutral species: 879—881, 886—1159
Cations: 1160—1321
Complexes: C1-C525

A

B

C

H

I

J

L

M

N

O

P

Q

R

S

T

SUBJECT INDEX

A

B

H

I

K

L

M

N

O

P

Q

R

S

T

U

Z